AF710358

WITHDRAWN

COGNITION AND LANGUAGE

A Series in Psycholinguistics • Series Editor: R. W. RIEBER

Recent Volumes in this Series:

AMERICAN AND CHINESE PERCEPTIONS AND BELIEF SYSTEMS:
A People's Republic of China–Taiwanese Comparison
Lorand B. Szalay, Jean B. Strohl, Liu Fu, and Pen-Shui Lao

THE COLLECTED WORKS OF L. S. VYGOTSKY
Volume 1: Problems of General Psychology
Volume 2: The Fundamentals of Defectology (Abnormal Psychology and Learning Disabilities)
Volume 3: Problems of the Theory and History of Psychology
Volume 4: The History of the Development of Higher Mental Functions
Volume 5: Child Psychology
Volume 6: Scientific Legacy

EXPERIMENTAL SLIPS AND HUMAN ERROR:
Exploring the Architecture of Volition
Edited by Bernard J. Baars

LANGUAGE, THOUGHT, AND THE BRAIN
Tatyana B. Glezerman and Victoria I. Balkoski

PSYCHOENVIRONMENTAL FORCES IN SUBSTANCE ABUSE PREVENTION
Lorand B. Szalay, Jean Bryson Strohl, and Kathleen T. Doherty

THE PSYCHOPATHOLOGY OF LANGUAGE AND COGNITION
Robert W. Rieber and Harold J. Vetter

TIME, WILL, AND MENTAL PROCESS
Jason W. Brown

UNDERSTANDING MEXICANS AND AMERICANS:
Cultural Perspectives in Conflict
Rogelio Diaz-Guerrero and Lorand B. Szalay

A Continuation Order Plan is available for this series. A continuation order will bring delivery of each new volume immediately upon publication. Volumes are billed only upon actual shipment. For further information please contact the publisher.

Language, Thought, and the Brain

Tatyana B. Glezerman and
Victoria I. Balkoski
Albany Medical College
Albany, New York

Kluwer Academic / Plenum Publishers
New York, Boston, Dordrecht, London, Moscow

QP
399
.G57
1999

Library of Congress Cataloging-in-Publication Data

Glezerman, T. B. (Tat′îana Borisovna)
Language, thought, and the brain / Tatyana Glezerman and Victoria Balkoski.
p. cm. -- (Cognition and language)
Includes bibliographical references and index.
ISBN 0-306-46096-3
1. Neurolinguistics. 2. Psycholinguistics. 3. Cognition.
4. Psychology, Pathological. I. Balkoski, Victoria. II. Title.
III. Series.
QP399.G57 1999
612.8'2--dc21 99-38463
CIP

ISBN: 0-306-46096-3

© 1999 Kluwer Academic / Plenum Publishers
233 Spring Street, New York, N.Y. 10013

10 9 8 7 6 5 4 3 2 1

All rights reserved

No part of this book may be reproduced, stored in a retrieval system, or transmitted in any form or by any means, electronic, mechanical, photocopying, microfilming, recording, or otherwise, without written permission from the Publisher

Printed in the United States of America

To our families

Preface

The purpose of this book is to present a novel, comprehensive hypothesis about the relationship of human language and thought to specialization of the brain. Drawing on data from a wide variety of modern and classical sources and multiple—sometimes disparate—disciplines, we offer an original attempt to relate these segments of information in a framework tracing the historical connections and common origins of language codes and aspects of thinking with their phylogenetic roots.

In this time of unparalleled technological capabilities and the resulting enormous accumulation of separate and generally uncoordinated facts, such attempts at synthesis and cohesive theorizing are essential if we are to make sense of what we already know, as well as to construct a model which we can, in turn, explore in a more directed fashion. We have revisited and integrated insights from the past and new findings from multiple disciplines, as well as our own clinical studies, as a base from which to speculate. Our model of cerebral organization of language provides a framework for greater understanding and future investigation not only of our stated areas of interest but also in a much broader context, with important ramifications for research and conceptualization in allied fields examining localization and organization of higher functions in the brain. This has tremendously exciting implications for the emerging appreciation of brain mechanisms in psychiatric disorders. The model has immediate practical applications as well, offering a theoretical basis for a new approach to the rehabilitation of various language disorders, such as aphasia and developmental language disorders.

Thus, although the purpose of this book is to present a model of the connections between language, thought, and the brain, its scope is much broader. It outlines an approach to the study of the cerebral basis and cerebral organization of self and of symbolic thinking, whose disturbances are at the core of psychiatric disorders.

In the book we review in detail aphasia—language disorder due to local brain damage. Even in this "simple" model, with one focus on the brain, the clinical picture does not just represent loss of function of the damaged area but results from interaction between damaged and intact areas as an attempt at spontaneous compensation. These interactions are predetermined by evolutionarily fixed patterns of cortical connectivity, functional systems underlying language behavior in humans. Evolutionarily determined patterns of connections in the brain underlie a broad repertoire of human behavior. Psychiatric disorders and schizophrenia in particular elude definitive localization in the brain because there is a change in the brain connectivity pattern itself (developmental disorder), rather than damage to one area. We propose a correlation between psychopathological patterns and brain connectivity patterns.

Finally, we do not intend to give an exhaustive description of points of view and underlying facts in modern neuropsychology, neurolinguistics, and neuropsychiatry. Rather, by touching on a vast number of topics from aphasia and thought disorder due to focal brain damage to psychiatric thought disorder (delusions), our overall goal is to present a gestalt—the whole picture, but not by any means the full picture.

Contents

PART I. INTRODUCTION TO THE PROBLEM AND APPROACH

1. Basic Factors in the Human Brain's Differentiation Underlying Cerebral Organization of Language Ability

PART II. CEREBRAL ORGANIZATION OF LANGUAGE AND THOUGHT

2. *Temporal-Occipital Region: Visual Object Perception, Thought and Word*

3. *Temporal Region and "Sound-Articulate" Speech*

4. Parietal-Occipital Region: Spatial Perception and Word Form

5. Frontal Region: Thought and Sentence

PART III. CONCLUSIONS, REFLECTIONS, PERSPECTIVES

6. Thought and Focal Brain Damage

7. *Perspectives for Psychiatry*

I

Introduction to the Problem and Approach

1

Basic Factors in the Human Brain's Differentiation Underlying Cerebral Organization of Language Ability

1.1. Background

The history of intrahemispheric "localizing" of language functions dates back more than a century. Basic data on this subject were obtained from patients with focal brain damage by correlating language deficits (aphasia) with locations of the lesions within the cerebral hemisphere found on autopsy. Numerous studies have shown that different and specific language disorders accompany damage to the particular cortical areas in the left hemisphere: frontal, temporal, and parietal. These areas have been called "speech zones" of the brain.

The localization of higher cortical functions in classical neurology was considered in terms of independent "brain centers," although some theorists warned against too narrow a localization of cortical functions. These concepts eventually gave way to the contemporary concept of functional systems (Luria, 1966/1980), which suggests that any complex function such as language is realized by the interaction of several cortical areas, each of them making specific contributions to the whole. In the norm, functional systems are highly integrated, and thus it is difficult to "extract" the contribution of discrete cortical areas.

It is pathology—in particular, aphasia—that is the natural experiment in which focal cortical lesions result in a "falling out" of that discrete component of language subserved by this area in the norm (Luria, 1947/1970; Luria, 1966/1980). Using his concept of functional systems, Luria attempted to connect specific language disorders with dysfunction of the corresponding cortical area (Luria, 1966/1980).

Modern progress in our understanding of cerebral organization of language is a result of new neuroimaging techniques that allow immediate correlation of observations *in vivo* with the location of brain damage (Damasio, 1992; Damasio & Damasio, 1989). The findings to date have been consistent with classical descriptions, but in addition, they showed a much higher degree of cortical differentiation for language, or "linguistic specificity" of certain cytoarchitectural fields, than was previously thought (Damasio & Damasio, 1992).

All these studies, both classical and modern, focused primarily on the localization of language functions within the left hemisphere. Thus, they considered brain mechanisms of language in the framework of intrahemispheric cortical differentiation. Another aspect of cortical differentiation is interhemispheric specialization. Since Broca's finding in 1861 of a left frontal lobe lesion in a patient with expressive speech impairment, language has been connected with the left, dominant, hemisphere. However, functions of the right, nondominant, hemisphere have remained unknown for a long period of time. Research in the field of interhemispheric specialization was stimulated by the famous experiments in "split-brain" patients in the 1960s (Springer & Deutsch, 1989). At present, there is a vast amount of data suggesting the importance for normal speech activity of not only the traditional, dominant left hemisphere, but also the right hemisphere. However, these data are not systematized in the context of cerebral organization of language ability as a whole. What remains unexplored are the specific contributions of discrete cortical regions within each hemisphere and the interactions between intra- and interhemispheric dimensions.

Another concept regarding organization of cerebral functions was introduced in the period of classical neurology. It was the idea that cerebral, and in particular, cortical functions are hierarchically organized. Applying this principle to language, such seminal figures as Jackson and Head made a distinction between the symbolic aspects of speech activity (language) and its sensory and motor components (Jackson, 1958; Head, 1926/1963). However, these ideas were not considered in the context of cortical cytoarchitectonics and localization of linguistic functions.

The Russian physiologist Nicolai Bernstein (1947, 1967) developed a theory and an elegant system of function levels based on the morphological vertical hierarchy in brain differentiation. He also was the only author to combine "horizontal" (intrahemispheric) and "vertical" (hierarchical) principles of brain differentiation. His model has nearly been forgotten. It was called to our attention by Russian linguist Vjacheslav Ivanov (1978) and American linguist Roman Jakobson (1970), who recognized the potential of Bernstein's ideas and named him a leading biologist of our time.

The model of cerebral organization of language ability proposed in this book is inspired by the insights of Bernstein (1947). In this connection, we will outline Bernstein's system of brain function vertical organization. Although we begin with

Bernstein, our chief concern is with contemporary issues regarding cerebral organization of language. We thus extended Bernstein's system to include a symbolic function level, which was not described by Bernstein; we considered structure–function vertical hierarchy together with intrahemispheric and interhemispheric specialization and applied it to language.

1.2. Bernstein's Model of Hierarchical Cerebral Organization of Movements

Bernstein studied the cerebral organization of motions, and using this as a model, developed a comprehensive theory in which he explained the connections between the vertical hierarchy of brain structures and the order of the function levels. According to Bernstein, cerebral organization of motions can be represented as a multistory building composed of hierarchically overlaid stories of different phylogenetic ages that correspond to certain function levels.

Bernstein's highly insightful work not only examined the vertical principle in movement's cerebral organization but also considered each level as a relatively autonomous functional system comprising two parts operating concurrently: posterior brain regions, associated with afferentation, and anterior brain regions, associated with efferent systems. Here Bernstein followed the chief principle of the brain's horizontal differentiation, the Bell and Magendie rule, which was first applied to higher cerebral functions by Jackson. The rule states that at all levels of CNS evolution and in all CNS parts (spinal cord, brain stem, subcortical areas, cortex), the afferent systems occupy the posterior side and the efferent systems the anterior. Another general rule of basic horizontal differentiation states that the mode of operation of the posterior brain is simultaneous or spatial synthesis, whereas the mode of operation of the anterior brain is successive synthesis of constituents in time—successive or temporal synthesis. Bernstein interpreted this rule regarding each function level separately, i.e., each level is characterized by its specific simultaneous (posterior brain) and successive synthesis (anterior brain). In Bernstein's terms, each function level operates in the frame of its own "synthetic space" and "synthetic time." Regarding movement formation in the brain, Bernstein postulated the presence of a "movement image" or engram: "It is clear that each of the variations of a movement (for example, drawing a circle large or small, directly in front of oneself or to one side, on a horizontal piece of paper or on a vertical blackboard, and so on) demands a quite different muscular formula; and even more than this, involves a completely different set of muscles in the action. The almost equal facility and accuracy with which all these variations can be performed is evidence for the fact that they are ultimately determined by one and the same higher directional engram.... [It is] structurally extremely far removed (and because of this also probably localizationally very distant) from any resemblance whatever to the joint–muscle schemata; it is extremely geometrical, representing a very abstract motor image of space" (Bernstein, 1967, p. 49). In movement formation, separate images are responsible for the semantic structure of movement (action), represented in the posterior brain, and its motor composition, represented in the anterior brain. The semantic structure of an action is here understood as "sensory or sensory-gnostic synthesis that is adequate to the assigned task and can provide a solution to it" (Bernstein, 1947, p. 34). This sensory synthesis Bernstein termed *leading afferentation*. By afferentation, Bernstein refers not just to "raw" sensations but sensory information of one or several modalities that is integrated according to a spatial framework that is unique for each level. Thus, at each level, afferentation is connected with spatial synthesis. It gives information about the field of space in which an action is organized; it "models" the spatial configuration of an action. The motor composition of an action is the realization in time of the action's spatial image. Bernstein (1967) postulated the presence in the brain of a guiding engram—a motor image of a movement (specific for each level): "It must contain within itself ... the entire scheme of the movement as it is expanded in time. It must also guarantee the order and the rhythm of the realization of this scheme" (p. 39). The motor image of a movement is incorporated, according to Bernstein, in the coordinational time axis at each level. Bernstein correlated the semantic structure of an action (afferentation) with spatial (simultaneous) synthesis, and its motor composition with temporal (successive) synthesis. Table 1 shows the hierarchy of the function levels, the relation of spatial synthesis to afferentation and of temporal synthesis to an action itself (Bernstein, 1947).

Emphasizing the multilayered cerebral organization of any complex movement, Bernstein indicates that the highest level participating in the movement's formation is the leading one. It is conscious and voluntary, directly responding to the action's task. The levels lower than the leading one do participate in the realization of the movement, but in an assimilated way, as background.

What follows are brief descriptions of Bernstein's model of function levels.

1.2.1. Level A

Level A is defined by Bernstein as a level of paleokinetic regulation. The leading afferentation of this level is of kinesthetic and vestibular origin, which give information about body position in the gravitational field. The brain substratum receiving this afferent input is not the chief collector of kinesthetic sense, the thalamus, but the cerebellum. The integration of the kinesthetic and vestibular senses creates the "synthetic space" of the A level: the vertical position of the body as a weight category in the gravitational field. At this level there is no division into "self-space" and "non-self-space." The effector center of this level is in the mesencephalon (red nucleus group).

TABLE 1. Function Levels

Function level	Anatomic base Post/Ant	"Synthetic Space"/ Leading afferentation	"Synthetic Time"/ Type of movement
A	Cerebellum/ Mesencephalon	*Body position against gravity: up-down/* Proprioceptive Vestibular	*Simple Rhythm/* Muscular tone
B	Thalamus/ Pallidum	*Space of one's own body/* Proprioceptive Tactile	*Complex rhythm, individual pattern/* Synergetic movements
C	Cortical primary sensory fields/	*External space/* Visual Auditory Vestibular Tactile Proprioceptive	*Moment, speed, duration/* Goal-directed moving in external space
C1	Striatum		
C2	Primary motor cortex		
D	Parietal cortex/ Premotor cortex	*Object topological scheme/* Visual Proprioceptive Tactile Auditory	*Semantic sequence/* Object action (praxis)

Note. Based on *The Construction of Movements* by N. Bernstein, 1947.

Normal functioning of the A level provides muscular tone. These "movements" of level A—muscle tone—serve two functions: (1) maintenance of the vertical position and posture and (2) background tone necessary for further muscle contractions. Although muscular tone is the background for all movements, Bernstein gives three situations in which the A level's motions in the pure form can be observed on their own. One is the quick oscillatory and vibratory movements of piano and violin performers, which are quite automatized but volitional. Level A's motions may also be seen when they come forward involuntarily in some physiological states, such as shivering when cold and shaking when fearful. Finally, these movements may become apparent as a result of hyperfunction of the effector center; the monotonous, rhythmical resting tremor of Parkinsonism is an example. Synthetic time of the A level (see Table 1) represents a simple rhythm that can be depicted as a pure sinusoid.

1.2.2. Level B

Level B is defined by Bernstein as a level of synergy. The leading afferentation of this level is kinesthetic and tactile. The brain substratum receiving this afferent input is the thalamus, the main subcortical sensory center. At the B level, all kinesthetic and tactile sensations are integrated according to the body's single coordinate system. In other words, the synthetic space of level B is defined by Bernstein as one's own body coordinate system formed ("filled") by kinesthetic and tactile sensations.

Bernstein points out that there is no influence of the vestibular system at level B. Cerebellar kinesthetic and vestibular afferentation of level A subserves the function of keeping balance (muscular tone, vertical position) and there is no division into inner and outer space at level A. At level B, a constant flow of kinesthetic information about changing body position builds one invariable image, space of one's own body, independent of outside space and body position at any given moment.

If we separated the B level from the lower and higher ones, we would obtain the space of one's own body, which we will refer to as "I-space," isolated from the external world. This is because afferentation of the B level includes kinesthetic and tactile sense without the admixture of distance modalities such as visual, auditory, and olfactory (Bernstein, 1947).

The effector center of the B level is the pallidum. Bernstein indicates that neither lower nor higher effector centers in the brain can get such complete primary information about body position and movements as the pallidum obtains from the thalamus. Level B is the level of possession of one's own body. Movements of the B level represent the extensive "synergetic chorus" of the simultaneous and coordinated contraction of numerous muscle groups, occasionally acting independently but mostly serving as a background for all complex movements and actions.

Because of the characteristic afferentation, movements of the B level are always completely introverted and tied with the body no matter the external surroundings. They are self-sufficient movements, changed only by their own inner harmony and organization. They are, in fact, a kind of propriomotor function. The lack of auditory and visual information, and as a consequence, nonpossession of the external world, restricts the number of independent movements implemented by the B level. A few examples of such "pure" B level movements would be the plastic, rhythmical movements of Eastern dance; pantomime; habitual, monotonous-mechanical motions; and certain half-voluntary movements such as stretching. The other instance in which the B level movements appear at the foreground is the pathological hyperfunction of level B secondary to the lesion of the upper extrapyramidal center, the corpus striatum. These are various hyperkineses—excessive synergies. Bernstein gives the following figurative description of these pathological movements: "It is as though disinhibited hyperfunction of the B level throws open the gates of the phylogenetic zoo ... and then, from the deep of the motor system, all these emerge: deformed, grotesque backgrounds without figures, without meaning and adequacy; various torsion spasms, fragments of ancient movements such as chorea and athetosis; involuntary growls and utterances; psychomotor chimeras, madness of the effector system" (1947, p. 79).

Aimed at one's own body, with no regard for the external world, the B level movements themselves have a cyclic, rhythmical nature. In contrast to the A level movements that present a simple sinusoid, the B level movements have a very complicated pattern that, however, asserts itself exactly and invariably whenever called into play. Bernstein emphasized the "chased reiteration" of these movements: "They are as alike as two coins" (p. 69). Bernstein also indicated that it is the B level that brings the inner propriomotor rhythm to all aperiodic movements of the higher levels. It is, in fact, an individual rhythm reflecting the uniqueness of one's own internal body space and is manifest in all kinds of motions and behaviors, such as gestures, mimicry, handwriting, dancing, playing music or sports, and individual speech characteristics. Correspondingly, synthetic time of the B level is an individual propriomotor rhythm.

1.2.3. Level C

Level C is defined by Bernstein as a level of the external spatial field. Although practically all sensations participate in the powerful afferent synthesis of the C level, the distance modalities of vision and hearing play the leading role, thus determining the external nature of level C's spatial field. The afferent centers of the C level are represented by the primary projective cortical fields. With regard to synthetic space, the C level is the polar opposite of the B level: whereas synthetic space of the B level is the body space, at the C level it is non-self-space.

Afferent synthesis of the C level allows evaluation of the physical parameters of external space and its objects: distance and depth; mass, size, and three-dimensional shape; object localization and position; movements and interacting forces. Bernstein points out level C's greater degree of objectivity compared with the other levels, indicating that it is level C's "metricity" and "geometricity" which constitute the basis of precise and accurate movements.

The kinesthetic and tactile sensations that take part in the formation of level C's afferent synthesis are quite different in their meaning from those destined for level B. Rather than focusing on and defining one's own body, as in level B, in level C they are projected on external objects as things or categories possessing mass, shape, consistency, texture, and so on. The vestibular analyzer is of great importance in the C level's afferentation, vestibular sensations being integrated with the kinesthetic and visual ones.

Bernstein indicates that the proprioceptive (kinesthetic) pathways terminate at the several "phylogenetic stories" of the brain corresponding to the A, the B, and the C functional levels. The meaning of kinesthetic sensation is different at each of the given levels and is determined by the context of the functional system. In the A level, the kinesthetic pathways terminate in ancient cerebellum, where part of the vestibular system's pathways also end. Thus afferentation of the A level, as well as afferentation of the C level, is characterized by the close connection between the kinesthetic and vestibular senses. On the other hand, at the "thalamic" B level, there is almost exclusively a refined, proprioceptive sensation, and there are no appreciable connections with the vestibular system. At the A level, the significance of the proprioceptive-vestibular afferentation lies in determination of the body position in the external force field. The external force field is estimated by its meaningfulness for muscle tone (maintaining vertical body position). At this level, proprioception is projected on the body and gives the sense of weight and position.

Kinesthetic pathways pertaining to the C level terminate in the postcentral cortical area, and here give the sense of weight and position of objects in the external force field.

The brain substratum of the effector side of the C level, according to Bernstein, consists of two different (by their phylogenetic age) formations: the corpus striatum and the pyramidal cortex, which have different sources of afferentation. The corpus striatum's afferentation is, to the greatest degree, provided by the proprioceptive-vestibular system, whereas the afferentation for the cortical motor center of the C level is mainly provided by the visual-vestibular system. Thus Bernstein distinguishes two sublevels within level C: C1, consisting of the corpus striatum and its corresponding afferentation; and C2, the pyramidal cortex and its afferentation.

Summarizing the numerous movements implemented at the C level, Bernstein differentiates several types, such as locomotion (walking, running, swimming, and so on); shifting things in space and "manipulation with space" (for example, typing). Sublevel C1 implements adjustments to the external field during moving; this is its contribution to locomotion. Manipulation-with-space motions, such as the finger movements of the typist or musician, as well as weight-shifting-in-space motions are also connected with sublevel C1.

Because of the prevalence of the visual afferentation at the C2 sublevel, it is concerned with the preciseness and accuracy of the motion and its final result. Sublevel C2 implements a movement's projection on its final point in the external space. Throwing, imitating, and take-aim motions are connected with the C2 level.

In general, Bernstein concludes that motions of the C level "carry, weigh on, pull, take, bear, throw over, and so on." Directed at the external world (extroverted), these motions have their beginning and their end; they have a distinctly expressed, goal-directed character.

Thus, contrary to the B level, whose movements are cyclic, movements of the C level are aperiodic; accordingly, whereas synthetic time of the B level is rhythmic, subjective time of the C level is speed and moment. On the other hand, the C level's movements are not yet the rational manipulation of the object. They represent manipulation of the external space and its geometric forms and forces.

1.2.4. Level D

Level D is defined by Bernstein as a level of object action. We will refer to it as a gnostic-praxic level, and to level C as a sensory-motor level. This level provides the transition to an increasingly meaningful ordering of the surrounding world. Level D constitutes the first step of this new stage of cognition of the external world and is characterized by distinguishing external objects for active use. The "rational perception" of this level is manifested by identifying an object as such within the external space. Thus, the synthetic space of the D level is the object. But it is "not the object in itself as geometric shape and as something with a definite mass and consistency, but the rational aspect of acting with it" (Bernstein, 1947, p. 123). Although afferentation of the D level uses all distance modalities, it differs substantially from the sensory message about the same object as perceived at the C level. Afferentation at the D level distinguishes in an object those features which determine how one should behave with it. To define the afferentation at the D level, Bernstein introduces the concept of *topological scheme*: "Topological scheme is the totality of the qualitative peculiarities of an object, independent of its size, shape, configuration, and so on. For instance, for a cup as an object of definite use, neither height nor width, nor roundness nor squareness have any substantial meaning; what is relevant to it is its having solid sides, an unbroken bottom, and a handle—all these signs are purely topological" (1947, p. 125).

The afferent center of the D level is the inferior parietal cortical region and the secondary cortical fields. The effector center of the D Level is the premotor cortex.

According to Bernstein, object actions are elementary behavioral acts, determined by the meaning of the task. Examples of object actions (praxis) range from simple acts such as putting on and buttoning a coat, sharpening a pencil, striking a match, putting a letter in an envelope and sealing it, and writing letters of the alphabet, to more complex, multistepped actions. The latter represent successive chains of movements—"kinetic melodies" (Luria, 1966/1980)—in which the discrete movements are united by semantic motives that cannot be reduced to just moving things in space and overcoming the external forces. "Semantics" of these complex chain processes are based on the D level afferentation that determines not only what to do with the object but also in which consecutive order (Bernstein, 1947). Driving a car, working with a tool, and cooking are examples of multistepped object actions. Bernstein concluded that synthetic time of the D level is also not metrical, but topological. It is the semantic (causal) sequence. Categories *post hoc* and *propter hoc* are crystallized at the D level.

Bernstein mentions the existence of function levels that are hierarchically above the level of object actions. He notes that symbolic actions do not occur at the level of object actions, and their impairment may be connected with focal brain lesions, which, by their localization, differ from the lesions causing apraxic disorders. Although Bernstein did not consider the symbolic function level in detail, he did outline his conception of the evolution of the function levels.

Using the dichotomous model of intrahemispheric, "horizontal" differentiation, Bernstein emphasized that evolution of the function levels should be understood as evolution of simultaneous and successive synthesis. He proposed an evolutionary chain—afferent system–metric space–topologic space (object)–concept—which is related to posterior brain, and an evolutionary chain—effector system–time–subject—which is related to anterior brain (see Table 2).

1.3. Cytoarchitectural View

Contemporary progress in disciplines related to the problem of "brain and behavior" made obvious that the brain mechanisms underlying basic blocks of human behavior are acting neither at the level of molecules nor the single neuron, but rather at the level of neuron aggregates organized according to definite principles—*brain architecture*. Architecture is understood as the manner in which cells and myelinated fibers are grouped together. The original cytoarchitectonic map of the cerebral cortex delineated at the turn of the century (see Brodmann, 1908, 1909; Campbell, 1905; Vogt & Vogt, 1919; C. von Economo & Koskines, 1925) remains valid at present.

The cytoarchitecture of the cerebral cortex includes its division into units—cytoarchitectural fields. The following parameters are usually used to distinguish one cytoarchitectural field from another: the layering in depth of the cortical sheet and, within each layer, cell qualitative and quantitative characteristics; their spatial distribution and organization; and their connections (myeloarchitectonics). Since electrophysiological studies have confirmed the borders of the cytoarchitectural

Table 2. Horizontal (Intrahemispheric) Differentiation and Evolution of Function Levels

Posterior brain	Anterior brain
Afferent systems/	Effector systems/
Synthetic space:	Synthetic time:
1. Metric space	1. Movements
2. Topologic space (object)	2. Meaningful, multilinked action
3. Concept	3. Behavior
	4. Subject
Simultaneous synthesis	Successive synthesis

Note: Based on *The Construction of Movements* by N. Bernstein, 1947.

fields outlined by anatomical methods, it has been axiomatic that differences in anatomical organization of cytoarchitectural fields reflect functional differences.

A new stage in brain cytoarchitectonic studies began as more refined and precise neuroanatomical methods became available. Most of these studies are conducted on primates, combining anatomical and physiological methods. Significant progress in the precise tracing of cortical connections was achieved using antero- and retrograde degeneration techniques in animal models (Pandya & Yeterian, 1985). New techniques, such as staining cortical areas for the mitochondrial enzyme cytochrome oxidase, allowed the study of cortical architecture at a subcellular level (Livingstone & Hubel, 1988). Much of these new data can be extrapolated to the human brain, to some degree. This is especially important for myeloarchitectonics, which has been much less able to be examined in the human brain.

In general, modern experimental studies have revealed further subdivisions of the cytoarchitectural fields into specialized structural-functional units, and the number of discovered specialized areas within the traditional cytoarchitectural fields is increasing. New data have shown that both high segregation and integration are characteristic of the cortical structural-functional organization.

Both classical and recent cyto- and myeloarchitectonic studies have postulated that the cytoarchitectural approach integrates the structure and the phylogenetic development of cortical areas rather than merely "parcellating" the cortex according to cellular physiognomy (Pandya & Yeterian, 1985).

One can suppose that language is subserved by the most recent (from a phylogenetic perspective), specific to human, cortical formations, which implement so-called symbolic functions. Concerning the brain cortex, the principle of vertical hierarchy is manifested by differentiation of the cortical cytoarchitectural fields according to their structure–function organization and phylogenetic age. Traditionally, three types of fields are distinguished in the human brain: primary (or projective); secondary (associative); and tertiary (Luria, 1966/1980).

With lesions of the primary fields, sensory and motor disorders appear; as pertains to speech, dysarthrias. In the function hierarchy of the cerebral cortex, the information processing implemented by the primary fields corresponds to the sensory-motor level, level C according to Bernstein.

The secondary, associative fields are formed above the primary ones. They process the information that is received from the primary fields, implementing the particular integration of the modality-specific material. In the function hierarchy of the cortical fields, these operations belong to the gnostic-praxic level, level D according to Bernstein. Speech agnosias and apraxias appear when the corresponding zones of the left hemisphere are affected (beginning from the stage of the secondary fields, cortical functions are lateralized).

The tertiary fields, which emerged latest of all in phylogenesis, are found only in humans. In the function hierarchy of the cerebral cortex, operations fulfilled by these fields might belong to the symbolic level. Formed on top of the secondary fields, the tertiary ones are rather supramodal than extramodal. It might be supposed that the information, coming to the tertiary fields from the secondary fields in the form of generalized physical parameters of modality-specific stimuli, becomes source material for forming symbols and operating with them (thinking). The cytoarchitectural map of the cerebral cortex of the human brain (see Figure 1) clearly demonstrates that the tertiary fields occupy a substantial part of the convex surface of the cortex. Also worth mentioning here is the fact that most of this ample territory—namely, the frontal-temporal-parietal region—belongs to the so-called speech zones of the brain.

Modern anatomical and physiological data obtained in primates have shown that the hierarchical chain of areas where integration progressively occurs on the transition from one area to the next is not the only strategy of the cerebral cortex. Indeed, there is an intricate combination of both hierarchical and parallel strategies (Zeki & Shipp, 1988).

Within the visual pathway, several serial pathways running in parallel were discovered. Each of these pathways was specialized for processing separate attributes of the visual image: color, form, or motion. These myeloarchitectural data, together with the cytoarchitectural findings of specialized cell populations and specialized areas within the visual cortex, shed a new light upon structural-functional cortical organization, demonstrating how profoundly the cerebral cortex is differentiated, even within one modality. Here we deal with cortical differentiation along the horizontal dimension—intrahemispheric.

In visual and auditory cortex of various mammal species, another type of parallel connections was revealed: parallel pathways to associative cortex that bypass the primary cortex.

Classical cytoarchitectonics considered fibers between cortical areas as a sequential, unidirectional flow of connections from the primary fields to the associative areas. Such understanding of corticocortical connections easily led to the conclusion that the associative areas depend on the sensory pathways to the primary areas and, moreover, that a higher level function of the associative areas is dependent in its development on the lower function of the primary areas. This concept has traditionally existed in neuropsychology for many years although it was contradicted by the clinical data. As was found, even in early brain damage to the primary fields before the association cortex fulfilled its ontogenetic development, no significant impairment of the higher cortical functions is usually observed.

The discovery that the pattern of cortical connectivity presents with both hierarchical and parallel organization has very important implications for understanding of information processing in the cortex. If associative cortex, though receiving input from the primary fields, has also its own access to information,

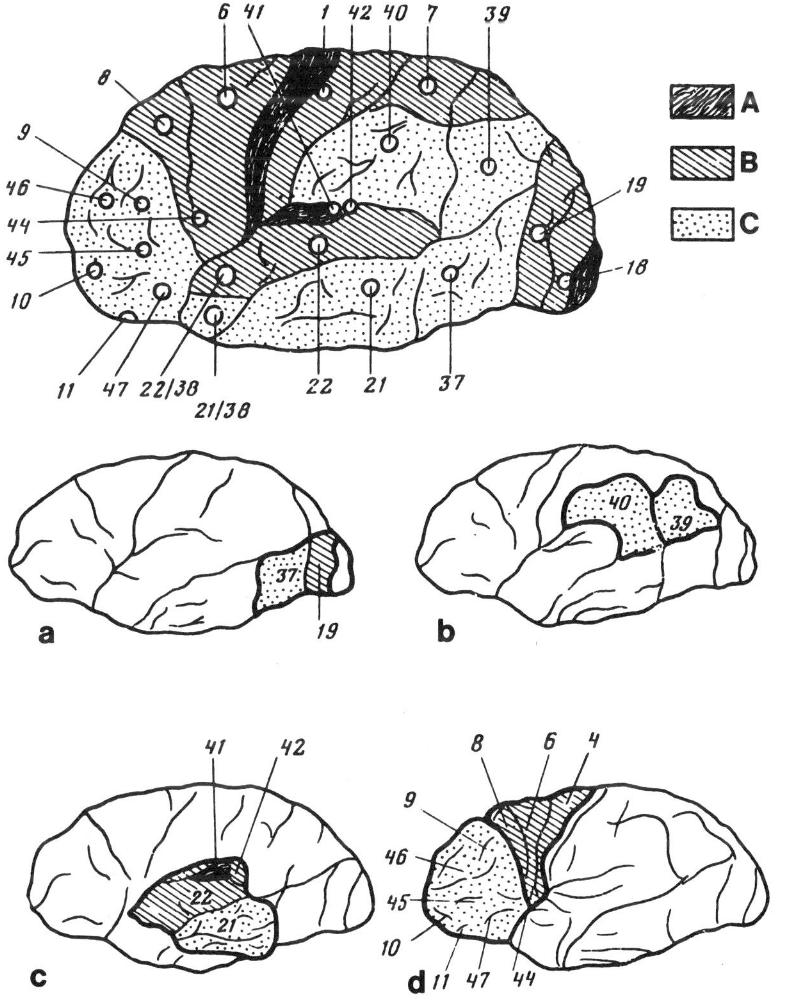

FIGURE 1. Map of the human cerebral cortex's cytoarchitectural fields. (Numbers designate Brodmann's cytoarchitectonic fields.) Top: Convex surface of the cortex: A—primary fields; B—secondary fields; C—tertiary fields. Bottom: Tertiary fields with the corresponding secondary field: a, Temporal-occipital region (tertiary field 37); b, Parietal-occipital region (tertiary fields 39, 40); c, Temporal region (tertiary field 21); d, Frontal region (tertiary fields 9, 10, 11, 45, 46, 47). From *Psychophysiological Base of Thinking Disorder in Aphasia*, by T. B. Glezerman, 1986, Moscow: Nauka. Reprinted by permission.

information may be processed simultaneously at the level of primary and at the level of associative cortex.

In recent studies, attempts have been made to identify specialized subdivisions of visual cortex (discovered initially in primates) in human brain. The first results are intriguing. Area V5, specialized for visual motion, was recognized within the secondary, associative visual cortex of the human brain (Watson, Myers, Frackowiak, Hajnal, Woods, Macciotta, Shopp, & Zeki, 1993). V5 receives its predominant visual input from the primary visual cortex, but there are other routes from the retina to V5 that bypass the primary visual cortex. They include the direct projection from the lateral geniculate nucleus and the input from a middle brain visual center, the superior colliculus. It was shown that visual signals may even reach V5 before reaching the primary visual cortex (Beckers & Homberg, 1992). Recent PET studies in patients with damage to the primary visual cortex have found that the direct subcortical input is sufficiently potent for V5 to preserve its function. It was proved by both conscious experience of visual motion (reported by the patient) and selective activity of V5 (registered in rCBF change) (Barbur, Watson, Frackowiak, & Zeki, 1993).

All these data are consistent with Bernstein's concept of hierarchical, "multistoried" functional organization of the human brain, and they give the anatomical base for the following assumptions: (1) information is processed simultaneously at the different function levels; (2) function levels are relatively independent of each other.

Modern experimental myeloarchitectural studies have found that the cortical connections of the associative areas of the cerebral cortex are organized in a systematic manner, and there is a close correlation between the connections and architecture of these regions. For example, each sector of the sensory association areas is connected with a frontal region that has basically similar architectonic features. In other words, each sector is connected with a portion of the frontal lobe that appears to occupy a similar stage of architectonic differentiation. This would imply that each sensory association sector may have developed in parallel with a specific frontal region, and with this region may constitute a functional subsystem within the cerebral cortex (Pandya & Yeterian, 1985). Independently from these studies, paleoneurologic data of brain macrostructure evolution in anthropogenesis suggest that two main focuses of intense growth—in the temporal-parietal-occipital region and in the frontal region—coincide in time with the formation of the superior longitudinal fascicle, which connects these brain areas in modern man (Kochetkova, 1973).

These data are consistent with Bernstein's concept that each functional level is a separate, relatively autonomous system represented by a "couple," posterior brain region and anterior brain region of the same phylogenetic age.

Regarding the symbolic function level, we may assume that although different tertiary fields might have been formed at different stages of ontogenesis, only after the process of their "assembly" into an integral functional system was completed did human thinking become possible. On the other hand, every tertiary field's symbolic function makes its own specific contribution to thinking and this specificity of contribution is preconditioned by the history of the given field—that is, by the modality-specific region over which it was formed during phylogenesis. Thus, every modality has its indirect representation in thought. This specificity of symbolic function of the different tertiary fields might be a premise for subserving different linguistic functions.

Indeed, the task for neurolinguistics is to understand how the linguistic concept *language* is connected with the brain's symbolic function. It is our assumption that in the triad *language–thought–brain*, language and thought are connected through the brain's symbolic function. More precisely, the roots of the language–thought connection go back into the history of the formation of the human brain and its tertiary cortical fields. Thus, data in the phylogenesis of language and thought must become an important source for neurolinguistics.

Data of modern linguistics may be another source. According to Jakobson (1970), the discovery of phonological and grammatical linguistic universals showed that there are definite general rules of thinking. According to these rules, the child masters the unknown-to-him language models. Concepts of contemporary linguistics, such as separation of language as a code system from speech, distinguishing the different language codes (phonological, semantic, syntactical, morphological), separation of phonology from phonetics, are very useful for neurolinguistics because they may reflect the "natural division" of language, its cerebral organization.

Jakobson emphasized the correspondence between the cerebral and linguistic bases in aphasia. He indicated that the pure linguistic typology of aphasia that was developed without consideration of anatomical data gave, nevertheless, a surprisingly coherent picture, very close to the topography of the cerebral lesions which underlie those language disorders (Jakobson, 1980). Regarding aphasia, the task for neurolinguistics is not only to establish the correlations between the patterns of language dissolution and the location of the lesion in the brain (these data have been accumulated for more than a century) but also to understand the connections between the type of language disorder and dysfunction of the corresponding "symbolic" area of the brain.

Trying to reconstruct language processing in the brain, we have in mind that language is a kind of two-faced Janus, represented by *language standard* and *language ability* (Leontjev, 1969). Language standard refers to the objectively existing system of any given language. It is the system of rules concerning language sounds (phonology) and language meanings (semantics, morphology, and syntax). The language standard is environmental and outside of the self. It is independent of one's brain development and pertains to the external, social, speaking community. Until speech begins, the language standard as a system of

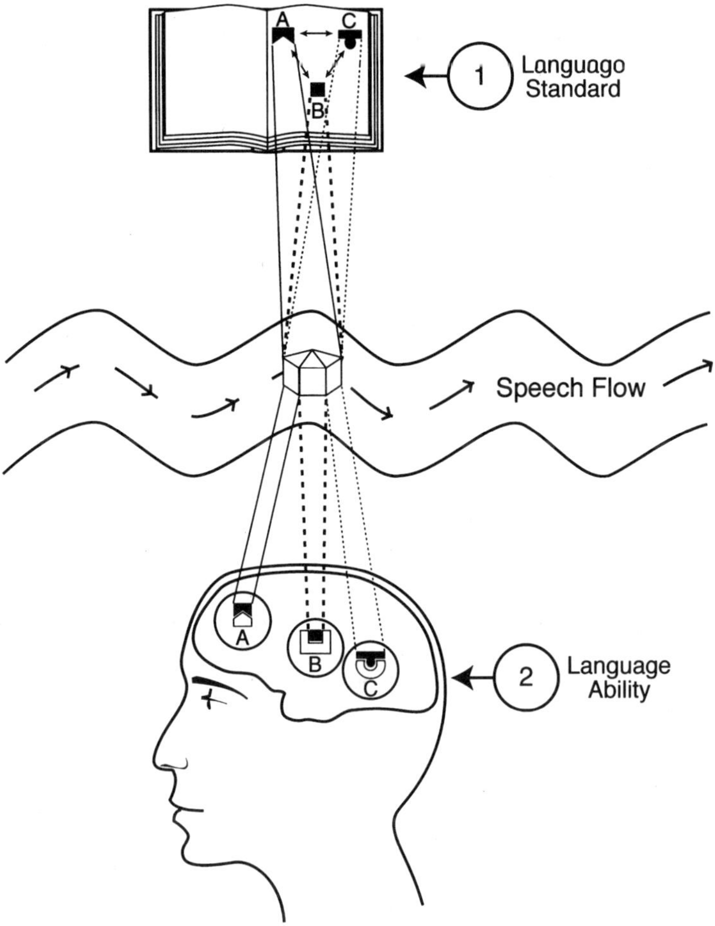

FIGURE 2. Language standard and language ability: A—syntactic code; B—phonological code; C—semantic code.

rules (codes) resides in the abstract. It is speech flow that is the tangible evidence of the language standard. With linguistic tools, one can analyze any segment of speech flow, this "river of sounds," and "extract" the rules governing it (Figure 2).

Language ability refers to the potential ability of the human brain to operate with symbols. Thanks to this ability, the child "extracts" language as a code system from the speech flow in which he is immersed at his speaking community and builds his own utterance according to the given language rules. Language ability is precisely the "junction" where the linguistic concept "language" can be correlated with the brain mechanisms of speech function.

Language ability has a complex cerebral organization. Processing any temporal segment of speech flow, different cortical speech zones have selective specificity for the language codes implicitly presented. Figure 2 illustrates both the double nature of language and the selective specificity of cortical speech zones using the metaphor of receptor specificity, in which the various "shapes" (language codes hidden in speech flow) will find matching puzzle pieces in cortical "receptor sites." As the speech flow bathes the developing child, it is due to the selective specificity of the speech cortical areas that the child "reinvents" the language within himself and begins to express back to his language speaking community. It is of interest to note that the child builds his own utterances according to the language rules but he does not "understand" these rules until he studies grammar (language standard) at a much later time. Thus, we may suppose that, in human speech activity, the language standard is realized through the "mastery" of the cerebral cortical zones subserving various symbolic functions.

1.4. Functional Asymmetry of the Brain, Intrahemispheric Specialization, and Function Level

Functional brain asymmetry, particularly the left hemispheric specialization for language, represents a final stage of brain evolution in anthropogenesis. At present, the published research on the functional asymmetry of animals does not contradict the concept of the human brain's functional asymmetry uniqueness, because the phenomena implied are qualitatively different.

Bianchi (1967) traced the history of functional asymmetry far into the depths of vertebrate phylogenesis. He showed that the morphological premise for functional asymmetry is the presence of double structures in the central nervous system. The morphological evolution from symmetric centers up to the double structures connected by commissures corresponds to the development of functional asymmetry. On the basis of comparative physiological study, Bianchi suggests that development of the double structured formations in the brain was connected with the refinement of visual-spatial discrimination. Bianchi conducted a series of experiments with animals of different species (including mammals) and found that, under the effect of a lateralized stimulus—a light source, for example—the potentials aroused in symmetric sectors of double brain regions are asymmetric. The author explains this functional asymmetry by the principle of physiological dominance, in which he postulates that the focus of activity in one hemisphere, which receives the input earlier, suppresses the symmetric centers of the opposite hemisphere and is simultaneously reinforced by impulses returning along the commissural paths from these centers. Depending on the localization of the stimulus in the external spatial field, one or the other hemisphere may dominate. Moreover, the localization of the stimulus determines the degree of the asymmetry as well. The less lateralized the stimulus, the less hemispheric asymmetry is manifested.

From our perspective, at this stage of the brain's evolution, functional *specialization* of the hemispheres is absent—either hemisphere may be dominant with regard to the same function. One hemisphere dominates only through the agency of an external stimulus, and it is the asymmetrical localization of the stimulus that is the immediate cause of the hemispheric functional asymmetry. Nevertheless, the very possibility of functional asymmetry appearing under the action of an external stimulus depends on internal factors: the duplication of the brain's centers, the contralateral connections with the periphery, and the mechanisms of physiological dominance.

Bianchi also demonstrated that the evolution of functional asymmetry is closely tied to the evolution of the sensory and motor centers. At every stage of the brain's evolution in vertebrates, functional asymmetry was most expressed where the basic analyzer centers were concentrated. The general process in evolution is the telencephalization and corticalization of functions; that is, moving the analyzer centers into neocortex. The latter process finds its highest expression in mammals, which have evolved cerebral hemispheres with extensive connections.

The concepts of intrahemispheric specialization and function level introduced earlier are both tied to the progressive evolutionary movement of the motor and sensory analyzer centers into the cerebral cortex. Thus, functional asymmetry of brain hemispheres is connected with both intrahemispheric specialization and leading functional level at the given phylogenetic stage. Applying Bernstein's theory we can say that the stage at which full transition of the analyzer centers into the neocortex occurs belongs to level C. There is a correspondence at this level between the "extraversion" of level C's spatial field and the dependence of functional asymmetry on external stimuli as described by Bianchi. "Adjusting" to spatial localization of stimuli, the hemisphere's dominance plays a role in providing absolute physical parameters of external objects: their location and extent in space, their direction of movements, distances between them, and so on. Thus, at this stage of evolution, functional asymmetry is in full harmony with the leading functional level, level C, the most objective level, the level of "metricity and geometricity."

The cardinal data about functional asymmetry of the brain in man were obtained from the studies done in brain-damaged and split-brain subjects. Later, brain lateralization was studied extensively in normal subjects using dichotic tests in which information was presented to one of the visual half fields or to one of the ears.

Before the studies in the split-brain patients and normal subjects were done, there was a finding that is of interest to us. In the early 1950s, Mishkin and Forgays demonstrated an advantage for the right visual field for recognition of English words and a slight advantage in favor of the left visual field when Yiddish words were presented to subjects who could read Yiddish (cited from Springer & Deutsch, 1989). The advantage for the right visual field with English words was, however, considerably greater than that for the left with Yiddish words. The authors' interpretation of the results was that acquired directional reading habits caused better processing of written English in the left hemisphere, while the right-to-left reading of Yiddish is processed more accurately in the right hemisphere. The unequal size of the visual field differences, however, was left unexplained (Springer & Deutsch, 1989). After the studies in split-brain subjects showed a dramatic advantage for the right visual field for reading words, it was suggested that Mishkin and Forgays' findings may have been due to two factors operating simultaneously: (1) left hemisphere dominance for language in general, and (2) acquired reading habits in a particular language, which is much less significant. The hypotheses were tested, with vertical presentation of both English and Yiddish words. In this study, where the second factor, directional scanning, was eliminated, a right visual field advantage was found for both English and Yiddish words (Springer & Deutsch, 1989).

We can interpret the second factor as a contribution of level C to the reading process. Indeed, directional scanning represents goal-directed movements based on spatial position of the external stimuli. Even if directional scanning is a higher function than discrimination of lateralized stimuli, it still belongs to the C level, resulting in a mental image of absolute physical parameters of letters, without their symbolic meaning. Thus, level C, with its particular type of functional asymmetry, does contribute to language processing in man.

The higher the functional level, the closer the connection between functional asymmetry and the peculiarities of intrahemispheric specialization of that level. At the gnostic-praxic level, level D according to Bernstein, the disparity of left and right hands appears: object actions are carried out mainly by the right hand (left hemisphere), and thus afferentation of object actions is provided by the left hemisphere, too.

At level D, the functional specialization of the hemispheres has developed. Although the rudiments of hemispheric specialization are to be found in animals, the fully developed form is displayed only in man (Bianchi, 1979). According to the literature (Springer & Deutsch, 1989; Ivanov, 1978), there are two commonly held viewpoints regarding interhemispheric differences in man: (1) the left hemisphere specializes in one set of functions (speech, praxis) and the right on the others (visual, visual-spatial perception), and (2) the hemispheres differ in their modes of information processing. The second point of view is currently recognized by most scholars. The principle of the "left brain" (the left hemispheric mode of information processing) consists of the breaking down of processed information (analysis) and successive sifting of the resulting variants, leading to synthesis.

It is supposed that the representations of the left hemisphere don't exist in a whole, integral form but rather are put together as a certain combination of discrete units (features), and as various combinations of these units tied together as members of a continual logical, series (categorical recognition). The right hemisphere, however, appears to process information in a holistic manner. Here, the emerging image of the whole is not dissected and parcelled: the integral image is represented in its nonrepeatable uniqueness (individualized recognition) (Bradshaw & Nettleton, 1981; Nebes, 1978; Bogen, 1969). Ivanov (1978) illustrates the possible differences in the two hemisphere's strategies by the picture in which each whole image in the right hemisphere is represented as a sequence of discrete features, signs, and qualities in the left hemisphere (Figure 3). In general, the right hemisphere operates with discrete combinations of whole continuous images, whereas the left hemisphere functions with continual combinations of discrete signs (Ivanov, 1978). Modern anatomical data regarding cortical connections have tended to support the concept "two hemispheres—two cognitive styles." For example, in the right hemisphere the dendritic overlap among cortical columns is greater than in the left hemisphere, allowing for the possibility of more joint (synchronous) responses, which may correspond to a more "holistic processing" style (Seldon, 1982). The much greater center–center distance between columns in the left hemisphere is consistent with a better segregation of input and more independence in responses on the left (Seldon, 1985). Because of their greater

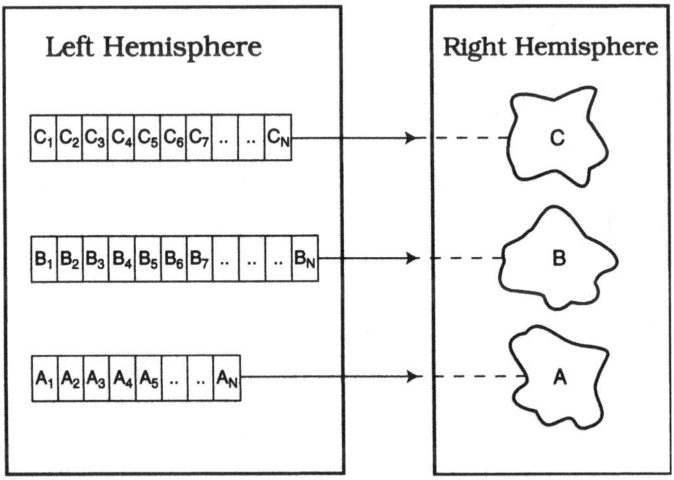

FIGURE 3. Interhemispheric specialization: analytic versus holistic information processing. Modification of Figure 26 from *Asymmetry of the Brain and Semiotic Systems* by V. Ivanov, 1978, Moscow: Sovetskkoje Radio. Reprinted by permssion.

tangential extent, dendrites of the left hemisphere columns interact with more afferents than their right hemisphere counterparts, suggesting that the columns of the left hemisphere should be more responsive to converging afferent signals and more capable of establishing the early categorical response.

Authors argued that the principle of information processing characteristic of each hemisphere determines its dominance with regard to the various functions (Levy, 1968). For example, the orientation of the left hemisphere to analytical, logical tasks determines its connection with "natural language." The difference between hemispheres is not what kind of stimuli they process but rather how they deal with stimuli (Springer & Deutsch, 1989). There is evidence that human-specific, "logical," left hemisphere information processing emerged earlier in evolution than articulated speech, supporting the notion of this cognitive functional specialization of the hemispheres. Further, this specific left-hemispheric strategy of information processing is apparent not only for language but also for nonverbal functions executed by the left hemisphere (Kock, 1967).

The term *function* is usually related to intrahemispheric specialization. However, functions connected with the certain secondary or tertiary cortical areas are, at the same time, "interpreted" through the strategy of the hemisphere involved. It results in the difference of the left and right hemisphere functions, connected with the same brain area within the hemisphere. For example, whereas the superior-posterior region of the left temporal lobe is responsible for so-called phonematic hearing, the symmetric zone in the right hemisphere subserves perception of complex symbolic nonverbal sounds such as applause, laughter, a child weeping, a dog barking, the sound of a passing train, and so on.

In man, both a new function level (level of object action) and a new cognitive style ("left brain") develops. Cognition that suggests an active attitude to the surrounding world and whose very first stage is distinguishing objects for use is inseparably connected with analysis, as in the left-hemispheric mode of information processing. Indeed, if we return to Bernstein's definition of topological scheme as a combination of the certain discrete features of an object, we will recognize "the left-hemispheric image." Thus, at the new function level (level D), "left-brain analysis" is intimately interwoven into processing of modal specific information (intrahemispheric specialization) and contributes substantially to the afferentation of praxis.

In general, at the gnostic-praxic function level, the difference in information processing by the hemispheres is displayed through their specialization: the left hemisphere provides for the "active" function (object praxis in general, speech praxis in particular), whereas the right hemisphere is concerned with the "contemplative" functions (visual, visuospatial perception).

We suggest that, at the symbolic function level, the difference in the hemisphere's cognitive mechanism presents as two opposite modes of thought. The duality of the human mind, discussed from ancient times, was attributed to the left-

TABLE 3. Thought Dichotomies

Price	Analytic, reductionist, simple, provable	Synthetic, concrete, complex, disorderly
Ruesch	Digital codification (discursive, verbal, logical)	Analogic codification (nondiscursive, nonverbal, eidetic)
Young	Abstract	Maplike
Spearman	Abstract reasoning—"eduction of relations"	Analogic reasoning—"eduction of correlates"
Hobbs	Directed	Free or unordered
Bruner	Rational	Metaphoric
Levi-Strauss	Positive	Mythic
James	Differential	Existential

Note: Adopted from *The Other Side of the Brain* by J. E. Bogen, 1969.

right hemisphere opposition since the experiments with split-brain patients. Table 3 lists some of these oppositions, collected by Bogen (1969) from various sources, which may be attributed to "left" versus "right" thought (Table 3).

The important component of interhemispheric specialization is the interaction between the hemispheres. It is manifested in two aspects: the activity of the hemispheres is complementary; the functioning of one hemisphere produces a suppressive effect on the activity of the other, so that they "act" in turn (reciprocal interaction) (Springer & Deutsch, 1989).

In evolution, reciprocal interaction tends to appear earlier than complementary interaction. It is, as a matter of fact, inseparable from functional asymmetry and its physiological mechanism in animals. It is useful to remember that, according to the principle of physiological dominance, the focus of increased activity in one hemisphere suppresses the symmetrical centers of the other hemisphere and simultaneously is strengthened through the commissural paths from the symmetric centers. Data showing the leveling of functional asymmetry after sectioning of the corpus callosum in animals gives one more piece of evidence that functional asymmetry at this stage of its evolution is sustained mainly through the reciprocal mechanism of hemispheric interaction.

The complementary interaction of the hemispheres in its full meaning, that is, cooperation of two qualitatively different but mutually supplementary cognitive mechanisms, may arise only with the appearance of the new, left hemispheric way of information processing (levels D and E, according to Bernstein). Hemispheric complementarity suggests the parallel development of both cognitive mechanisms in the process of phylogenesis, the "left" as well as the "right." The mutual supplementation of these two diametrically opposite mechanisms can be implemented only if the hemispheres are working in alternating order (the reciprocal interaction).

1.5. Summary

In summary, three basic principles in cerebral cortex differentiation reflect the main directions of human brain evolution:

1. Differentiation of the cerebral cortex by the level of structure-function organization ("vertical")
2. Differentiation within the brain hemisphere ("horizontal")
3. Differentiation between the hemispheres ("functional asymmetry")

Horizontal differentiation is based on the difference in modality and phylogenetically came from the movement of the analyzer's centers into the new cortex. Vertical differentiation arose from progressive evolution of the sensory and motor regions, which led to the higher-level function. Differentiation between the hemispheres came from the evolution of the double structures in the brain, and it represents the opposition between "left" and "right" cognitive mechanisms.

During historical development of the human brain, the aforementioned three fundamental factors underwent a complex interrelated and interdetermined evolutionary process. Every stage of the brain's evolution is characterized by its own peculiar combination of these factors and represents a complicated knot in which they are closely interwoven. Every stage of brain evolution represents that congruence in the brain's vertical, horizontal, and interhemispheric differentiation which is necessary to attain the highest (for the evolutionary stage) functional level—the leading functional level, according to Bernstein.

These three types of human brain specialization, their historically determined interconnection and their interdependence are fundamental for neurolinguistics. This applies both to language ability in the norm and to aphasia, as well as to the development of new rehabilitative methods in speech-language disorders.

Usually, the contribution of each of the presented factors of brain differentiation to localization of mental processes is considered separately from the others. Analyzing cerebral organization of language ability in man, we will connect each discrete component of language (language code) with the functional properties of the corresponding cortical area. In turn, functional specialization of the cortical areas will be considered in the frame of three dimensional cortical differentiation: vertical—symbolic versus gnostic-praxic functional level; horizontal—anterior versus posterior cortex with the further subspecialization within each of them; interhemispheric—left versus right cognitive mechanisms.

II

Cerebral Organization of Language and Thought

2

Temporal-Occipital Region: Visual Object Perception, Thought and Word

2.1. Delineation of Anatomical Region

2.1.1. Cytoarchitectural Data and the Implication for Functional Properties of the Temporal-Occipital Region

The temporal-occipital region, or inferior-posterior part of the temporal lobe, corresponds cytoarchitecturally to field 37 (Brodmann's classification). A detailed and comprehensive cytoarchitectural study of field 37, including comparative cytoarchitectonics and phylogenesis, was conducted by Blinkov at the Moscow Brain Institute (Blinkov, 1938, 1955). Field 37 was described as a phylogenetically young, specific for human, tertiary field that was formed by means of differentiation of the transitional structures located in primates between auditory and visual cortex. "In the process of human brain development," Blinkov stated, "the cortical zones, in which auditory, visual, and proprioceptive-tactile pathways are terminated, move apart and the area of the regions located between them enlarges" (Blinkov, 1955). Based on his studies of the phylogenesis and ontogenesis of the human brain, Blinkov determined that a certain part of field 37 was evidently formed from secondary field 19 of the occipital region, having driven the latter to the rear. He found that the convolutions and fissures that in monkeys correspond to field 19 are partially occupied by field 37 in the human brain. He also concluded that, at the early stages of the ontogenetic development of the human brain, the occipital cortex took part in the formation of field 37. The architecture of field 37 has revealed that certain features that are characteristic of temporal (auditory) and occipital (visual) areas are combined. In addition, field 37 also has its own cytoarchitectural peculiarities, and these features are unique to the human brain. Blinkov divided field 37 into six subfields according to their cytoarchitectural properties (Figure 4), and he indicated that it is not so much by size as by differentiation into subfields that the temporal-occipital region in man differs from the comparable region in monkeys. The proportion of phylogenetically new cytoarchitecture unique to field 37 is not equal in each of the six subfields, and is greatest in the central subfield. The five peripheral subfields, however, include transitional features in addition to those unique to field 37. Of all six subfields, only the central one is asymmetrical: its size is larger in the left hemisphere than in the right.

Based on the cytoarchitectural studies of field 37, we offer the following speculations about a corresponding functional organization of this area:

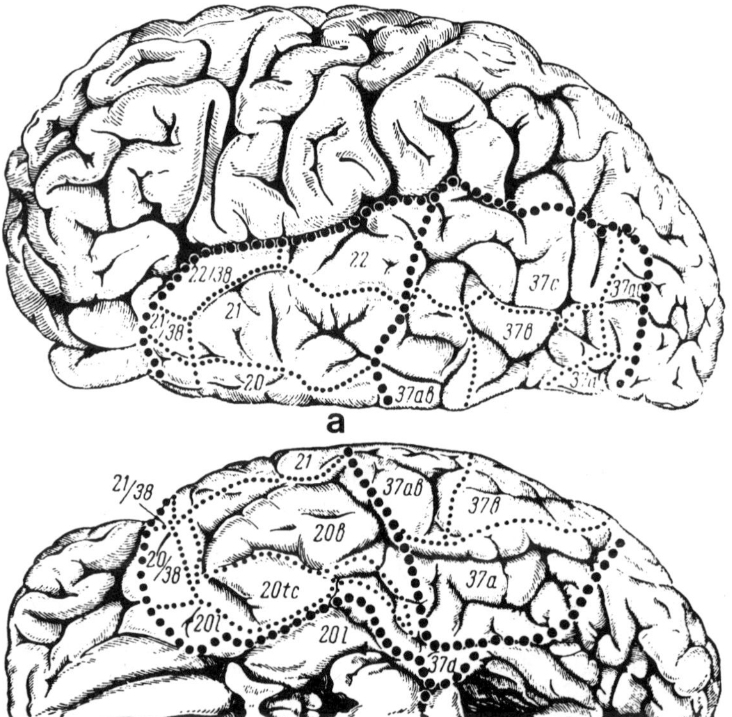

FIGURE 4. Cytoarchitectural field's border according to Brodmann with addition on lateral (a) and basal (b) cortical surface of the temporal lobe by S. Blinkov, 1955. Fields are indicated by numbers, subfields by letters. Bold dots—field borders; regular dots—subfield borders. From *Visual Agnosias* by E. Kock, 1967. Reprinted by permission.

1. The differentiation of field 37 into six subfields may indicate a functional specialization within the field (differentiation along the horizontal dimension).
2. The different phylogenetic age and the different level of structural organization of the separate subfields may indicate that field 37 is also nonhomogeneous with regard to the level of the functions provided by it (vertical differentiation).
3. The distinctions between the central subfield and the five peripheral subfields may suggest that different functional levels exist within field 37 for implementing operations. It seems likely that the highest level of function, the symbolic one, is connected with the phylogenetically youngest central subfield, a kind of "nucleus" of field 37.
4. Field 37 is considered in the literature as a tertiary cytoarchitectural field implementing supramodal information processing. The cytoarchitectural investigation conducted by Blinkov (1955) has shown that this field contains both new and phylogenetically older formations. The latter are characteristic of secondary, not tertiary, fields. Thus, in general, field 37 may be considered a complex one: on the one hand, it is the integrative visual-auditory field implementing cross-modality processing, which pertains not to the symbolic level but to the gnostic-praxic level (or a transitional level between them); on the other hand, it provides a symbolic level (supramodal) that cannot be reduced to a modality-specific level, even though it is integrative.

2.1.2. Myeloarchitectural Data and the Implication for Functional Properties of the Temporal-Occipital Region

Modern cytoarchitectural studies conducted on mammals have provided more detailed data about myeloarchitectonics of the temporal-occipital region. Although there is no complete functional equivalent to field 37 in mammals, including primates, the inferotemporal region considered as a continuation of the visual system in primates (Mishkin, 1972) may be homologous to field 37 in man.

Studies have confirmed the classical concept that there is a sequential flow of connections within the visually related areas of the occipital and inferotemporal cortices. The primary visual area (cytoarchitectural field 17, the "striate" visual cortex) receiving its visual input from the subcortical visual center, the lateral geniculate body, projects to the prestriate visual area (cytoarchitectural field 18). Field 18 is connected with field 19, which in turn projects to the inferotemporal cortex (see review by Pandya & Yeterian, 1985). The visual association area in mammals, thus, includes field 18 and 19, which surround the striate cortex, and it extends into the inferior temporal region.

Besides these sequential pathways from the primary to the association visual cortex, a number of parallel pathways have more recently been found. The lateral geniculate body sends projections in parallel to both striate and extrastriate (association) visual cortex. Also, in a wide variety of mammals including primates, the visual functions of the temporal lobe may depend, at least partially, on a parallel tectopulvinar pathway that bypasses both the lateral geniculate body and the striate cortex (Pandya & Yeterian, 1985).

Studies of intercortical connections have also showed that information from the visual system to the frontal lobe is distributed in a stepwise manner. The visual association area that corresponds to secondary fields 18 and 19 projects to the premotor region, secondary field 8. The inferior temporal area, however, projects to prefrontal and orbitofrontal cortex fields 46 and 11, which are tertiary fields. In other words, each stage of visual association areas is related to a frontal region which appears to occupy a similar level of architectonic organization. The same is characteristic of the connectivity pattern between the visual system and the temporal (auditory) region. Visual association areas project to those auditory association regions which appear to have analogous architectonic features and appear to occupy a comparable stage within their own architectonic lines (Pandya & Yeterian, 1985).

What can we infer from the presented anatomical data regarding functional organization of the temporal-occipital region?

1. The sequential connectivity pattern within the visual system suggests that there is a hierarchy of levels in the visually related areas of the occipital and inferotemporal cortices with a sequential flow of information from the lower to the higher levels.
2. The parallel pathways to the association cortex allow the simultaneous processing of information at the different levels of the visual system.
3. Connections with the distinct areas of the auditory-temporal region might reflect cross-modality properties of the inferotemporal region.
4. Connections between corresponding (by stage of architectonic organization) regions of the visual system and the frontal lobe might indicate that each successive level represents a functional system, which is consistent with Bernstein's theory of functional levels.

2.1.3. New Data about Horizontal Differentiation within the Visual Cortex at the Lower Functional Levels and the Implications for Functional Properties of the Temporal-Occipital Region

Within the primary visual cortex—area V1 in primates, corresponding to field 17 in human—specialized cells selectively responding to different attributes of the visual scene (such as color, shape, and motion) have been found (see Zeki & Shipp, 1988; Zeki, 1993; Zeki & Lamb, 1994; Livingstone & Hubell, 1988). There is also a characteristic architecture in the distribution of the specialized cells in area V1. Cells responding to wavelength (color selective) and cells responding to lines of a particular orientation (form selective) are located in two separate sets of cells in layers 2 and 3 of V1. Cells responding to both line orientation and direction (form and movement selective) are found in layer 4B of V1. These functionally distinct groups of cells are segregated by the pathways to which they belong. Color and form selective cells in layers 2 and 3 receive their input from the Parvo system, the parvocellular subdivision of the lateral geniculate body; form and movement selective cells in layer 4B receive their input from the Magno system, the magnocellular subdivision of the lateral geniculate body. Cells of the parvocellular layers of the lateral geniculate body are sensitive to difference in wavelength, whereas cells of the magnocellular layers are sensitive to temporal aspects of visual stimuli. Parvo and Magno pathways running in parallel within the visual pathway remain segregated through the whole visual system. From layers 2 and 3 of V1, the Parvo system projects to prestriate visual association cortex, namely, to area V4. The Magno pathway from layer 4B of V1 projects to another area of prestriate cortex, V5 (MT). These two areas of visual association cortex are functionally specialized: V4 for color and static form recognition, V5 (MT) for movement and dynamic form detection. Livingstone and Hubel (1988) indicated that the segregation of functions beginning in the visual system at the earliest levels became more and more pronounced at each successive level, changing from a specialized grouping of cells to specialized areas. These authors pointed out that the different subdivisions of the extrastriate visual cortex in the primate be related to different characteristics of the cells of the striate area.

Segregation of functions with continuation of Parvo and Magno systems extends beyond the extrastriate occipital cortex. V4 projects to the temporal-occipital region, and V5 projects to the parietal-occipital region. Livingstone and Hubel (1988) supposed that temporal visual areas may represent the continuation of the Parvo system, and the parietal areas the continuation of the Magno pathway.

These new anatomophysiological data allow us to draw the following conclusions and to make the following extrapolations regarding the human temporal-occipital region:

1. An intimate interconnection exists between vertical and horizontal differentiation in the visual cortical system. The higher the functional level, the more horizontal differentiation (functional segregation) is expressed in the visual cortex. Although at the level of the primary visual cortex there is a functional specialization for different visual attributes (color, form, movement, depth, and so on), these visual attributes are perceived in the context of the whole visual scene. At the higher levels, they are perceived, however, in the context of *form* (area V4 specialized for static form and area V5 specialized for dynamic form).

2. The transition from specialized cells within area V1 to specialized areas in the secondary visual cortex corresponds to the re-organization of synthetic space in Bernstein's system—the transition from the C level (external spatial field) to the D level (object).
3. The divergence of *static form* (Parvo system) and *dynamic form* (Magno system) with their corresponding projections to the temporal-occipital and parietal-occipital regions suggests that form—static versus dynamic—is used for different purposes at the highest levels. Indeed, it has been suggested that the temporal-occipital region, receiving its input from the Parvo system (which is phylogenetically younger and well developed only in primates), might be necessary for identification of objects by their appearance (Mishkin, Ungerleider, & Macko, 1983; Livingstone & Hubel, 1988).

2.2. Neurobehavioral Correlates: Visual Object Gnosis Connected with the Temporal-Occipital Region

Having studied visual agnosias in brain-damaged patients, Kock (1967) concluded that the zone involved in visual gnosis disorder far exceeds the limits of the visual occipital cortex proper (fields 18 and 19). In addition, data obtained by Kock suggest the possibility of selective visual gnosis disturbances. Disorders in object recognition (visual object agnosia), in letter recognition (visual letter agnosia), and in face recognition (visual facial agnosia) were found as independent, separate phenomena. Deficits in color recognition were often combined with either letter agnosia or facial agnosia, but not with object recognition deficits.

The selective visual gnosis defects revealed by Kock were observed in patients with lesions in different locations in the posterior cortex. All kinds of visual agnosia, except object recognition deficit, were found accompanying lesions of the occipital visual cortex proper. Letter, number, and color agnosias occurred with damage to the occipital region of the left hemisphere. Facial agnosia resulted when the lesion involved the symmetrical zone in the right hemisphere. According to Kock (1967), visual object agnosia occurs with lesions not in the occipital but the temporal region; specifically, the inferior-posterior section of the convex temporal cortex, which corresponds to cytoarchitectural field 37.

Subsequent studies have also found that the ability to identify objects can be altered with temporal damage (Newcombe & Russel, 1969). In monkeys, lesions in the inferior temporal area appear to result in deficits in object discrimination (Ungerleider & Mishkin, 1982). As mentioned in section 2.1, new anatomophysiological data showing that the temporal-occipital region may be a continuation of the phylogenetically younger Parvo visual pathway, well developed only in primates, suggest that this parvotemporal lobe system is especially suited for the detailed analysis of multiple visual attributes needed for identification of objects (Livingstone & Hubel, 1988).

It was accepted in the literature that visual object agnosia is found more often and presents with more severe forms when the right hemisphere is damaged rather than corresponding sections of the left (DeRenzi & Spinnler, 1966). Kock was able to demonstrate by detailed neuropsychological analysis that, in actuality, there are qualitative differences in the syndrome of visual object agnosia in patients with lesions in symmetrical areas of the different hemispheres. Kock found that damage to the right hemisphere results in a disorder of direct recognition of the whole. When presented with object pictures, patients with lesions in the right inferior-posterior temporal region frequently misrecognize the objects. The errors (paragnosias), according to Kock, are based on fragmentation of visual perception in these patients, and their tendency to use the perceived separate pieces or fragments to construct a whole image. This leads at times to bizarre responses: for example, a patient presented with a picture of a comb might state that it is a bench, responding to the salient features of parallel lines. Kock noted that visual gnosis disorder is selective for objects in these patients: it does not involve face recognition or understanding of complex pictures of whole scenes (Kock, 1967).

Damage to the left inferior-posterior temporal region, Kock found, is manifested by a disorder of object naming (object anomic aphasia) and by a delay in recognition response. Patients with lesions in this area were able to recognize colored, realistic pictures of objects, but they had difficulties in those tasks requiring distinguishing a figure from the background or recognizing images under difficult conditions (visual mask picture tests, overlapping figure tests, and so on). Thus, although on the surface the manifestations of unilateral lesions of the right or left inferior-posterior temporal area appear to be quantitatively different, Kock's research has shown that they are qualitatively different neuropsychological syndromes (Kock, 1967).

More recent neuropsychological and neurophysiological studies further elucidate the qualitative differences of these two syndromes (Delis, Robertson, & Efron, 1986). In these studies, patients with unilateral brain damage were presented stimuli with larger forms (global) containing smaller forms (local) (Figure 5). It was found that lesions of the right temporal area disrupted the organization of a global form, whereas lesions of the left temporal area were more likely to affect the ability to respond to local forms. These findings are consistent with the current understanding of basic differences in "left" and "right" types of information processing. As presented in the introduction, it is felt that the right hemisphere operates according to a *gestalthaft* holistic processing mode and implements individualized recognition, whereas the left hemisphere operates in an analytic mode, implementing categorical recognition.

In Kock's work, when patients with left temporal lesions were presented with object pictures, they not only had delayed recognition and difficulties naming

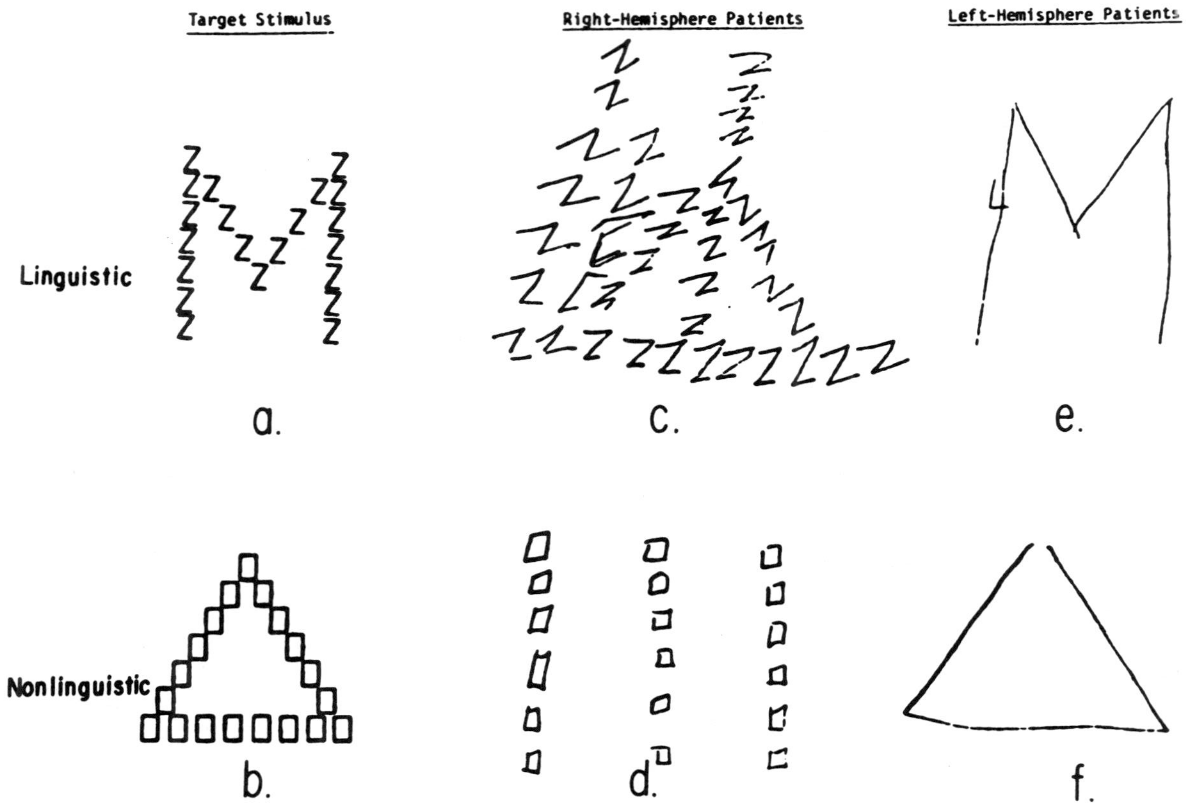

FIGURE 5. Recalled drawings of patients with unilateral brain lesions: (a) linguistic symbols; (b) nonlinguistic symbols; (c) and (d) drawings of RHD patients illustrating correct recall of the smaller forms but not the larger forms; (e) and (f) drawings of LHD patients illustrating correct recall of the larger forms but not the smaller forms. From "Hemispheric Specialization of Memory," by D. C. Delis, L. C. Robertson, and R. Efron, 1986, *Neuropsychologia*, *24*, pp. 205–214. Reprinted by permission.

objects but also failed to recognize objects as members of definite categories. Kock (1967) distinguished a "temporal-according-to-dominant-type syndrome," a discrete syndrome complex that results from damage to the inferior-posterior temporal area of the left, dominant, hemisphere. It includes: 1) visual object agnosia expressed by delay in visual recognition and failure to distinguish the figure from the background; 2) object anomic aphasia; and 3) disorder of categorical recognition selective for objects and forms.

More recent studies have confirmed the existence of selective visual-gnostic disorders (Hecaen & Albert, 1978; Warrington & Shallice, 1984; Goldenberg, 1992; Feinberg, Schindler, Ochoa, Kwan, & Farah, 1994). We dwell on Kock's work in more detail because she was the only author to notice some connection ("triple association") among visual object gnostic disorder, partial categorical thinking disorder, and linguistic naming disorder. Anomic aphasia per se (whose localization in field 37 has been recognized for more than a century) is usually considered as a separate entity, a selective naming disorder in an otherwise intact individual (Benson & Geschwind, 1971). However, Pick (1931/1973), Head (1926/1963), and Goldstein (1948) suggested that anomic aphasia is a particular manifestation of a more general impairment of categorical thinking. Goldstein took the most extreme position. According to Goldstein, the difficulty in finding words as names for objects is the consequence of a change in the total personality in brain-damaged patients: specifically, an impairment of so-called categorical or abstract attitude, in which patients lose the ability to understand the "categorical aspect" of things (Goldstein, 1948).

Describing "temporal-according-to-dominant-type syndrome," Kock emphasized that the disorder of "abstract attitude" in this syndrome was partial and selective. It did not concern letters, colors, and spatial relationships. Moreover, Kock described another selective disorder within the visual modality—impairment of categorical recognition of letters and colors in patients with lesions in the left occipital region. Kock indicates that, besides occurring with temporal-occipital lesions, a disorder of "abstract attitude" is observed only when the frontal lobe of the left hemisphere is damaged. However, in "frontal lobe syndrome," disorder of "abstract attitude" is global, not selective. Kock concluded that impairments in "temporal-according-to-dominant-type syndrome" involved object gnosis, object thinking, and object naming. Kock believed that the left inferior-posterior temporal region represents the separate anatomical system that implements the particular type of visual gnosis, and that it is the selective and specific gnostic disorder which is the cause for anomic aphasia. Thus, whereas Goldstein inferred anomic aphasia results from the loss of abstract attitude, Kock took another extreme position, assuming that anomic aphasia (language disorder) is secondary to visual gnostic disorder.

Based on these considerations, the following questions are raised:

1. If various kinds of visual gnosis are related specifically to the occipital region, why is visual object gnosis related to the integrative region, placed between auditory (temporal) and visual (occipital) cortex?
2. Why is visual object gnosis related to a functionally higher level (tertiary field 37) than other kinds of visual gnosis, connected with secondary occipital fields?
3. What is *categorical recognition* and how is it related to language?

In answering these questions, we will consider the hierarchical structure of visual object discrimination based on function levels and the left hemispheric mode of information processing.

2.3. Left Temporal-Occipital Region: Multilevel Visual Object Processing, Categorical Classification, and Logical Grammatical Language Code

> "Objects and concepts are man-made, or at least nervous system–made." [Whitefield, 1985]

As we mentioned earlier, the heterogeneity of the inferior-posterior temporal region (cytoarchitectural field 37) and its complex hierarchical organization including the separate subfields of different phylogenetic age and origin give an indication that this region might implement functions at both the gnostic-praxic and symbolic levels. Clinical data regarding dysfunctions of this region, ranging from visual object agnosia to anomic aphasia and selective, "modality-specific" disorder of categorical thinking, give further support to this supposition.

In 1986, one of us offered the hypothesis that in the inferior-posterior temporal region of the left hemisphere, visual images are represented as certain combinations of discrete visual features and that this "left-hemispheric" analysis and subsequent synthesis proceed simultaneously at (at least) three functional levels (Glezerman, 1986). The first level roughly corresponds to a transitional level between the C and D levels in Bernstein's system; the second level corresponds to the gnostic-praxic level (or the D level, according to Bernstein); and the third level corresponds to the symbolic, or the E level.

According to our hypothesis, the same visual image is multiplied in the inferior-posterior temporal region, being represented at three functional levels. However, multiple representations are not identical, because reorganization of subjective space at each successive level leads to a reconstruction of the discrete signs. As a result, the image has acquired different meanings, appearing as a thing with visual parameters of size, form, color, and so on; as an object of use; and as a concept. "Product of output," or what we perceive, is complex, incorporating in itself all levels of conceptualization; but, depending on the task, we become most conscious of the leading functional level. We can draw a parallel between the new anatomophysiological data regarding visual cortex in primates that were introduced in chapter 1 and in previous sections of this chapter and our hypothesis of multilevel visual object recognition connected with the left inferior-posterior temporal region. Discovery of parallel pathways to extrastriate cortex that bypass striate, primary cortex, is in concordance with an assumption that information is processed at different function levels in parallel (simultaneously). Along the hierarchically organized pathways in visual cortex, at each stage of the hierarchical chain, cells have larger receptive fields than their counterparts at the previous stage, which might be the basis for a higher level of conceptualization at each successive functional level (Zeki & Shipp, 1988; Zeki & Lamb, 1994). Backward connections from associative to primary visual cortex are diffuse and nonspecific (Zeki & Shipp, 1988). It is supposed that return input from specialized areas may modulate activity in the various cells in the primary visual cortex, depending on associative cortex's needs: "An area performing a specialized higher function will tap any source of information that is useful" (Zeki & Shipp, 1988). Organization of backward connections in visual pathways gives support to Bernstein's theory that lower functional levels might be used in an assimilated way, as a background. Connections existing between specialized pathways at each level of the visual system (Zeki & Shipp, 1988) might be an anatomical basis for having a coherent percept, or, in our terms, synthetic image as a certain combination of signs, at each level. Zeki (1993) and Zeki and Lamb (1994) indicate that the human brain does not just analyze or imitate the visual world but constructs it, and interpretation is an inextricable part of sensation from the early stage in visual processing in visual cortex. It might be supposed that the human brain "constructs" the visual world at every functional level of the visual system, and every level has its own interpretation, its own "version," of the visual world.

At the first level one distinguishes (or analyzes) concrete and concrete-situational features and signs. Signs at this level are related to objective features of things, such as size, shape, color, and the like, but are contextually bound. They characterize an object as it occurs in its particular situation, rather than the object in itself. The left hemisphere synthesizes or structures the "image," which, at this level, is the totality of the concrete signs of the given object-situation. Consider the example of a cup (Figure 6, I). Each cup is seen and recognized by its physical characteristics in each particular situation, but no generalization is made regarding *cup* as the same thing. To put it differently, we have here the most objective, physical visual image of a cup in its situation, which parallels the metricity and geometricity of Bernstein's level C. At this stage, one does not yet separate the image of the object/action-which-is-to-be-performed-with-it-situationally from the situation itself. Thus, at this level, the meaning of an object is limited by manipulation with it in its given situation.

The gnostic-praxic level follows—it is built upon the sensory-motor level. At

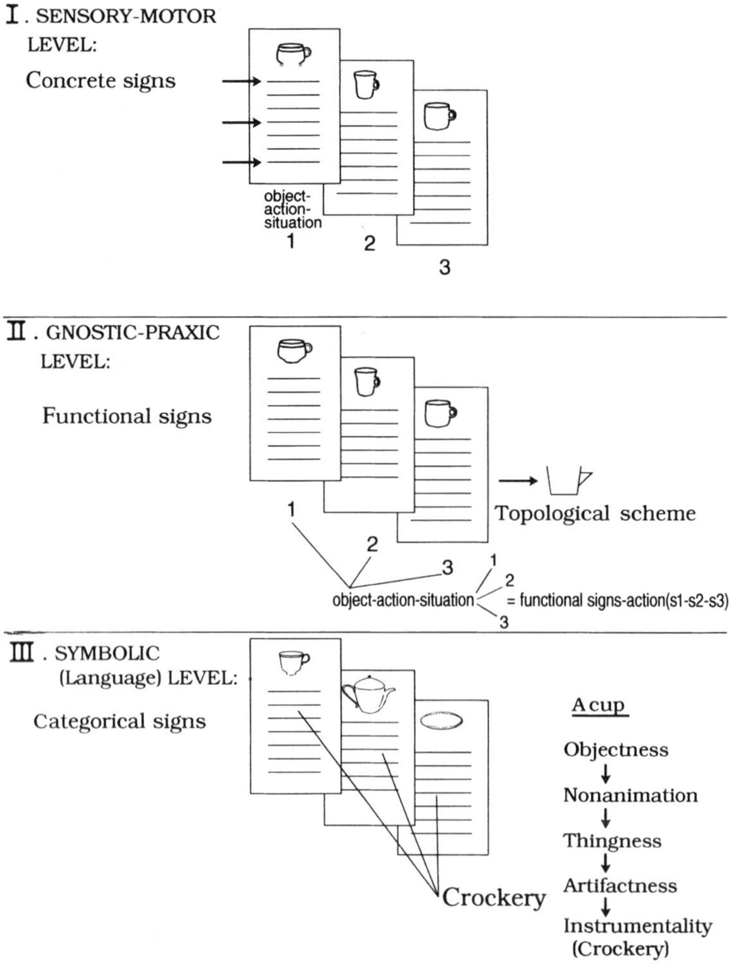

FIGURE 6. Model of the hierarchical multilevel information processing in the left inferior-posterior temporal cortical region.

this level, the distinguishing (analysis) and the following structuring (synthesis) of functional signs take place. The functional signs of the object are such signs that are necessary to act with it. Thus, the left-hemispheric synthetic visual image at the gnostic-praxic level presents, in essence, Bernstein's topological scheme of an object, which is, in turn, afferentation for praxis. For example, the topological scheme of a cup is formed by the combination of the following topological features: its having a bottom, a handle, solid walls, and an open upper part, independent of its actual size, shape, color, and so on. What Bernstein called topological features we will call *functional signs*, which take into consideration level of abstraction at the gnostic-praxic stage. Generalization on a functional level characterizes objects as such but does not characterize their relations with other objects. The functional signs of an object determine the action with it in several situations, so they determine usefulness of the object in these situations. Thus, at this level, the connection with situation is indirectly preserved (Figure 6, II).

To understand better generalization connected with the gnostic-praxic level, we will compare synthetic space/visual afferentation, or the semantic ground of an action, at levels C and D, according to Bernstein. Bernstein writes: "The spatial field of the C level represents the most objective space;... at the following D level, space evolves toward schematization, which results in a gaining of semantic organization and 'order' but a loss of the strict objective, photographic correspondence with real metrical parameters." For example, "the normal adult would use functional level D [as the leading one] when drawing objects; he would draw schemes, not shapes. As a result, a person who is learning the art of drawing has to learn how to look at the factual world, 'switching on' optic control, free from higher level interpretation" (Bernstein, 1947, p. 124). The goal-directed, precise movements at level C represent manipulation with the external space filled with forms and forces, but they are not yet the rational manipulation with objects (tools). "Movements" of the D level are object actions, and topological scheme—being, as we understand it, the combination of functional signs of an object—is the spatial (visual) configuration of action. This "fixed mold" of object action is an internal scheme acquired in ontogenesis by training. When the internal afferent base of object action is impaired, as occurs in patients with focal lesions in the left posterior cortical areas, comprehension of the object as a tool is lost (ideatory apraxia). The patients do not know *what* to do with matches, scissors, needle, and so on.*

At present, evidence exists that before articulate speech appeared, communication was performed by means of gestures (Ivanov, 1978). Among the great number of various gestures, the following may be distinguished: imitative (gesture as an imitation of action with an object), descriptive (gesture as demonstrative depiction of object features), and indicative. As Blonsky (1935) remarks, at the stage preceding the development of articulated speech, human communication

*Because our interest here is directed to the cortical field connected with the visual modality, so far we have only spoken about topological scheme in this modality. However, because the gnostic-praxic level is provided by the secondary, modality-specific cortical fields, it is understandable that the topological scheme of the same object exists in different modalities. In other words, afferentation for praxis is provided by several modalities including visual and kinesthetic. For example, the topologcial scheme of a *cup* in the kinesthetic modality is the combination of those hand "poses" and positions that are complementary to functional signs of *cup* (handle, solid walls, open upper part, bottom) and give the kinesthetic plan for cup use.

took the form of action-representation, including imitative and indicative gestures. Gestures are closely related to object action (praxis).

Imitative gestures are symbolic actions without the object, based on the internal topological scheme of the object, mostly in the kinesthetic modality. The connection between imitative gestures and kinesthetic afferentation is illustrated by the fact that patients with lesions in the postcentral parietal region suffering from kinesthetic agnosia experience special difficulties in the reproduction of object action in the absence of the object. These patients cannot show how to stir sugar in a glass of tea with a spoon, how to hammer, how to pour a cup of tea, and so on.

The topological scheme in the visual modality (visual object gnosis) is a base for indicative gestures. To indicate a separate object means to distinguish it in consciousness. Such a function cannot be performed merely by means of the cognitive mechanism of the right hemisphere in which nondividable, whole situations are stored. It becomes possible only while analyzing complex situations with functional signs. It is just the topological scheme of the object, that is, left hemispheric analysis at the gnostic-praxic level, which might "develop" an image of a separate object in the right hemisphere (Figure 7).

Historically, when articulated words emerged, they replaced indicative gesture. At the same time, sound words were combined with imitative gesture to form a message (Blonsky, 1935), which was, in general, the whole situation represented in the right hemisphere. However, at that stage, although word meaning was undifferentiated, it included reference to a separate object within, a left-hemispheric "nucleus." For example, one word *nu* in a series of African nonwritten languages, denotes the following: *this*, *I*, *eye*, *to look*, *to know*, *nose*, *mouth*, *to drink*, *water*, *tooth*, *to bite*, *to eat*, *to speak*, *to listen*, *ear*, *hand*, *five*, *two* (Blonsky, 1935). Although word meaning is diffuse, the image of the whole visual, actional-demonstrative situation ("man and his activities") and the emerging separate

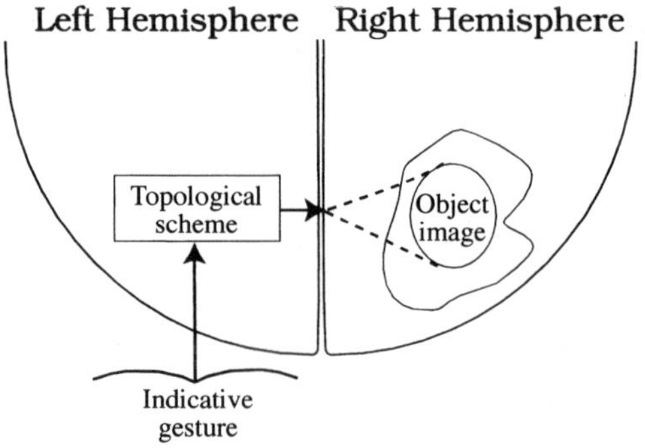

FIGURE 7. Indicative gesture and topological scheme of an object.

object coexist in it. It is "this"—a sound replacement of the indicating hand that gives the opportunity to refer to any one object depending on context. Together with the first words to appear, the word "this" evolved. Many languages have the same word roots to denote "hand" and "speech," and many more languages have the same or very similar words to denote "to speak" and "to show" (Blonsky, 1935). Thus, while the uttering of a word (nomination) replaces the indicative gesture, that which was indicated is not the external object but its left hemispheric image, namely, the topological scheme (Figure 7). Nomination at that stage might be implemented by activating the connections between the auditory temporal region (word sound) and the temporal-occipital region, most likely the peripheral "visual" subfields of field 37.

At the third functional level, the symbolic, the distinguishing (analysis) and structuring (synthesis) of categorical signs are realized. From the successive series of numerous signs and features of objects obtained during left hemispheric analysis, one sign common for the given group of objects and differing this group from the other groups is distinguished. Such a sign is called a categorical one. It is by this sign that the belonging of an object to a corresponding group (category) is defined. Within each of the categorical groups, subgroups are distinguished that possess their own distinctive signs: by a progressive distinguishing of the specific categorical signs, more specific categories are formed from the more general. For example, *a cup* belongs both to the general category of objects and to the following subordinate categories: inanimate objects, things, man-made things (artifacts), things made for certain needs (instruments) (Figure 6, III).

It is obvious that, from functional level C to functional level E, the visual construct in the left hemisphere is transformed from concrete shape to more and more abstract schemes. As we attempt to understand how visual images are interpreted at the symbolic level, we suggest some representation of category. For example, categorical representation of *cup* may include the generalized perception of the objects categorized as "dinnerware," the representation of animation may be connected with perception of a multitude of animated objects, and in the continuous series of their signs the subcategories of persons, agents of action, and so on, may be distinguished.

Linguists suggest that the "inventory" of categorical signs is the base of the so-called hidden or covert grammar that forms the "inner logical framework of language" (Katznelson, 1972). The fact that grammatical distinctions ascend from categorical ones is illustrated by the general division of words into grammatical classes (that is, parts of speech). This classification, in the opinion of linguists, reflects the human capacity for categorical classification of the outer world (division into substances, actions-states-relations, qualities-properties).

Katznelson indicates that there are also more specific hidden grammatical categories embodied in lexical meanings. In the words "tablecloth" and "waiter," for example, *objectness* is a common categorical sign. In the word "tablecloth," the following subcategories of more specific order stand behind the category *objectness*: *nonanimation* → *thingness* → *artificial thing* (artifact, product) → *instrumentality* (meant for certain needs). In the word "waiter," the subordinate categories are: *animation* → *person* → *agentivity* (acting person) → *occupationness* (the person has a certain kind of occupation) (Katznelson, 1972).

According to Katznelson, certain combinations of categorical signs form the categorical component of word meaning. However, he emphasized that this categorical component of word meaning at the same time represents the grammatical, or rather the logicogrammatical component in the word. The categorical signs perform the grammatical functions that determine the semantic "combinability" of words. This hidden grammar also plays a role in the formation of particular contexts; the polysemy of one word in a sentence may be eliminated by certain categorical signs of the others.

For example, in the sentences "The table is covered with the tablecloth" and "The table is covered by the waiter," the verb "to cover" does not have the same meaning. In the English language, the prepositions *with* and *by* can help clarify the meaning of the verb. In the Russian language, however, there are no prepositions in this sentence and the lexical ambiguity of the verb is dismissed by means of the categorical signs of the nouns, namely, *instrumentality* (tablecloth) and *agentivity* (waiter).

Thus, reorganization of synthetic space/visual afferentation at the symbolic level, "refracting" through the left-hemispheric type of information processing, leads to transformation of visual representation into categorical classification of the external world. At present, there are neuropsychological data indicating that the semantic (categorical) system underlying word meaning is related to the temporal area of the left hemisphere, more precisely: areas within the temporal lobe that do not belong to the classical speech-connected temporal cortex (Coughlan & Warrington, 1978; Hodges, Patterson, Oxbury, & Funnel, 1992; Damasio & Tranel, 1993).

It is with this region that our ability to categorize is likely connected, and categorical signs appear here as discrete units of categorical thinking. It should be noted, however, that the left temporal region performs for categorical thinking the specific role of a "storehouse," an "arsenal," a "deliverer" of the units; whereas the operations with these units are related to the frontal region of the left hemisphere. In general, categorical signs represent units for categorical thinking. On the other hand, they also form the categorical component of word meaning, and, finally, they are units of so-called hidden grammar.

What is the relationship between categorical system, or categorical representations in general, and categorical component of word meaning?

Categorical representations as such are nonverbal and multidimensional. For example, the continuous series of categorical signs representing the general category "objectness" at each subcategorical level always imply alternatives: animation versus nonanimation; persons versus nonpersons; agents versus instruments; artifacts versus phenomena. In a word, the categories necessary just for the given notion get actualized. It is the categorical component of word meaning in which categorical representations obtain their specific expression in language. The categorical component of word meaning is not whole but, as any left-hemispheric representation, is a certain sequence of signs. For example, the categorical component of the word "cup" includes the following linear sequence of categorical signs: *objectness → nonanimation → thingness → artifactness → instrumentality (dinnerware)*. This hierarchical system is only implicitly contained within the categorical system. Indeed, the notion does not exist beyond its phonological shape, or, to be more exact, it gets actualized at the very moment when it is joined with the corresponding phonological shape of the word. Word sound is that organizing factor which can "extract" the particular sequence (characteristic for the given notion) from the continuous categorical sign series. In the historical development of language, categorical component formation and phonological shape formation of a word were reciprocal processes: the definite sequence of categorical signs was consolidated by sound designations characteristic for the given language. In different languages, the notions that correspond to similar objects or actions may differ in the number of categorical signs. For example, the categorical component of the English word "to come" includes the following categorical signs: movement → active movement → direction of movement. In the Russian language, several notions correspond to this verb. The categorical component of these Russian verbs also includes the more specific sign—*mode of movement* (that is, to come by car, by walking, and so on). Direction of movement is lacking in the meaning itself in the Russian word. It is expressed not by categorical signs embodied in word meaning but by overt grammar, which has its sound expression in language, the prefix (Figure 8).

Thus, the differences between the categorical classification of the external world and the categorical component of word meaning are these:

1. Categorical representations in themselves are multidimensional and volumetric, whereas the categorical component of word meaning has a linear, hierarchical structure.
2. Categorical representations are nonverbal, whereas the categorical component of word meaning does not exist beyond its phonological shape.
3. Categorical representations are universal, whereas the categorical component of word meaning is idioethnical.

Generally speaking, highly important for neurolinguistics is the fact that the categorical representations get "grown" from visual object perception and that the units of those representations, that is, categorical signs, form the categorical component of word meaning.

The level hierarchy to be found in the function of visual object perception and

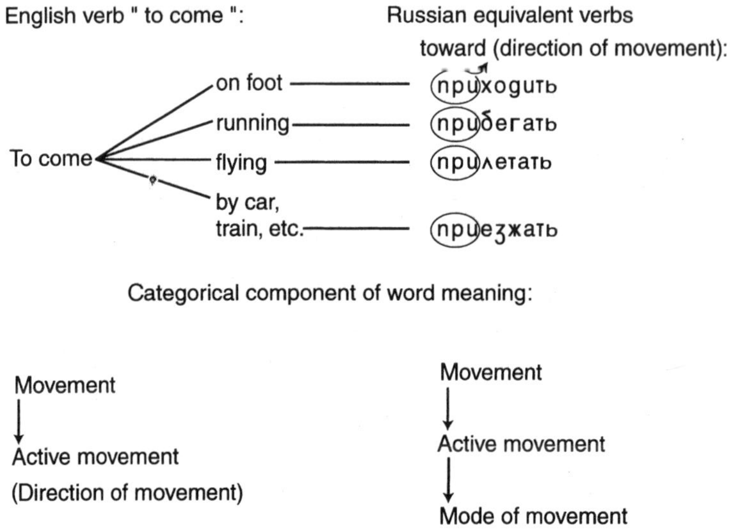

FIGURE 8. Idioethnicity of the categorical meaning of a word.

its connection with categorical thinking and language allow us to explain its particular situation as compared to other forms of visual gnosis, namely, its "moving" into the tertiary cortical field.

2.4. Right Temporal-Occipital Region: Visual Object Perception and Visual Symbolic Thinking

In this section, we will consider visual object perception connected with the right inferior-posterior temporal region and its contribution to word meaning. Data from numerous neuropsychological studies support the existence of this contribution. Summarizing research results on hemispheric differences in normal people, Moscovitch (1983) indicates: "Hemispheric differences are large in processing syntactic or phonological properties of language; however, ... in processing semantic properties of language—hemispheric differences are likely to be smaller because each hemisphere contributes in its own way to the task" (p. 94). A gross defect in understanding speech usually requires, in addition to a left hemispheric lesion, an associated disconnection of the right temporal lobe. Studies of regional cerebral blood flow patterns in subjects who solved cognitive tasks of word recognition or word memorizing showed an increase in regional activity in both left and right inferior temporal regions (Goldenberg, Podreka, Steiner, & Willness, 1987).

Authors agree that the semantic system of the right inferior temporal region is represented by "visual imagery" (Goldenberg et al., 1987). There is also an assumption that the generation and storage of mental visual images is a right hemisphere contribution to verbal memory (Denis, 1979; Paivio, 1979). Jones (1976) reported that imagery mnemonics improve the performance of patients with left temporal lobectomy but not of those with right temporal lobectomy in verbal paired associated tasks. Jones-Gotman and Milner (1974) and Jones-Gotman (1979) established that an intact right temporal lobe is crucial for normal performance on verbal tasks that have an imagery component.

What kind of visual images are characteristic for the right hemisphere?

As we discussed in the introductory chapter, according to an accepted assumption about hemispheric difference in information processing, whereas the left hemisphere is supposed to operate in an analytic mode, the right hemisphere operates in a *gestalthaft*, holistic processing mode. We conceive of the left hemispheric representation as a combination of separate signs and the right hemispheric representation as a single whole. In the right hemisphere, the visual object images are perceived and stored as included within situations (visual scenes). Each visual situation is, in fact, an integral global entity. It is not subdivided into component parts and not connected with other situations. The right hemispheric representation includes not only the visual picture as such but also emotions and affects that both determine the picture and are determined by it. This incorporates one's individual experience as memories that are directly stored as unchangeable, stationary wholes.

Drews (1987) studied word interpretation in normal individuals using a divided visual field technique; it was shown that the left hemisphere prefers logical, categorical semantic relations whereas the right hemisphere prefers situational semantic relations. In this study, sets of word pairs were presented to the left visual field (right hemisphere) or right visual field (left hemisphere). These sets were: 40 word pairs in which the two words belonged to the same level of a certain category (bus–train, saw–axe); 40 pairs in which each target word described local context for the prime word (coffin–earth, shepherd–pasture); and 80 word pairs in which the two words were semantically unrelated (avenue-grip). Normal, right-handed subjects were asked to decide whether the two words presented in quick succession "had something to do with each other." Results of this experiment showed that the left hemisphere performed better on words connected categorically, whereas the right hemisphere did better on words connected situationally. The level of left hemisphere performance on situationally connected words was the same as with semantically unrelated words; the right hemisphere performance on categorically connected words was the same as with semantically unrelated words. Other studies, which examined patients with a temporarily inhibited left hemisphere showed that the right hemisphere interprets words by means of visual-situational associations. For example, one such patient responded to the word "water" with the following associations: "summer–to bathe–competition–swimming–fishing," which reproduced the whole visual scene (Ivanov, 1978). Thus, a word seems to get visual-situational decoding in the right hemisphere. Individual interpretations of words depend on the peculiarities of the right hemispheric visual scenes.

According to the right hemisphere gestalthaft mode of information processing, the images of one and the same object represented in different situations are not connected with each other: the object images in the right hemisphere are singular. The meaning of some words, probably right-hemispheric in their origin, in so-called primitive languages gives us an example of uniqueness of the visual image separated from the images of the same object by its situational context. For example, in the Australian language Aranta there is no general word denoting "a leaf"; instead, there are several words: *kanta*—round leaves; *ibala*—oval and fleecy leaves; *iana*—fleshy leaves. There is no word "hair" in Aranta but the following cluster of words: *panga*—long hair; *pantja*—long, trailing hair; *aratja*—straight hair standing upright. On the other hand, one word may mean more than one thing, association through a visual situational context: for example, *inta* in Aranta means at the same time "stone" and "recumbent." The word *inka* means "foot," "footprint," and "steep" (cliff or mountain path) (examples taken from Katznelson, 1986, pp. 94–95).

Based on the right hemisphere preference for forming situational associations, Drews (1987) supposed that visual scenes–situations represent the organizational structure of the right hemisphere's lexical knowledge. However, data indicate that the right hemisphere's semantic ability goes beyond just visual situational understanding of words. In the case of patients with right hemisphere damage, it was found that the comprehension of metaphor was impaired. When asked to match verbal expression of metaphor with pictures, patients with damage in the right hemisphere usually choose a literal picture (Winner & Gardner, 1977). There is also a suggestion that so-called visual-imaginative thinking is connected with the right hemisphere (Ivanov, 1978).

Considering the interdependence between interhemispheric specialization and vertical differentiation, one can suppose that each functional level, as a stage in left hemisphere development, is related to the definite pattern in the right hemisphere's organizational structure. Taking into account the cytoarchitectural heterogeneity of field 37 and its division into subfields of different phylogenetic age, it seems probable that within the right temporal region functions can vary in their degree of complexity, ranging from visual perception per se to visual-figurative thinking, the latter being a necessary component of creative thinking.

The situational visual picture is phylogenetically the most ancient layer of human visual representations. Were one to be able to examine this layer on its own, one would find the absence of subject-object division; absence of one's own "I" as separate from the environment; and the projection of emotions onto external objects. Kretschmer (1927) indicates that primitive man projected his affects to the outside, perceiving them in a similar way as visual and acoustic input. For example, whereas modern people may say: "I experience fear of the dead body," primitive people would say: "Taboo (that is, fear) is sitting in the dead body."

Because of the indivisibility of emotions and visual pictures in the right hemisphere, very different situations, through their association with the same emotions, can be united into a symbolic system. In consequence, while retaining its own identity, a single situation becomes identified with other situations in a symbolic system. A symbol will result from the reduction of a series of situations into compressed form. We will refer to this process as "symbolic-situational thought." Symbolic-situational thought may be recognized in the word meaning of primitive languages. For example, the word *ota* in one Native American language means "moving along a surface, performed with effort, slowly" (for example, crawling). The word *u-mani* means the possessive pronoun "his." The word *utotama* literally means "his to move with effort." In reality, the meaning of the word is "his older brother" (example taken from Blonsky, 1935, p. 96). As Blonsky remarks, the older brother in this Native American culture is the guardian of the child, his teacher and tutor in the clan. It is the idea of the continuous responsibility and care of the older brother that is represented in the visual-situational rendering of the word.

Another example of situational symbolic thinking, when subjective experience unites different situation into a symbolic system, can be drawn from word meaning in Aranta. So, one and the same word *ngu* denotes: water lily roots, hidden under the water, sleeping men, sleep, man's bones [unseen, like underwater roots], and an interrogative pronoun related to a man unseen by the speaker (this example taken from Ivanov, 1978, p. 45). In this case the objects and phenomena are united on the subjective connection, that is, by their relation to the subject ("unseen"). The inseparability of subject and object is displayed in that under the general word meaning not only "unseen" objects are grouped but also those which "don't see," e.g., sleeping men.

In symbolic-situational thought, situations have no objective properties in common. It is subjective factors—affects and emotions—that unites them into a symbolic system. However, the right temporal region can also provide an "objective" factor for uniting situations into a symbolic system: resemblance of holistic forms. The left hemisphere establishes connections between objects based on kinships of their properties, features, and signs. The right hemisphere is oriented toward perceiving the whole, and its gestalthaft cognitive mechanism allows identification of visual objects according to resemblance of their holistic form, even if their content is different. Moscovitch (1976) presented patients with sixteen randomly ordered drawings of common objects each of which belonged to one of four taxonomic or shape categories (Figure 9). In recalling the names of these drawings, normal subjects usually cluster their responses according to both lexical

FIGURE 9. Test of categorical and holistic form associations. From "Stages of Processing and Hemispheric Differences," by M. Moscovitch, 1983, Cambridge, MA: MIT Press. Adapted by permission.

and shape categories. Patients with right temporal lobectomy, however, clustered only by taxonomic category, whereas patients with left temporal lobectomy clustered primarily by shape.

Another example can be drawn from studies of split-brain patients. Pictures that can be matched by either their function (such as cake on a plate and a knife and fork) or their appearance (such as cake on a plate and a hat with brim) (Figure 10) were handled differently by the two hemispheres. With ambiguous instructions simply to match similar stimuli, the left hemisphere of the split-brain patients matches by function, and the right hemisphere matches by appearance (example and picture from Levy and Trevarthen, 1976).

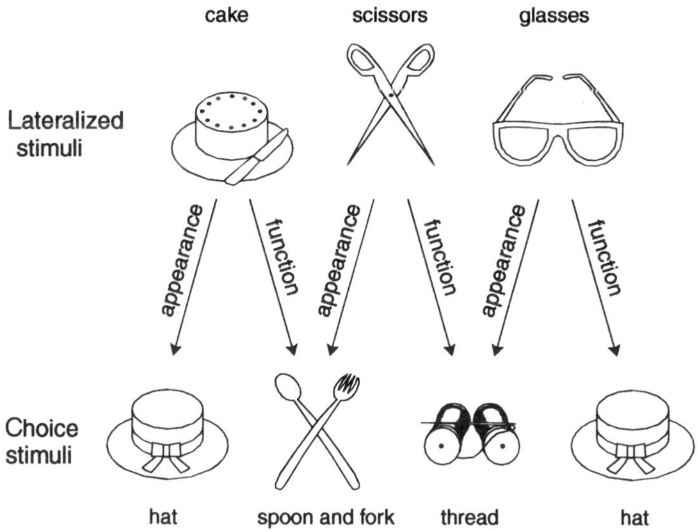

FIGURE 10. Test of function and holistic form associations. By J. Levy & C. Trevarthen, 1976, *Journal of Experimental Psychology: Human Perception and Performance*, 2, pp. 299–312. Copyright © 1976 by the American Psychological Association. Adapted with permission.

Let us return to examples from the language Aranta. As mentioned earlier, several words in Aranta correspond to our word "hair." On the other hand, each of these words in Aranta corresponds to several words in our language. For example, *panga* means not only long hair but also cave above water, brushwood of trees before entrance to a cave, and solar eclipse; *pantja* means not only long, trailing hair, but also black night and the deep; *aratia* means not only straight hair standing upright but also the straight road (examples are from Katznelson, 1986, p. 96). Thus, we can see two cognitive strategies that both reflect, in our opinion, the organizational structure of the right hemisphere:

1. Single visual scenes (situations)—images of the same object enclosed in different situations—are not identified with each other. Different names (words) denote long hair, long trailing hair, and straight hair standing upright.
2. On the other hand, each of these singular images is identified with similar-in-appearance images from alien domains, forming the symbolic system. Here the symbolic system is represented in language: one word designates long trailing hair, black night, the deep.

Let us give two more examples from the language Aranta demonstrating right-hemispheric classification: *ilbala* means not only the oval or angular leaf but also bird's feather, bird's wing, and fin; one word designates knee, curved bone, bend of the river and earthworm. Here we can again see that a visual holistic form, identified within different situations, unites them into a symbolic system (examples were taken from Katznelson, 1986, p. 95).

Blonsky (1935) examined the types of visual associations emerging in normal individuals during the relaxed drowsy state in response to visual, verbal, or kinesthetic stimuli. Here are some examples of the responses Blonsky recorded in his experiments: *matches*—"something hot in the hand ... white teeth ... the dents against background"; *coin*—"round cat's muzzle ... a tree with round crown"; *strip of paper*—"white color, from the white very clearly a yellow brass tube protrudes"; *little stick*—"wooden ramrod ... a spear is flying toward a tree ... a river ... a raft of sticks on the river" (p. 52). In the last example we can see a "flowing" of one holistic form through singular situations. Situational context itself was recognized by some but not by all individuals. Here we see how a symbol can be formed through reduction of a series of object images in which the images from different situations are identified based solely on similarities of holistic form. We believe that this process of symbol formation was possible only at a level of "left brain" development at which the ability to form topological schemes had evolved. Indeed, although the whole images of objects as such are represented in the right hemisphere, the right-hemispheric cognitive mechanism alone cannot distinguish the separate image out of a situation. The topological scheme in the left hemisphere corresponds to the whole object image in the right, and it is the topological scheme in the left hemisphere that forces the separate image to be developed within the situation (Figure 7). Once the separate image has developed in the right hemisphere, similar holistic visual forms, even with different contents, are simultaneously identified in different situations. This is the way we form the polysemantic individual right hemispheric symbol. We will refer to this process as "symbolic-object thought." Figure 11 illustrates the increasingly complex steps in right hemispheric cognition.

Kretchmer (1927) studied the historical development of human visual object images and analyzed the mechanisms of visual symbol formation. At that time, although the facts of hemispheric specialization for speech and perception were already being discussed, little was known about the interhemispheric differences in cognition. This assumed, it is more interesting yet that the formation of symbols described by Kretchmer is in correspondence with what we now know about the right-hemispheric type of information processing. Indeed, Kretchmer showed how object symbol formation may proceed further. According to Kretchmer, visual-imaginative thinking developed through the reduction of the image series into a compressed form. Kretchmer postulated the following three principles of symbol formation:

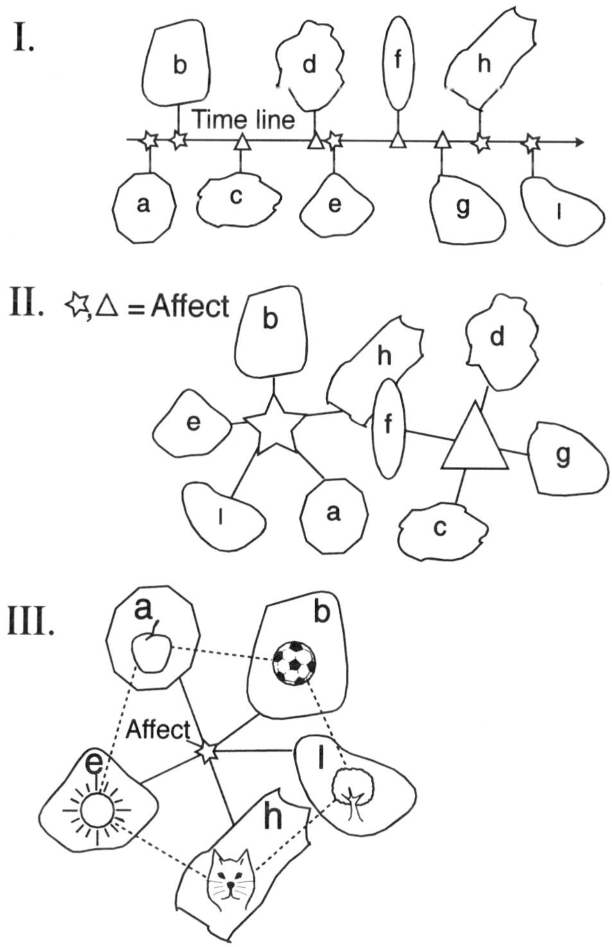

FIGURE 11. Steps in right hemispheric cognition: I, Situational thought; II, Symbolic-situational thought; III, Symbolic-object thought.

1. *Agglutination* (condensing). Kretchmer illustrates this type of symbol formation with diverse examples from mythology and primeval art (sphinxes, centaurs, fawns, angels, griffins, and so on). "The forms of man and animal, lion and eagle, animal and plant are interwoven in the closest way, or are integrally composed in the body of the fantastic creature" (1927, p. 82). In syntheses of this kind, images are condensed into a single total image of human and animal figures, resulting in a

symbol possessing substantial emotional (affectual) saturation, uniting superhuman reason and power.

2. *Displacement.* Here, the process of reduction of the image series is expressed through displacement of emotional effect onto one part of the whole, which therefore becomes a symbol.
3. *Stylization.* In this type of symbolization, images are reduced to simple geometric forms through what Kretchmer believed was a primary tendency of the psychic apparatus itself. Kretchmer felt that there was an emotional component to this tendency to reduce a complicated surrounding world into stylized forms.*

At this point, we will attempt to put forth our conceptualization of the right hemispheric symbol, contrasting in with left hemisphere analytic processing.

The right hemispheric symbol is the bearer of the integral meaning, simultaneously and indivisibly incorporating "content" within the "form." In the left hemisphere, it is the certain sequence of discrete units that corresponds to the meaning. In the right hemispheric symbol, the image content is inseparable from affects—subject and object are inseparable. As expressed by Shilder (cited by Blonsky, 1935), "Symbolization produced by the power of imagination permeates images and perceptions with another sense, different from what they directly express but nevertheless connected, and then presents these images in a new symbolic expression.... Yet the visual image (form) preserves its own value. There are no monosemantic relations between 'content' and 'form.' "

In the complex cerebral organization of symbol formation in the right hemisphere, one may, at least theoretically, distinguish a component of visual form association, an ability of the cortex; and an emotional, subjective component contributed by deep, subcortical structures. It is likely that the inseparability of these components in the right hemisphere may be explained through the more diffuse excitation involving both cortical and subcortical areas in the right hemisphere than in the left (Maslov, 1983). This is probably why the content of a symbol and the visual image through which it is expressed form a single integrated representation. The symbol is, in itself, a result of thought process that cannot be realized without the participation of the frontal cortical area. However, as our goal in this chapter is limited by the role of field 37 of the right hemisphere in demonstrative-imaginative thinking, the cognitive mechanisms determining right hemispheric associations will be considered in the corresponding chapter.

The individual symbol, which, figuratively speaking, is passing through the

*There is a fundamental difference between Kretchmer's theory regarding stylization and our concept of topological scheme. Topological scheme is nothing but synthesis following analysis; that is, the distinction of functional signs. Stylization, as we understand it, is form extraction in its immediate sense; it is based on the associations of holistic forms. Because this process is always subjective, the separate parts of the form can be exaggerated. The same applies to displacement: the choice of the image-replacing detail is subjective and does not correspond to analysis of the image by its features and signs.

emotional constituent of personality during its formation, is closely connected with individual personality. One's unique symbolic system formation is dependent on the early period of ontogenesis in childhood, when cerebral dominance belongs to the right hemisphere. Both genotype (and individual differences in the development of separate brain areas) and early life experience determine the influence of the prespeech period on symbolic system formation. It is worth noting that researchers in the field of brain cytoarchitectonics, using any cytoarchitectural parameters, have found that there is significant variability among homologous brain areas of different people (Blinkov & Glezer, 1964). So, the individuality and nonreproducibility of the right hemispheric symbol have two sources: the uniqueness of the object and the uniqueness of the subject.

Summing up the possible contribution of field 37 of the right hemisphere to visual figurative thinking, we propose that the basic, increasingly complex, steps are these:

1. Situational thought
2. Symbolic-situational thought
3. Symbolic-object thought

Thus, while the left hemisphere extracts typical features of objects and phenomena, abstracting from the "whole," the right hemisphere, on the contrary, "deepens" into the single, having the individual symbol as its highest cognitive step.

2.5. Cerebral Organization of Word Meaning

In this section, building on what we have said thus far about visual cognition connected with the temporal-occipital area, we will propose our model of the organization and representation of word meaning in the cerebral cortex.

The "meaning" is one side of a word as a language unit, the other side is the sound (phonological code). In linguistics, a word is considered an integral unit of its two sides: "signifier" (sound) and "signified" (meaning). The double-sided nature of words may become apparent in patients with focal brain lesions. Indeed, although each of the sides implies the existence of the other, and the word is only realized in their unity, in pathology (namely, in patients with different focal brain lesions) selective impairment of one side (word sound or word meaning) can be observed, the other side being preserved. This suggests that the cerebral cortical representation of the signifier and the signified differs topographically. It is known that word sound is impaired with damage to the left temporal region (cytoarchitectural fields 21, 22, 42). Word meaning, on the other hand, is probably connected with the functioning of the left temporal-occipital region (cytoarchitectural field 37). Our model of the cerebral organization of word meaning is illustrated in Figure 12.

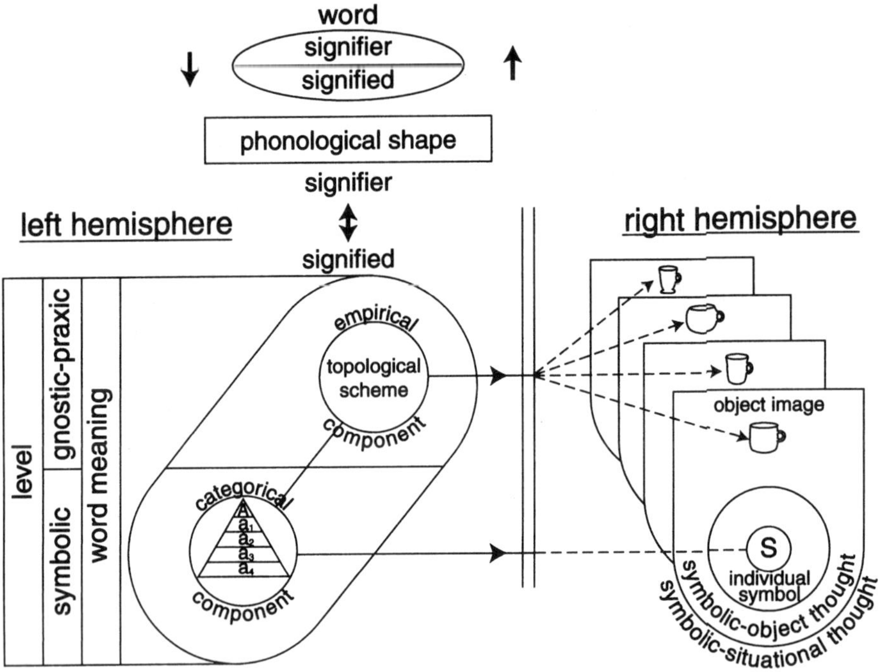

FIGURE 12. Model of cerebral organization of word meaning.

We distinguish the left hemispheric word meaning and its right hemispheric equivalent. There are two components in the meaning of a word: empirical and categorical. The empirical component of word meaning is represented by the certain combination of functional signs (topological scheme); it corresponds to the psychological term *object reference*. The categorical component of word meaning is represented by the definite hierarchical sequence of categorical signs; in general, it corresponds to psychological term *concept*. "Categorical" and "empirical" appear to be the most appropriate linguistic terms for the search for correlates with the function of the temporal-occipital region. The right hemispheric equivalent of the empirical component of word meaning is the object image. The image of the object is being "developed" within the existing right hemispheric visual scenes/situations. The equivalent of the categorical component of word meaning in the right hemisphere is the symbol, which is formed by the reduction of image series (see Figure 12). As one can see, this model is hierarchical and three-dimensional. It considers the role of three basic factors of brain differentiation in word meaning formation:

1. The contribution of the visual modality (horizontal differentiation)
2. The contribution of the gnostic-praxic and the symbolic function levels (vertical differentiation)
3. The contribution of the left and right hemispheres (interhemispheric differentiation)

The model shows that the "fullness" and depth ("volumeness") of word meaning is determined by the combination of the discrete components, which are subserved by the different areas and hemispheres working as one functional system. The functional system determines both direct and indirect connections among the components of word meaning's complex structure. In normal individuals, word sound (phonological code) is directly connected with the empirical and categorical components (left hemisphere) and through them with the right-hemispheric equivalents (see Figure 12). Considering the left hemispheric dominance, one can suppose that the direction of interhemispheric connections is primarily from the left to the right hemisphere. The right hemispheric associations related to the particular word are evoked by their left hemispheric counterparts: the object image (and situations) stands behind the empirical component (topological scheme), and the individual symbol stands behind the categorical component. Degree of awareness of this "out-of-language" right hemispheric content of words varies in the different individuals, but in general is marginal. Based on experimental studies, Moscovitch (1983) notes: "On verbal tasks, special techniques are required to free the right hemisphere from the dominance of the left to reveal its contribution to normal performance" (p. 103). The right hemispheric representations are driven away to the periphery of consciousness parallel to the formation of the left hemispheric dominance connected with development of the language system. "The hard nucleus of an uttered word is spiritually accompanied with something like a halo of evaporation from images and strong affects merged together" (Kretschmer, 1927, p. 123).

The relationship of the right hemispheric representations to word meaning becomes more apparent when categorical meaning is diminished; in the situations listed in the following section, we give some examples of "exposed" right hemispheric representations:

1. *The historical development of language* (*phylogenesis*). At a certain stage of brain (and language) development, the left hemisphere has not yet the abstract notions to correspond to the aggregations of images that pertain to the right hemisphere. We may consider again the example of the word *ngu* in the nonwritten language Aranta, which means: this, I, eye, to look, to know, nose, mouth, to drink, water, tooth, to bite, to eat, to speak, to listen, ear, hand, two, five. It should be noted that the objects and actions included in the meaning of the word *ngu* do not contain common objective features. They are united on the basis of their common presence within an integral visual-"actional" situation. The uniting factors serve

not objective connections but an integrity of perception ("man in action"). Here we see word meaning formation is based on visual-situational thinking. Another illustration from language history is the Indian word *utotama* (older brother; literally translated as his-to-move-with-effort), an example that we earlier cited from Blonsky's work. We deal in this case with word meaning formed by symbolic situational associations in the right hemisphere (symbolic-situational thinking).

In general, the given examples illustrate word formation based on right hemispheric representations. The complex visual images occupy the key position here, because categorical word meaning at this stage is absent. In modern man, these right hemispheric representations persist, submerged in subconsciousness.

2. *Individual development* (*ontogenesis*). Prespeech and early speech development is characterized by the predominance of the right hemisphere. This is manifested in egocentric, affect-laden word meaning, and vague, nondifferentiated concepts. This is considered in detail by L. S. Vygotsky (1960). In his work, Vygotsky gives an example of a child who called a duck swimming in a pond *kwa*, a sound-imitating word; the same name was then applied to every liquid, including milk from his bottle. Later, after seeing a picture of an eagle on a coin, the coin acquired the same name, *kwa*, which then became the word for all round, coin-reminding objects. In this example, word meaning is formed on the basis of the whole, integral perception—the visual scene-situation. One original visual scene and images included within the scene are identified, and subsequent visual images that are similar in their appearance become united in the integral perception.

In this illustration (ontogenesis of language), right hemispheric word meaning based on complex visual images appears at the foreground, because the left hemispheric ability to construct categorical word meaning has not yet developed. The traces of early right hemispheric associations may play a role in individual differences in the "sense" of a word in adults.

3. *Schizophrenia*. Patients with schizophrenia demonstrate so-called autistic thinking, in which the prevalence of their own affectual experiences over reality may be accompanied by a predominance of right hemispheric symbolic associations in word meaning. Kretchmer (1927) gives an example of a schizophrenic patient's description of his mental experience connected with the word *infinity*: "The images of a tower came crowding upon me, circle after circle; a cylinder gets indirectly driven into the common picture. All this is moving and in a state of growth, the circles acquiring depth and through this turning into cylinders, the towers growing higher and higher. All this is quite spontaneous, like an expressionistic picture or a dream" (p. 133). In this case, the uncovering of the symbolic system standing "behind" the word occurs because of an imbalance between the "individual sense" of a word and its categorical meaning (thought disorder).

4. *Extraordinarily high ability to form visual associations*. A case of an individual with extraordinary demonstrative-visual thinking was extensively studied and described by Luria (1968). This man, S, a well-known mnemonist, reported

that the visual object associations related to every word were always present in his consciousness: "When I hear the word *green*, I see a green flowerpot with flowers; *red*, a man in a red shirt approaching; *blue*, somebody waves a blue flag from a window," and so on (Luria, 1968, p. 20). Thus, for S, the visual association induced by a word was a single situation. Often it was an individual scene from his early childhood: *thief*—"this is a half-dark room when it is evening, when one doesn't yet put on the lights, and you can hear a rustle, and he takes a piece of bread from the shelf.... This I heard when I was small—bread from the shelf—but where?... Thus in our small pantry" (p. 52). The visual image, having arisen from a single visual situation, was "fixed" to the word, which could not be "thought" without the paired image. The image thus became a sign of the word and its second "meaning." Here are examples of "visual meanings" of words for S: *rider*—"foot with a spur"; *America*—"Uncle Sam"; *eternity*—"old sage, God from the Bible"; *oppositions*—"two dark clouds driving over each other"; *something*—"a cloud of vapor"; *nothing*—"a liquid cloud."

When given a word-remembering task, S turned the words into visual images, which were taken from single situations, and then just "read" these images. He could remember an almost limitless number of words for years, but at the same time was unable to extract from the list those words belonging to one category—for example, "names of birds" or "liquids"—a task requiring left hemispheric classification ability.

S also experienced tremendous difficulties understanding lengthy texts. Every word gave rise to a visual situation; single visual situations would pass in a train through his mind, forming a "second text" that interfered with his understanding of the original.

In this case, we see the uneven development of cognitive abilities, with an extraordinarily high development and prevalence of the right hemisphere, which in turn is represented by its most ancient component: visual-situational thought.

Another example of unusual right hemispheric ability was observed by one of us (Glezerman, 1986) in an individual, N, who gave the following visual associations in response to the word "remember": "How I perceive remembering? I am holding a thin glass goblet with a very long, thin stem; with lines etched in marking every year, full of far-off but distinct remembrances, mixed with rattling ice bits. The glass becomes dim, but I touch it with my fingertips and small drops limply flow down, leaving paths through which one can see something. So I am often sitting in my armchair, holding in my hand a little curved glass screen, turning it lightly in my fingers, stroking it with my finger, and I see there inside, something clinging and knocking between the ice bits—I am mixing the contents with a cocktail stick and drinking it lazily." In this case we deal with the individual symbol (symbolic-object thought). In these last two examples, right hemispheric associations prevail and come to the surface, probably because of unusually high or peculiar development of the corresponding cortical regions.

The different parts of word meaning represent the "layers" in language history. The most ancient are the visual-situational and the symbolic-situational content of words, followed by the empirical component. More "young" historically are the categorical component and the individual symbol (see Figure 12). Corresponding to this, we can assume that the cortical representation of word meaning includes regions of different phylogenetic age. We may suppose that the empirical component may be related to the peripheral subfields of field 37, whereas the categorical component is connected with the central subfield. In connection with this, the following data may be of interest. Goldenberg et al. (1987) investigated patterns of cerebral blood flow in normal individuals who memorized abstract and concrete nouns. It was found that the task "switched on" a functional system consisting of several areas rather than a single region. The inferior temporal areas of both hemispheres, although with the left hemisphere's superiority, were involved in memorizing abstract nouns. When concrete nouns were being memorized, the occipital region was involved as well. The authors noted that the association between the occipital and inferior temporal regions is specific to concrete nouns.

The meaning of concrete nouns includes empirical and categorical components, whereas in abstract nouns the empirical component is reduced. Based on this, we may suppose that the difference in localization between abstract and concrete words may be attributed to the empirical component that is related to the more peripheral part of the inferior temporal region (field 37), bordered with the occipital region.

The connection and interdependence of the categorical component and the individual symbol in word meaning (see Figure 12) can be illustrated by the history of words in modern language. The specific set of categorical signs characteristic for a concept in a given language (idioethnicity) might have been formed under the influence of the right hemispheric symbolic associations. For instance, the Russian word *sutki* (meaning the 24-hour day) originally meant *seam*, the place where two fabric pieces are connected. It came to denote any joint, a corner in the Russian *izba* (house), the place of two walls coming together; later, in a figurative sense, the place where day and night join together, and afterward embracing the time from twilight to twilight (Vygotsky, 1934/1962).

Based on the preceding example, we can also assume that there are two types of symbolic associations connected with a word: (1) those that had been fixed to the word during its formation in language history (and, therefore, common for the given language); and (2) individual symbols (the individual sense of the word). In general, the impregnation of words with symbolic associations ("fullness" of symbols), as well as one's degree of awareness of the "symbolic sense," depend on the individual peculiarities of right hemisphere function. Development of the right temporal-occipital area varies significantly from individual to individual. We have mentioned already that genotype and both prespeech experience and verbal environment contribute to the brain's different region variability. Prespeech experience may determine which words in one's vocabulary will be more saturated with symbolic sense, and to what degree.

Interaction of categorical and individual sense "develop" word meaning in opposite directions: toward typification (monosemy) and individualization (polysemy). Individual symbolic associations are nothing but that "nutrient medium" for language which adds shades of meaning and, as a result, paves the way for distinguishing more and more specific categorical signs and, thus, further differentiation of existing concepts.

In the case of focal brain lesions, the left hemispheric word meaning components and the right hemispheric word meaning equivalents that in the norm are closely interwoven and united into a functional system may be selectively impaired. In the next part of this chapter we will consider selective disorders of word meaning (anomic aphasia, visual anomia) that are a result of focal lesions in the left temporal-occipital region. We also mention the complex direct and indirect connections between the discrete component parts within the functional system of word meaning. The system may be disarranged by abnormal interactions of primarily intact components, too.

An example of a disordered functional system of word meaning occurs in sensory aphasia, which we feel results in a reversal of direction in the connections between left hemispheric word meaning components and the right hemispheric equivalents. It is the phonological shape—the word sound code—which is primarily damaged in this type of aphasia (dysfunction of the left superior-posterior temporal area). In this pathology, word sound becomes unstable and undifferentiated and is unable, as in the norm, to trigger the system that goes from phonological code through empirical component to an object image (see Figure 14). Without a clear word sound, the categorical component also cannot be actualized. The right hemispheric representations become dominant, and now they may direct the choice of a sound word form. Here is an example (taken from Glezerman, 1986, p. 199). Patient M at the Speech Center in Moscow had mild sensory aphasia following a stroke a year earlier. He was asked to explain the meaning of the word *tirade*, which in Russian includes elements of a pompous speech, ironically referred to as solemn. His response was, "By the way, I'm sorry, this is somewhere upstairs, *terrace*, so, that is, where I was climbing ... the gallery, on which story I was sitting ... near the theater." We have a very interesting paraphasia here: *tirade* is replaced by *terrace*. First of all, this is a replacement by sound resemblance: in Russian, these two words sound very similar. But here there is a double association, by meaning as well as sound. However, it is not the left hemispheric meaning that ties the words *tirade* and *terrace*; it is actually the right hemispheric symbolic association. We can propose the following mechanism of this paraphasia. In patient M, the phonological code of the word *tirade* exists but is very unstable. It evoked a few categorical signs connected with "lofty style," and "speech-making." However, the categorical component as a certain hierarchical sequence of categorical signs is not actualized because word sound is already lost. Instead, vague concepts of "lofty style" and "speech-making" induced in the right hemisphere the "physical" visual equivalents: "upstairs," "story," "theater," feeling of ascent ("I was climbing"). The right hemispheric associations are now leading in a search for the lost word sound. As a result, the new word *terrace* emerges, which is similar to the presented word *tirade* in both sound and right hemispheric associations, although these words have nothing in common in their meaning in language.

In sensory aphasia, one may also observe the peculiar narrowing of word meaning where literal meaning, both concrete and abstract, is lost, while the figurative meaning is preserved. Here are a few examples of word meaning explanations by patients with sensory aphasia: *pipe*—"peace pipe"; *dwarf*—"dwarf's thought ... diminutive, pygmean soul"; *sharp*—"well, sharp question, it means complex, unpleasant, and also tongue, sharp tongue, everybody is afraid of it" (examples taken from Bein, 1961, pp. 117–139). Assuming that the figurative word meaning is connected with visual-symbolic associations, and using our model of cerebral organization of word meaning, we can explain preservation of figurative word meaning in sensory aphasia by the fact that the right hemisphere symbolic equivalent of word meaning is most distant from the defective phonology of the word.

Here is another example of using individual symbols in word explanations by a patient, S, with sensory aphasia: *resist*—"Spartacus"; *enormous*—"Gulliver" (Glezerman, unpublished data). We can see that patient's response is far from the direct explanation of word meaning, but it conveys emotionally saturated visual images that are equivalent to the meaning of these words. This example is also interesting because it shows that in the formation of the individual symbol not only idiosyncratic images but also cultural symbols can be used.

Disorder in word meaning's functional system may be due to disconnection (ranging from weakening to a complete block) among its component parts. We may also suppose that different variants of selective disconnections within the functional system of word meaning may exist. To illustrate this statement, we will consider word meaning in patients with schizophrenia. Word meaning in schizophrenia is impaired, but the impairment does not correspond to any pattern of selective disorder in one component of word meaning. It presents with a variety of patterns, each of which can be explained by an increased contribution of one component into word meaning. The disproportionately increased may be any component (or equivalent) of word meaning. Here are a few examples.*

1. Patient R, when asked to classify pictures of objects, united into one group car, spoon, and wagon, explaining that they "can be united accord-

*Examples are taken from *Pathological Psychology* by B. Zeigarnick, 1976, Moscow: Moscow University Press, pp. 143, 144, 165, 170.

ing the principle of movement: one also moves the spoon to the mouth." In this example, symbolic-situational thought prevails.

2. In the same test patient S put into one group the following objects: flower, spoon, and shovel, saying "all of them are elongated objects." Here we may see right hemispheric classification according the resemblance of holistic forms.
3. In the same classification test patient N united into one group the following objects: spoon, bed, car, plane, and shovel. The patient's explanation was: "Iron, objects which testify to the power of human thought." Here we can see that the symbolic association of human power with iron leads in defining the word. It results in uniting objects from alien domains into one group.

These three examples have in common that word meaning is determined by its right hemispheric equivalents (1—visual-actional situation; 2—the holistic forms; 3—symbolic object associations) dissociated from its objective left hemispheric features.

4. Patient M put together pictures of a tree and a cockroach, replying "nobody knows where cockroaches came from, and nobody knows where trees came from." Here we observe so-called empty reasoning. Classification is based on categorical principle (patient used categorical sign origin to unite the objects), but estranged from its object reference, the categorical component became defective.
5. Patient S was asked to draw a picture that would remind him of the word "doubt." The patient drew a catfish (in the Russian language, the word doubt and catfish sound similar: *somneniye* and *som*, correspondingly. Here we observe so-called clang associations, associations according to word sound. Formal phonological operations of the left temporal region are disconnected from the word meaning functional system (left and right inferior temporal-occipital region). Thus, in this example, the signifier (phonological shape of the word) is pathologically increased and acquires its own idiosyncratic meaning for the patient.

The last two examples illustrate two variants of distorted word meaning due to the increased left hemispheric component that is dissociated from other parts of the functional system.

Considering both reciprocal and complementary interhemispheric relationship, we can suppose that at least one of the causes of word meaning impairment in schizophrenia is a weakening of connections between the left and right temporal-occipital regions. One hemisphere, released from the censorial control of the opposite hemisphere, takes over, its function becoming grotesquely exaggerated and distorting the whole functional system (word meaning).

Returning from pathology to the norm, we may expect uneven development of the separate components and equivalents (and connections between them) of word meaning in the frame of individual variability of word meaning neuropsychological profile within the normal population. The neuropsychological profile of any given individual is connected with the relative development of the separate cortical regions. Indeed, all researchers in the field of human brain cytoarchitectonics indicate that there is a great variability among the homologous brain areas of different people. On the other hand, there is no correlation in the level of development among the different cortical fields in one brain. "The uniqueness of each individual's brain is characterized by the relative development of the different cortical fields" (Blinkov & Glezer, 1964). There is also a suggestion that genetic determination of the different cortical cytoarchitectural fields is relatively independent in the late stages of their morphogenesis (Glezerman, 1983). Thus, because the development of any cortical field (cytoarchitectural and myeloarchitectural parameters) varies significantly in different individuals, and because the different cortical fields vary independently from one another, a tremendous number of combinations appear that underlie individual differences in functional systems. This all may be applied to cerebral organization of word meaning: prevalence of empirical or categorical meaning, visual-situational context or symbolic sense, and peculiarities of their connections will determine individual cognitive style.

2.6. Anomic Aphasia: What Is It?—Visual Anomia and Lexical Logico-Grammatical Aphasia

I have forgotten the word I wanted to say—
On severed wings, to play with the transparent ones,
The blind swallow flies back to her palace of shadows; ...

But I have forgotten what I wanted to say,
And a thought without flesh flies back to its
palace of shadows. [O. E. Mandelshtam, *Selected Poems*. Selected and translated by James Greene, Penguin Books, 1991, p. 34]

2.6.1. *Definition of Anomic Aphasia in the Literature*

Based on the functional properties of field 37 that have been described—in particular, its relation to word meaning—we may return to the analysis of anomic aphasia. Anomic aphasia is defined as a fluent aphasia in which the patient has a naming deficit in the absence of phonological, articulation, and any other language defects. The patient with anomic aphasia has word-finding difficulties in various forms of speech activity but especially in confrontation naming (when asked to name objects) (Geschwind, 1971; Heilman, Safran, & Geschwind, 1971; Benson &

Geschwind, 1971; Damasio, 1992). Word search and word substitutions are characteristic of anomic aphasia, and erroneous words produced by the patient are related in meaning to the target word—verbal semantic paraphasias (Head, 1926/1963; Goldstein, 1948; Caramazza & Berndt, 1978). Although syntax is intact, many authors indicate that spontaneous speech of patients with anomic aphasia is poor, deficient of content words (Caramazza & Berndt, 1978).

Early clinical-anatomical investigations indicated that anomic aphasia was most often associated with lesions in the left inferior-posterior temporal region, corresponding to cytoarchitectural field 37 (Nielson, 1936/1948). More modern studies confirmed the crucial role of the temporal lobe in naming (Newcombe et al., 1971; Coughlan & Warrington, 1978; Cappa, Cavallotti, & Vignolo, 1981). Neuroimaging techniques corroborate the involvement of Brodman area 37 in anomic aphasia (Alexander & Benson, 1991; Benson, 1994).

Although the symptom complex of anomic aphasia is well accepted, its pathogenesis is controversial. Initially, in classic neurology, this symptom complex was interpreted within the framework of associative psychology of the time to be the result of a disconnection between the center of auditory word images and the center of notions. However, others (Pick, 1931/1973; Head, 1926/1963; Goldstein, 1948) believed that the basis for anomic aphasia is a global disorder of categorical thinking. According to Goldstein, brain-damaged patients lose the "categorical or abstract attitude." Naming disorder, in his opinion, is one expression of this impairment of abstract attitude, because the naming of an object in the norm requires that one consider the object as a member of a category. In opposition to this, Luria (1947/1970) and Benson (1984) considered anomic aphasia as a pure speech deficit not accompanied by any thinking disorder. Kock (1967) suggested that there was a connection between anomic aphasia and visual object agnosia "of the left hemispheric type." More recent studies emphasized the connection between naming and word meaning. Caramazza and Berndt (1978) suggest that there is "some form of semantic representation of the informational elements that make up a particular conceptual entity" and that this is involved in naming disorder. Goodglass and Baker (1976) point to a connection between the ability of aphasic patients to find and understand words and their ability to trace conceptual relations among words. Progressive fluent aphasia with focal left temporal atrophy was described recently in five patients (Hodges et al., 1992), and characterized by anomia, impairment of single word comprehension, and marked reduction in the ability to generate exemplars from semantic categories (e.g., animals, vehicles, and so on). In contrast, phonology, syntax of spoken language, and comprehension of complex syntactic commands were preserved. Visuospatial ability was also intact. Thus, although it is becoming increasingly clear that naming is related to word meaning, the essence of the anomic syndrome as a language disorder and its interrelationship with nonverbal functions at different levels—categorical thinking (symbolic level) and visual object perception (gnostic-praxic level)—remains controversial.

2.6.2. *Naming in the Norm*

We will analyze anomic aphasia using our model of cerebral organization of word meaning (Figure 12). In the norm, naming is actualized by the connections between the empirical and categorical components of word meaning (field 37, left) and word sound (auditory temporal region, left). The named word is a realization of the unification of the signified and signifier (Figure 12). Considering the structural heterogeneity of field 37, we proposed, as mentioned previously, that the cerebral base for the categorical component of word meaning is the phylogenetically younger central subfield, whereas the empirical component is subserved by the peripheral subfields. As is known regarding the inferior temporal region in primates, the pattern of connectivity between the inferior temporal (visual) and auditory temporal regions follows a "phylogenetic order": areas comparable in their stage of cytoarchitectural differentiation are connected. It seems likely that each functional level participating in the cerebral organization of word meaning—gnostic-praxic (empirical component) and symbolic (categorical component)—has an independent approach to word sound. Applying the concept of a functional logic of cortical connectivity proposed by Zeki based on his discoveries at lower levels of the visual system (Zeki & Shipp, 1988), we assume that naming includes both sequential and parallel information processing. In general, naming should be understood as a complex process that is realized simultaneously at two (at least) functional levels: gnostic-praxic and symbolic.

Naming at the gnostic-praxic level is based on the connection between word sound and the empirical component of word meaning. As already suggested, the topological scheme of an object (Bernstein's term), or the left hemispheric synthetic image composed of the functional signs of the object, is the visual correlate of the empirical component of word meaning in the brain. Knowing that topological scheme represents the afferentation for praxis, we conclude that when the topological scheme is "named," the word sound supplants object action (praxis): thus, naming at the gnostic-praxic level represents a kind of replacement of praxis (Figure 7).

Naming at the symbolic level is based on the connection between word sound and the categorical component of word meaning. According to our model, the categorical component is "born" at the very moment of its reunion with word sound: this process includes distinguishing certain categorical signs from their continual series and a structuring of these categorical signs into a linear hierarchy.

2.6.3. *Naming Disorders Due to Selective Impairment of Gnostic-Praxic Level (Visual Anomia)*

If, as we proposed, the categorical and empirical components of word meaning have different locations within field 37, a limited lesion would affect one of the components in an isolated fashion. Indeed, impairment in recognition of schematic pictures and disturbance of figure-ground discrimination accompanied by visual object anomia (Kock, 1967) might be an example of selective naming disorder at the gnostic-praxic level. Praxis is provided by multimodal afferentation, and thus other modality-specific syndromes might be expected. In fact, such syndromes have been described in the literature: the inability to name objects presented by touch accompanied by an impairment of kinesthetic and tactile gnosis is one example; selective forgetting of names of smells accompanied by olfactory agnosia is another (Markova, 1961).

Beauvois, Sallient, Meninger, & L'Hermitte (1978) described a patient with a left parietal-occipital lesion who was unable to name objects presented to him by touch but could "act" with them correctly. The authors argued that this was a "pure" tactile anomia, unaccompanied by tactile-kinesthetic agnosia. However, when objects were placed in this patient's hand and he was asked not to name them or to act with them but to explain their use, he had difficulties with this task. Thus, the patient could not describe functional properties of objects, and we suppose that the topological scheme in the kinesthetic modality, or the afferent base for kinesthetic praxis, was impaired. That the patient could still "act" with objects correctly may be due to the habitual character of some actions that become automatisms, and their implementation is transferred to lower functional levels.

It is of interest to present here an observation of H. Goodglass (1983). Aphasic and nonaphasic brain-damaged patients and normal controls were presented with 48 pictures: 16 were of objects that would be identifiable also by smell (e.g., chocolate, gasoline); 16 were identifiable also by touch (e.g., a spoon, scissors); and the remaining 16 were identifiable by the sound of the action upon them (e.g., a bell, a typewriter). The subjects were asked to name each of the 48 objects in response to both its picture and presentation by its other input modality. For all subjects, naming was slower for smell and sound stimulation than for picture and touch. These findings may reflect the fact that the topological scheme of an object, fundamental for naming at the gnostic-praxic level, is represented mainly in visual and kinesthetic-tactile modalities.

Warrington (1975) described a patient who had a striking deficit in the comprehension of concrete words compared with abstract words (the same dissociation was later documented by the author in a second patient). Although the patient's comprehension of concrete words was exceptionally poor, he was not observed to have difficulty with any particular taxonomic category. The latter finding suggests that the categorical component of word meaning for the concrete words was intact, whereas the empirical component was impaired, a selective disorder at the gnostic-praxic level. In a later study, Warrington and Shallice (1984) described four post-encephalitic patients with bilateral temporal lobe lesions who showed significant discrepancies in their abilities with inanimate objects versus living things and food: although they could identify and name visually presented inanimate object pictures and define inanimate object words, they could not do so with pictures or words of living things and food. Here is an illustration of striking discrepancy in patients' responses to inanimate object words and living things words demonstrated in this work: *briefcase*—"small case used by students used to carry papers"; *compass*—"tools for telling the direction you are going"; *torch*—"handheld light"; *submarine*—"ship that goes underneath the sea"; *umbrella*—"object used to protect you from water that comes"; *parrot*—"don't know"; *snail*—"an insect animal"; *eel*—"not well"; *ostrich*—"unusual"; *wasp*—"bird that flies"; *crocus*—"rubbish material"; *holly*—"what you drink" (Warrington & Shallice, 1984, p. 838). The authors indicated that inanimate things were mostly artifacts, tools, and objects for use; consequently, functional attributes provided the definite characteristics for their identification. On the contrary, sensory features such as size, shape, color, texture, and so on, are more important for the identification of living things. They suggested that for precise identification of food and living things, a semantic system based on sensory features is required, whereas a semantic system based on functional specifications might have evolved for the identification of inanimate objects. This is in correspondence with our understanding, although the authors do not consider level hierarchy and assume that visual and verbal semantic systems are separate entities. Using our model, at the symbolic level, inanimate objects, living things, and food are supramodal concepts representing different categories. At the same time, however, at the lower functional levels (the gnostic-praxic level, or level D, according to Bernstein, and the sensory-motor, or level C, according to Bernstein) they are classified by their visual features. Because these two sets of objects are identified by different types of signs—functional signs for inanimate objects and sensory (concrete) signs with situational context for living things and food—they therefore may be unequally represented (and identified) at the lower levels, inanimate objects being mostly represented at the gnostic-praxic level and living things plus food at the sensory-motor. In Warrington and Shallice's patients, we hypothesize a selective impairment of the sensory-motor level that affects the empirical component of word meaning, leaving the categorical component intact (symbolic function level). These patients were unable to define words of living things and food but showed an excellent comprehension of abstract words: *debate*—"discussion between people, open discussion between groups"; *malice*—"to show bad will against somebody"; *caution*—"to be careful how you do something." But compare: *cabbage*—"used for eating; material, it's usually made from an animal"; *tobacco*—"one of the foods you eat" (Warrington & Shallice, 1984, p. 842). That "separate layers" which "supply" the empirical component of word meaning may be disrupted selectively is illustrated by the finding of *converse dissociation*, in which inanimate object names were impaired while food, animal, and plant were preserved (Warrington & McCarthy, 1983).

Thus, several variants of naming disorder that were described in the literature we believe can be attributed to the selective impairment of the gnostic-praxic level and even to a highly selective partial impairment within the gnostic-praxic level. We understand this as a selective disorder of the empirical component of word meaning (topological scheme of object), which is expressed by both left hemispheric type visual object agnosia and visual object anomia. In this disorder, visual object agnosia and visual object anomia constitute a single syndrome. We will call it *visual anomia.*

2.6.4. Naming Disorder Due to Selective Impairment at the Symbolic (Language) Level

There are neuropsychological findings that not only support the possibility of naming disorder due to selective impairment of the symbolic functional level but have revealed category-specific naming disorders suggesting the possibility of highly selective dysfunctions within the symbolic level (Mehta, Newcombe, & DeHaan, 1992; Farah & Wallace, 1992). There is also a suggestion that in the left temporal region, "relative system segregation exists for different categories of nouns" (Damasio & Tranel, 1993). To illustrate a category-specific disorder, we give an example of a patient described in the literature (Hart, Berndt, & Caramazza, 1985). The patient experienced considerable difficulty in naming individual fruits and vegetables, but was easily able to name objects from a large range of other categories (vehicles, tools, animals, clothing, and so on). The same selective naming disorder was observed when objects were presented to the patient by touch. The authors gave an example of a patient's striking inability to name such common items as a peach and an orange, while he could easily name less common items such an abacus and a sphinx. Performing on a classification test, the patient experienced difficulty only in categorizing pictures of fruits and vegetables; his performance on a large battery of neuropsychological tests was also unremarkable, except for this circumscribed area (Hart et al., 1985). The findings in this case suggest a selective disorder at the symbolic functional level based on the following: (1) the naming disorder was not modality-specific; (2) the naming disorder involved a set of objects that were related by their categorical signs rather than functional or situational context; (3) the patient was unable to correctly categorize these objects. The existence of such specificity suggests that the horizontal dimension within the symbolic level is highly differentiated in the norm.

According to our model, the selective disorder of word meaning at the symbolic level might be characterized by the depletion of categorical signs. (Hierarchical structuring of categorical signs per se, or left hemispheric action at the symbolic level, is associated with the left frontal region, which we will consider in the corresponding chapter.) The categorical signs that are, on the one hand, the units of the categorical component of word meaning are, on the other hand, the units of categorical thought. Depletion in the number of categorical signs forming the categorical component of word meaning will lead to name-finding difficulties. Depletion in the units, the operations with which are the essence of categorical thought, will lead to the specific thinking disorder. An example of how a patient with anomic aphasia performed on categorization tests follows (Glezerman, 1986). The patient was a 50-year-old man with high pre-morbid intellectual and professional functioning; he was seen the year after he suffered a stroke in the left temporal-parietal region, and at that point had mild-to-moderate naming disorder. The patient performed an analogy test in the following way: dress–coat—"fabric"; chair–table—"wood"; axe–saw—"iron." Analysis of the categorical component deficiency in this case is presented in Figure 13. It shows that the categorical component of the word pair axe–saw includes the following hierarchy of signs: "objectness" → "inanimateness" → "artifactness" (artificial thing) → instrumentality → "toolness." Each of the signs, being a member of the vertical hierarchy of one category, at the same time may enter other categories (continuity of the left-hemispheric information processing). However, the position of the same categorical sign on the "scale of ranks" of the different categories may be not equal. Implicit in the sign of "artificial thing" is that it is made from some material; material as a categorical sign, on the other hand, belongs to the category "substances." The patient was able to distinguish the categorical sign "artifactness," which signifies actualization of connections: "objectness" → "nonanimateness" → "artifactness." However, the specific categorical sign of instrumentality is lost, and the patient switched to a side association, "material," which belongs to the category "substances." As a result, the categorical component of word meaning was deficient, and its linear hierarchical structure was deformed.

In an object classification test, the patient united objects according to the categorical principle but the level of abstraction fluctuated. For example, the patient united into one group pictures of an elephant, a fox, a bug, a swan, a fish, and a swallow, calling this group "animals." At the same time, he united pictures of a dog, a horse, a cock, a goat, a cat, a pig, and a goose, naming this group "domestic animals" and refusing to merge the two groups under one name "animals." The peculiar thinking disorder observed here might be explained by a "falling out" of either more general or more specific categorical signs (supraordinate or subordinate), which resulted in the different kinds of categorical component narrowing—in the direction of concreteness or in the direction of excessive abstractness. Thus, there was no reduction to the concrete attitude, as Goldstein had postulated; the categorical attitude was retained.

We suppose that dysfunction of the left inferior-posterior temporal region at the symbolic level might be characterized by a partial and specific thinking disorder, due to the impoverishment of categorical signs but an intact ability to operate with them. What is the proposed mechanism for naming disorder? It was mentioned earlier that in the history of any given language, the categorical component of word meaning is formed in close connection with the phonological code. As a result, in different languages, concepts that correspond to the same

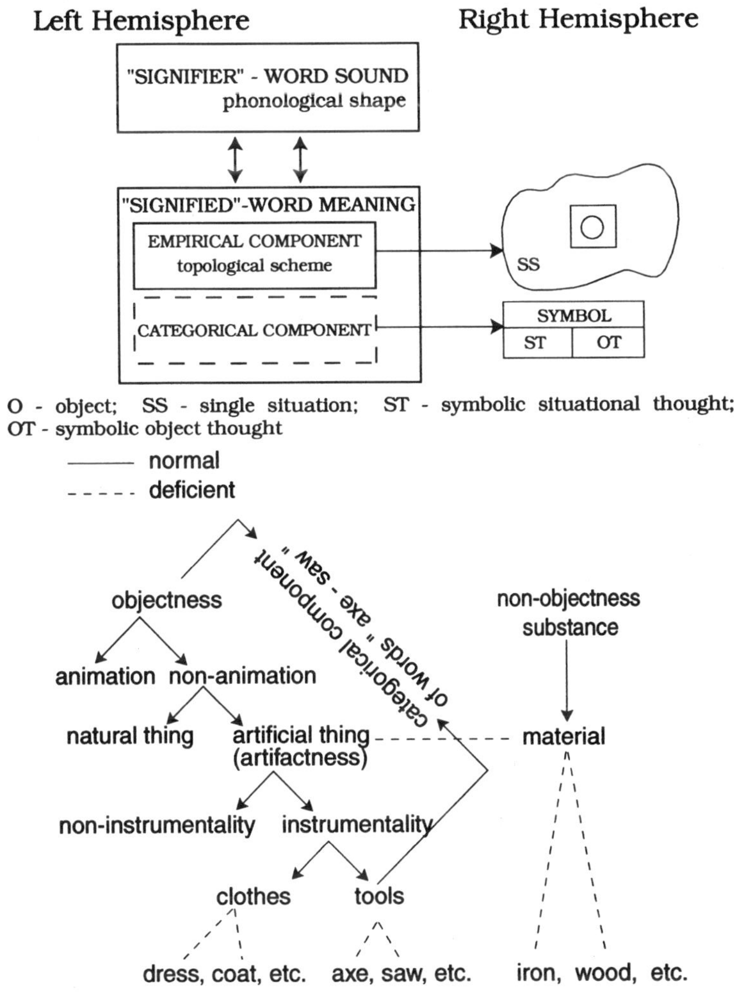

FIGURE 13. Deficiency of categorical word meaning in the patient with anomic aphasia.

object may differ in "size" (number of categorical signs). Within a given language, alteration of a concept's size may be observed under pathological conditions. The patient with anomic aphasia experiences a deficit of categorical signs; the categorical component of word meaning formed from these signs may therefore be deficient. The new sequence of categorical signs obtained does not fit the sound shape established in the given language, and this causes the difficulties in finding the word.

As was discussed earlier, the categorical signs in the categorical component of word meaning also represent so-called hidden grammar, which determines the semantic "combinability" of words (Katznelson, 1972). For example, verbs containing the categorical sign of instantaneous action cannot be combined with words having the categorical sign of duration: one cannot say "he dropped a glass for a long time." Such defining words as "purposely," "diligently," "lazily," "negligently," and so on, are compatible only with verbs that have the categorical signs of "voluntariness" and "intentionality": one cannot say "he was sleeping negligently" or "he was sleeping diligently" (Katznelson, 1972, 1986). Also, polysemy of one word in a word combination may be eliminated through the categorical signs of the other word (Katznelson, 1972). For example, the verb "force" has different meaning in the following two sentences: "The father forced her to change a decision" and "Rain forced her to return home." The categorical component of the verb "force" includes the categorical signs: causality → action against will → intentional action. For the causative meaning of the verb "force" to be expressed, there should be two partners in context: an initiator (agent) and a performer. The difference in the meaning of the verb "force" in the first and second sentences is determined by the categorical signs of the words *father* (agentivity) and *rain* (non-agentivity). In the second sentence, the verb does not have causative meaning (Katznelson, 1972).

In this light, we may suppose that depletion of categorical signs, the mechanism of partial categorical thinking disorder and word-finding difficulties, may also lead to impairment in understanding contextual connections and to the production of semantically incompatible word combinations in spontaneous speech.

As units of hidden grammar, categorical signs provide increasingly fine semantic discrimination, so that the presence or absence of one categorical sign may change the general meaning of a word. Katznelson gives an example of two verbs in the Russian language—*to throw* and *to drop*—which differ in that the categorical component of the verb to throw includes the categorical signs of voluntariness and deliberateness, whereas in the verb to drop, the categorical signs of nonvoluntariness and nondeliberateness are present. Thus, patients with deficient categorical components of word meaning may have significant difficulties in understanding slight difference (nuances) of word meaning, above all with synonyms.

As outlined, we distinguish a selective disorder of word meaning at the symbolic (language) level. In this disorder, anomia, partial and specific disorder of categorical thinking and disorder of semantic grammar constitute a single syndrome. We will call this symptom complex *lexical* (*logico-grammatical*) *aphasia*.

It is of interest here to touch on the dispute in the literature regarding aphasic disorder in patients with closed head injury (CHI). The most frequent type of aphasia observed after CHI is an anomic aphasia (Heilman, Safran, & Geschwind, 1971; Thomsen, 1975; Levin, Grossman, & Kelly, 1976; Knopman et al., 1984).

This is consistent with data that the dorsolateral surface of the temporal lobe is a frequent site of contusion in patients with CHI, whereas other regions of the speech area are rarely contused (Courville, 1942). Based on the "coup-contrecoup" mechanism of craniocerebral injuries, the orbitofrontal and contralateral temporal regions are frequently damaged at the same time (Courville, 1942; Heilman et al., 1971). Anomic aphasia was seen in those patients with CHI who had right orbitofrontal plus left temporal lesions (Heilman et al., 1971). Thomsen (1975) indicated that patients with anomic aphasia after CHI had, in addition to naming disorder, difficulty understanding synonyms, antonyms and metaphors, and describing a series of thematic pictures. Other authors had noted what they termed "non-aphasic" language disturbances in CHI patients that could not be attributed to dementia and that consisted of unusual and peculiar phraseology and using words and phrases in combinations that made their meaning difficult to understand (Weinstein et al., 1952). In many cases these difficulties in linguistic function persisted after the naming disorder had partially recovered. Some authors argued that CHI patients have a disorder of "communicative competence," with inappropriate communicative behavior that is qualitatively different from aphasic disorder. It is our hypothesis that the difficulties observed in CHI patients—the disorder in synonym–antonym understanding together with use of semantically incompatible word combinations and other difficulties in narrative speech—represent a semantic grammar disorder that is part of the lexical (logicogrammatical) aphasia syndrome we have already described. In anomic aphasia in patients with damage only to the left temporal area, semantic-grammar deficits will not stand forward, covered by more obvious word-finding difficulties. The problem in CHI patients with anomic aphasia is that they also have damage to the right orbitofrontal area. In the context of the resulting "right orbitofrontal syndrome" symptoms, such as impulsiveness, disinhibition, increased speech activity, and difficulty in self-monitoring, the language disorder in these patients is modified. In particular, the semantic-grammar deficits are present in exaggerated, grotesque form.

In summary, contradictory data regarding perceptual disorders and thinking disorders accompanying anomic aphasia might be attributed to the fact that what is called anomic aphasia in the literature is a heterogeneous condition including two independent and topographically separate syndromes: disorder of word meaning at the gnostic-praxic level, which we call *visual anomia*, and disorder of word meaning at the symbolic (language) level, which we have termed *lexical (logicogrammatical) aphasia.*

There are descriptions in the literature of highly selective naming disorders that indicate the possibility of even more limited impairments within both the gnostic-praxic level (Warrington, 1975; Warrington & Shallice, 1984) and the symbolic level (Hart et al., 1985).

In clinical practice, small focal lesions involving separate areas within field

37, which would lead to highly selective naming disorders, are rare. Most often one encounters more extended lesions of the left temporal region that result in disorders affecting both function levels. An example of such a complex combined disorder is described by Kock (1967) as the "temporal syndrome according to dominant type," in which disorders of language, visual gnosis and thought are closely interwoven. However, distinguishing those patients suffering from selective naming disorders at the gnostic-praxic or the symbolic level is important, because these two groups require very different strategies in rehabilitation work. What follows is a detailed description of a patient with logico-grammatical aphasia (Glezerman, 1986).

Patient R, a 50-year-old right-handed college graduate, presented with mild-to-moderate naming deficit one year after suffering a stroke in the left temporal-parietal region. Upon examination, his full-scale (FS) IQ was 108 with verbal IQ (V-IQ) of 102 and performance IQ (P-IQ) of 116. He had no difficulties in distinguishing oppositional phonemes and in comprehending words and sentences. In R's expressive speech, what was most striking was the marked paucity of spontaneous speech production; when he did speak, his speech was characterized by a decrease in content words and an exaggerated reliance on grammatical words and syntactical structure. For example, when repeating a story told to him in the indicative mood, he produced sentences in the subjunctive mood. There were no difficulties with word articulations, but there were pauses, word-finding difficulties, and occasional paraphasias based on semantic similarities of words. Similar difficulties were observed in naming: increased latent period and a few paraphasias (for example, instead of arm, he said "forearm"; instead of brown, "violet"; vase, "plate"; side, "part").

On the vocabulary subtest, R performed in the below average range (8). His level of concept formation fluctuated from situational to functional to categorical, although categorical responses were relatively rare. There were verbal paraphasias based on similarity of one or two categorical signs. For example, *fortitude*—"goal directedness."

On the similarities subtest, his scaled score was 9, in the low average range. There was again an unevenness of responses that indicated different levels of abstraction. There were answers that corresponded to the categorical level of abstraction: *dog–lion*—"animals"; *north–south*—"parts of the world." Other responses were based on the common functional sign: *orange–banana*—"one can eat them, edible ... taste." Some responses we interpreted as resulting from his specific deficit—loss of some categorical signs leading to incomplete categorical component of word: *dress–coat*—"fabric"; *axe–saw*—"iron"; *chair–table*—"wood" (see detailed analysis of these responses in the previous section, Figure 13).

In object classification, there was a deficiency due to the peculiar narrowing of the categorical component of word meaning, a falling out of either more general or more specific categorical signs. However, the so-called categorical attitude was preserved because the patient used the categorical principle in sorting out objects. When he was presented with cards of pictures of geometrical figures differing in shape, size, and color and asked to classify them, he united all the rectangle figures into one group and all others into a second group and named these groups: "rectangles" and "not rectangles." When asked to do it differently, he then put all the triangles in one group and all other shapes into another and labeled them "triangles" and "not triangles." After grouping by every shape, he went on to use the same "strategy" using color and size, showing no difficulty in switching from one sign to another. Thus, although the patient was able to use the categorical principle of classification, as in his use of the sign of geometric form, he had lost the more general categorical sign "form" and operated with more specific forms: triangle, square, rectangle, and so on.

In the three subtests that assess verbal logical thinking (Vocabulary, Similarities, and Arithmetic), R's Vocabulary and Similarities scaled scores were below and low average, whereas his score in arithmetic was significantly higher than average (14 versus 10). Analysis of patient R's profile and performance of subtests that include common factors will help explain this discrepancy. The common factor assessed in both Vocabulary and Similarities subtests is the level of concept formation. R's performance in these subtests demonstrated an unevenness in his level of concept formation, a very strong indicator of decreased ability compared to his premorbid level. The low score in Vocabulary is also not due to long-term memory deficit or lack of general knowledge: the patient's scaled score on the Information subtest, which assesses long-term memory of learned general knowledge (which in turn depends on education and cultural environment) was 14 versus the average of 10.

Although Arithmetic has the verbal logical thinking component in common with Vocabulary and Similarities, it also includes spatial analysis, measured in more pure form in the Block Design subtest. On this subtest, patient R received an extraordinary score of 19, near the highest possible 20. We assume, therefore, that his high Arithmetic subtest score is due to his exceptional spatial ability, which we see as a feature of his premorbid neuropsychological profile.

In the other nonverbal subtests, patient R scored in the average or low average range (picture completion, 9; object assembly, 10; and picture arrangement, 9). Although Object Assembly has spatial ability in common with Block Design, its emphasis is on spatial synthesis at the object level, which in turn overlaps with visual object gnosis and visual attention, measured by Picture Completion. Patient R had a very mild deficit in visual object gnosis, which may influence performance on Object Assembly and Picture Completion. Regarding Picture Arrangement, which includes ability to organize situations, emotional orientation to the situation, and visual gestalt thinking, we think that his low average score reflects features of his premorbid neuropsychological profile, which did not rely on a great deal of right hemispheric visual situational or right hemispheric symbolic thought. In conclusion, patient R's FS IQ of 108, although in the normal range, can be considered as decreased from his premorbid level. His major deficit is in the categorical component of word meaning, and unfortunately, he will not derive a great deal of assistance from the right hemispheric equivalents of word meaning because right hemispheric thinking is not one of his premorbid assets.

3

Temporal Region and "Sound-Articulate" Speech

3.1. Phylogenetic Connection between Phonological Language Code and Thought

In Chapter 2, we have given a detailed analysis of the cerebral organization of word meaning, what is signified (in linguistic terms). In this chapter, we will consider the cerebral basis of another side of the word as a language unit—word sound, the signifier. Human *sound articulate* speech as a new ability (compared with the sound communication of animals) was connected with the formation in phylogenesis of the left hemispheric cognitive mechanism, which determined a new form of thought. We will attempt to analyze the sources for human sound articulate speech and those historical stages of anthropogenesis that immediately preceded it. Such analysis will allow an understanding of how thought in modern man might be impaired when the sound (phonological) language code is damaged, and what phylogenetically more ancient "layers" would be exposed. The latter may play an important role in the theory and design of rehabilitation techniques in aphasia.

The complex topics of the origins of language and thought, and their interrelationship, has intrigued scientists and philosophers alike for centuries. We are interested solely in the neurolinguistic aspects: the structure of language and thought in connection with brain phylogenesis (cerebral organization of symbolic functions from an historical view). Paleoneurological data (obtained by examination of plastic casts—endocrane molds—of skulls of fossil man) about the transformation of brain macrostructure in the process of anthropogenesis have shown that brain "hominization" had been going on unevenly, by stages. At the early stages of anthropogenesis, as the analysis of endocrane molds has shown, the brain of fossil man did not substantially differ from that of anthropoid apes. The prehominid (australopithecine) brain was characterized by the "sphericity of the surface of all lobes that is nowhere disturbed by epicenters of more intensive growth; very insignificant frontal and lower parietal lobes and relatively large occipital lobes" (Kochetkova, 1973). The Australopithecus is believed to have been an erect creature who quite systematically acted with natural objects as tools and could even improve upon these objects. According to Kochetkova (1973), systematic action with natural objects indicates that, at the given stage of anthropogenesis, the concept of an object's usefulness (instrumentality) has developed. Anthropoid apes, on the other hand, can use objects as tools only when the objects are within the zone of vision (Vygotsky, 1934/1962). One may assume that the very first achievement in the evolution of mental processes was internalization of the visual situation in which objects were used. Comparing the hypothetical brain structure of prehominids with the character of their instrumental activity and trying to determine the highest brain functional level of this stage of evolution (the leading one, according to Bernstein), we arrive at a level intermediate between levels C and D. It should be noted that object action at its initial stage in evolution has considerable differences from praxis in modern man.

According to Bernstein's definition, praxis corresponds to that functional level in the organization of voluntary actions where their coordination is determined by the object (afferentation of this level). It is not current afferentation, input of information at the moment of action, but inner topological scheme determined by the semantic essence of the object. The semantic essence, in turn, represents a combination of the functional signs of the object and is closely connected with the analysis of the object's features, i.e., left hemispheric mode of information processing. The hypothetical brain structure of prehominids gives no evidence that at the early stages of anthropogenesis, the interhemispheric specialization, characteristic for gnostic-praxic level of modern man, took place. The representation consisted of a unity of the whole object image with the action that should be performed with it in the situation, and not yet separated from the situation itself (visual-action situation). The semantic essence of the object, thus, was limited by its usefulness in the singular situation (singular action). As to the modal-specificity of afferentation of these rudiments of praxis, we suppose that the leading and prevailing one was kinesthetic. Elaboration of kinesthetic gnosis does not require a higher order functional level because its morphological base in modern man is the secondary cytoarchitectural fields of the postcentral parietal region, which was well developed in the anthropoid apes.

The liberation of the hand resulting from erect posture and the subsequent constant manipulation with objects increase tremendously the volume of kinesthetic information supplied to the cortex. This, in turn, promotes the development of kinesthetic praxis (a great number of new hand postures and positions). Development of kinesthetic praxis results in elaboration of hand motor activity, which is expressed not only in systematic use of natural objects but in their reconstruction for more optimal use (Kochetkova, 1973). The presence of inner representation of "object-action-situation" makes possible the imitation of action without the object, or gesture. The connection of imitating gesture with kinesthetic analysis is illustrated by disorder of gesture activity in patients with lesions in the left postcentral parietal region (see examples in chapter 2).

Authors who had studied language phylogenesis had come to the conclusion that gesture was the major means of communication before sound articulate speech developed (Levy-Bruhl, 1930; Vygotsky, 1962; Blonsky, 1935; Ivanov, 1978), emerging early in anthropogenesis yet retaining considerable importance in some cultures up until this century. Gesture language is important in the prespeech period of ontogenesis, when visual-action thinking predominates (Piaget, 1977). The close relationship between kinesthetic gnosis, action-situational thinking, and gesture language is illustrated in the complex gesture language of some American Indian tribes studied by 19th century ethnologist Cushing (1892) who reported "they can think with their hands as modern man sometimes can think aloud" (cited by Ivanov, 1978, p. 62). In primates and other animals, body poses, expressions and sound signals can be behavioral manifestations of emotional reactions in response to situations; that is, the animal's signals are expressions of emotion caused by the particular visual situation. These signals then indirectly become the sign of this situation, performing a communicative function. Expressive movements and sound signals are typical right hemisphere language, in which expression of emotion and information about the situation are inseparable. The prehominids maintained these methods of communication conveying emotional state, which grew to play a complementary role as gesture language developed conveying messages about how to act with the object in the particular situation. Gesture as a communicative instrument represents in the early stages of anthropogenesis a right hemispheric language in which there is no division into signified (the meaning; here, action with the object in the situation) and signifier (here, gesture or imitation of action with the object in the situation).

We think that the emergence of situational thinking accompanied by kinesthetic analysis as the leading afferentation and gesture as a means of communication distinguishes prehominids from the animal world. Situational thinking may be considered the most ancient phylogenetic "layer" in modern man's psyche.

The major changes in brain structure organization that characterize the next stages of anthropogenesis make possible further developments in praxis, thinking, and language. Based on examination of endocrane molds of fossil man, the first focus of intensive growth emerges in Pithecanthropus in brain regions comparable to tertiary cytoarchitectural fields 37, 39, and 40 in modern man. The cultural remnants and tools of Pithecanthropus lead anthropologists to suggest that, at this stage, an internal representation of the form of a tool which they will manufacture in the external world emerges (Kochetkova, 1973). We think that the internal representation of form is possible if features and signs of the object are distinguished and analyzed; this may reflect cortical differentiation "along the horizontal and along the vertical" provided through the formation of the new fields 37 (analysis in visual-object perception), 39 (analysis in visual-spatial perception), and 40 (analysis in kinesthetic-spatial perception). In fact, analysis of features and signs of objects may indicate the beginnings of information processing characteristic of the left hemispheric cognitive mechanism. Here we may trace the mutual, interrelated, and interdependent phylogenetic development of *intra-* and *inter*hemispheric differentiation with the leading function level (vertical differentiation).

Morphological asymmetry of the skull of fossil man, as indirect evidence of functional hemispheric lateralization, was found first at the later stage of Neanderthal (Kochetkova, 1973; Abler, 1976). The focus of growth in fields 37, 39, and 40 discovered in the Pithecanthropus, continues to develop in subsequent stages of anthropogenesis, widening and dividing into two epicenters; the range of individual variability of these regions also increases, which Kochetkova (1973) connected with the process of further differentiation of these fields. Structural differentiation within the cytoarchitectural field might indicate functional differentiation "along the vertical." The interconnection and interdependence of hemispheric lateralization (mode of information processing) and vertical differentiation (the leading functional level) is expressed in stages or phylogenetic steps, each characterized by its own leading functional level and corresponding extent of interhemispheric specialization, and culminating with formation of the left hemispheric cognitive mechanism (categorical thinking).

At earlier stages of anthropogenesis, the visual-action situation became an internal representation, thus not needing the presence of actual situation in outer visual field. The internal representation is of the whole image, in which the object itself is not isolated from the action related with it in a single situation.

The first step in analysis of the visual-action situation is, we think, distinguishing of the concrete and concrete-situational signs. These signs characterize belonging of the object to visual-action situation: they do not connect with each other directly but through the situation; they are markers of the situation, giving certainty to this singular image. Thus, recalling the statement that the left hemisphere operates with continuous combinations of discrete units, we may say that at early stages of the left hemisphere cognitive mechanism formation, there is a discreteness in information structuring (distinguishing of signs), but continuity of sign series is not yet developed. One may suggest divergence with further phylogenetic development, when cognition became more and more dual: on the one hand, there is a "condensation" of situation-series, figurative image of action with the object, or image symbol of object function. On the other hand, there is analysis of situations with distinguishing of object signs, which are useful for action with it in several situations (situation series); that is, generalization based on the functional signs emerges. In connection with the functional sign, new classification of the surrounding world emerges. It characterizes objects as such but does not contain the object's "inner" connections and relationships; the indirect relation to the situation is still preserved. It is only by means of distinguishing an object's signs, that is, abstraction from the whole (object in situation), that man in phylogenesis is able to represent the object as a separate image. The distinguishing of functional signs as "left hemispheric" discrete units paves the way for the formation of the left hemispheric synthetic image, Bernstein's topological scheme of the object, which, in turn, allows actualization within the right hemispheric, condensed visual-action situation series—object image. The image of the separate object, on the other hand, is the basis for the analysis of the group of objects, distinguishing the signs that are common for the group and singling them out from objects of other groups (categorical signs). Because of this, at the next stages, categorical classification arises to serve as a system of rules ordering the object world, containing inner connections and interrelationships of objects. Thus, categorical and concrete thought have a common origin: they both are connected with the formation in phylogenesis of the left hemispheric cognitive mechanism. The difference between these two kinds of thought reflects the hierarchy of functional levels provided by cortical structures of different phylogenetic age.

Let us now return to that stage of brain phylogenesis at which fields 37, 39, and 40 start forming, and examine what new developments emerge in praxis, thinking, and language at this stage. In praxis, the functional role and usage of the object has become the semantic essence. Praxis, consequently, becomes similar to that of modern man. In thinking, "sprouts" of two new types of thought emerge, mostly related to field 37 formation: categorical and symbolic-situational. In language (gesture), spatial configuration of gesture becomes more complicated, and, probably, elements of conditionality emerge.

We think that there is no hemispheric specialization at this stage; both sprouts of thought are developing within each hemisphere. In general, it is the right hemisphere–like cognition within which elements of left type analysis emerge. Both sprouts are not separated yet and in whole thought might be characterized as syncretic and undifferentiated. Clinical data regarding consequences of brain damage occurring at different stages of ontogenesis give some evidence that at the early stages of ontogenesis, analogous to our understanding of phylogenesis, both hemispheres have the potential for left or right types of cognition.

This can be illustrated by data regarding speech function, the product of left hemispheric cognition in the norm, after unilateral brain damage in children (Basser, cited by Lenneberg, 1967). The data showed that at earlier stages of development (before age 2), both the left and right hemispheres have similar potential for subserving language. At later ages (2 to 10 years), left hemisphere lesions cause far more speech disorders than right, although the percentage of language disorder in this age group following right hemisphere lesions is still far greater than in adults, in which right hemisphere damage results in aphasia in only 3% of cases, the majority of whom are left-handed (Basser, cited by Lenneberg, 1967). It appears that at earlier stages of ontogenesis, the left and right hemispheres have similar potential to subserve language, whereas at later stages this function becomes more and more lateralized and connected with the left hemisphere. Thus, we suppose that division of "right" and "left" cognition starts within each hemisphere, and it is only at later stages that hemispheric specialization (for different type of thoughts) takes place.

The development of the two polar ways of cognition deepens the distance between them. This, in turn, facilitates further development of each one: reducing of situation-series (situational symbolic thought) and analysis of object image with distinguishing of categorical signs (category of "objectness" emerges). This stage cannot be thought of without sound articulate speech. As Katznelson (1972) aptly notes, "The world in reality is in no way a warehouse containing on its shelves classified objects and signs that only wait for us to hang an outer tag, a word–name, onto them.... The view that the world is a huge collection of random objects and signs is, in fact, brought about by the structure of language" (p. 141). While an object is named, the signs characteristic for it and distinguishing it from other objects are "gathered into a bundle," becoming actualized. On the other hand, connections between signs common for a group of objects are built (continuity of the left hemisphere cognitive mechanism). Thus, distinguishing of objects in sound articulate speech (naming) allows the establishing of their inner (by signs) connections and relations; that is, a new type of outer world classification—categorical classification—emerges. Again, in other words, at the stage of their phylogenetic formation, categorical signs and sound articulate speech were closely connected. This unity—"language–thinking" was preceded by the unity "perception–thinking." Thinking presented a "superstructure" over the corresponding perception type, which is most clearly expressed at the initial stages of anthropogenesis: kinesthetic gnosis-praxis—action-situational thinking.

It is important to emphasize that the first focus of intensive growth in the direction of brain "hominization" has been observed in the region corresponding to field 37, which, in modern man, is related to word meaning, that is, the signified. We have no data regarding emergence at that stage of a cortical region that is related in modern man with signifier (sound language code).

The transition from syncretic-perceptory representations to functional signs allows classification of the surrounding world—creates and broadens the "world of meanings." This broadening of the world of meanings requires an increase in word number. We assume that language, at the stage that we discuss, was represented by gestures and sound signals. It is the right brain principle of information processing that underlies this language: both depicting gesture and sound signal are not separable from the meaning being conducted thereby. However, the number of different sounds in stock, sounds that are suitable to convey the message, are not large. This limitation is confirmed by the low number of elementary sounds that are present in modern languages (Ivanov, 1978). Hocket, Ascher (cited by Cheif, 1975) indicated that the articulatory-acoustic space available was getting more and more densely packed; some sounds became so similar to each other that to distinguish them in both pronunciation and perception presented great difficulties for the human mouth, ear, and brain. Something needed to happen, or the whole system would have collapsed under its own burden.

The vocabulary of gesture language was also limited. The limitation is set by the quantity of depicting gestures to convey a message (imitating a single visual-action situation). This is confirmed by the extremely poor lexicon of gesture language in deaf individuals who have not learned special sign language (Ivanov, 1978).

One more fact should be noted. The number of sound signals used by different animal species for communication is approximately the same and corresponds to the number of phonemes in man; this latter is also similar, varying only within a small range, in different languages of the world (Ivanov, 1978).

Thus, at a certain stage of anthropogenesis, the expanding world of meanings might have encountered an obstacle to further development in the limited capacity to communicate these meanings. Before we attempt to speculate as to *how* this possible obstacle was "removed" with continued brain development, let us examine the structure of the signifier in modern language, which gives inexhaustible possibilities for transmission of information. Linguists indicate the *double discreteness* of language: "Both the world of concepts and the world of sounds are organized discretely" (Cheif, 1975). Both the semantic and sound plans of language have their own "division." This is supported by the cerebral organization of language ability. Indeed, as we understand it, left hemispheric representations do not exist as a whole but as a definite combination of discrete units that can be equally applied to both signified and signifier. The "divisions" of both language codes (semantic and sound) follow the basic rules of the left hemisphere cognitive mechanism but operate with different units. The units that constitute the signified are formed as a result of categorical recognition of the object world. They represent a logical inner framework of language to serve as a base for the classification of words. Units of the signifier are *formal* (conventional): they do not mean anything in and of themselves. The sound of the word is, quite literally, its code, the definite number and sequence of the conventional signs that code the meaning of the word. Although the number of conditional sound signs in the given language is limited (the alphabet), the number of their combinations is astronomically high.

According to modern linguistic concepts, it is the phonological system that underlies the sound aspect of language. It is represented for each language by a characteristic set of discrete signs of speech sounds, so-called distinctive features.

Certain combinations, bundles of distinctive features, form a phoneme, which is nothing but the constructing and defining unit of the signifier (Jakobson & Halle, 1956).

The description of the sound aspect of language in terms of distinctive features corresponds to the left brain principle of information processing. Definition of the phoneme as a bundle of distinctive features is compatible with the statement that left hemispheric representations do not exist in a whole, integral form, but merely as combinations of units–signs.

Signifier and signified are related to different regions in the left hemisphere, the temporal and temporal-occipital areas, respectively. In phylogenesis, cerebral functions of the supramodal symbolic level connected with the signifier and signified must have been built upon different modality-specific functions: auditory and visual. Besides these differences, there is a difference in principle of abstraction of processed information within the symbolic functional level while signifier and signified are formed. Processing of visual information (field 37 of the left hemisphere), we believe, takes place in parallel at two levels: generalization according to functional signs (gnostic-praxic level) and generalization according to categorical signs (symbolic level). Categorical classification of the surrounding world corresponds to the symbolic level of visual information processing. Categorical classification is, on the one hand, an autonomous parallel line of abstraction; on the other hand, it is a superstructure that is built upon the previous level—gnostic-praxic. Both levels reflect the surrounding world (left hemispheric semantic units).

In processing of acoustic information at the gnostic praxic level, we deal with the functional signs of speech sounds that are physical, acoustic/articulatory signs. The generalization of acoustic information at the symbolic level (phonological language code) has certain peculiarities that arise from the specificity of its relations with the lower, gnostic-praxic level of abstraction. According to Jakobson (1976), language uses the world of sound, sound matter, but selects in it some elements, dividing and sorting out according to its needs.

Sound, for language, is the form, the conventional sign that in itself does not mean anything but serves to differ meaningful units of language (morphemes and words). According to Trubetskoj (1960), the signifier in language consists of a certain number of elements whose essence is that they differ from each other. Every word must differ somehow from all other words in the same language. Applying this "signifier task" to the cerebral symbolic level, it might be hypothesized that the symbolic (language) level "selects" distinctive features from speech sounds, the discreteness of which is defined by their differences in relation to each other. Jakobson emphasized that, in phonological code, "It is not things that matter but relations between them; what is important in sound for language is not absolute signs but relative signs, distinguishing sounds from one another" (Jakobson, 1971a, p. 225). The task of distinguishing determines the structure of the phonological system as a combination or series of distinctive features of speech sound. The selection of phonological signs, the task of distinction of word sounds, is dictated by the language system (or, in terms of cerebral correlates, by the symbolic level), which will "extract" these features from "sound raw material" (Jakobson, 1976). The terminology used for the distinctive features of sounds in the phonological system is that of their physical counterparts, their correlates at the gnostic-praxic level—that is, acoustic and articulatory signs of speech sound. For example, the consonant *b* is distinguished by two distinctive features: combinations of place of articulation (labial) and voicing (acoustic characteristics). One of the models of phonological systems proposed by Jakobson states that the phonological system of a given language consists of binary oppositions, in which each phonological sign is a member of an opposition—that is, one variant of a two-choice selection. In binary opposition, one member represents a neutral, or unmarked (−), variant, and the other is characterized by the adding of some quality, the marked variant (+). For example, binary oppositions can be characterized by voicing(+)/voicelessness(−); compactness(+)/ diffuseness(−); acuteness(+) versus gravity(−); stridency(+) versus mellowness(−) (Jakobson's examples of binary oppositions, taken from the phonological system of the Polish language, Jacobson, 1971b, p. 312).

Thus, the divisions of signifier and signified in language differ significantly. The division of signified is connected with classification of the surrounding world. The signifier is the form for the signified. To be form, the signifier should not have its own meaning in language; therefore, features of speech sound do not have meaning in themselves, and the sounds of words are purely conventional.

In our attempt to reproduce phylogenesis of sound articulate speech, we may trace two parallel, relatively independent lines of its development, lines that have different roots. The first line is concerned with distinguishing of functional signs. As we have already discussed, many researchers believe that at the stage preceding sound articulate speech, communication took the form of action–performance, including imitative and indicative gestures. Analysis according to functional signs followed by synthesis (topological scheme according to Bernstein) are not merely afferentation of object action but became the basis for imitative gesture (topological scheme in the kinesthetic modality) and indicative gesture (topological scheme in the visual modality). At a certain stage, the gesture to imitate action with the object and the gesture to indicate the object might have been supplemented with "voice gesture," a sound expression that is similar to that of the specific sound connected with the object action (Blonsky, 1935; Levy-Bruhl, 1930). The similarity between speech sounds and natural sounds gives additional indication that speech could have originated from the imitation of sounds of the environment. It has been noted that the imitating activity of anthropoid apes, although embracing the movements of body parts, does not include speech apparatus, or imitative sounds (Bunak, 1966). There is some evidence that at the latest stage of anthropogenesis, beginning with Neanderthal man, changes in the organ systems of the larynx and mouth cavity arose that brought their structure nearer to the speech apparatus of modern man (Ivanov, 1978). In addition to the hypothesis of imitation, Bunak also proposed the idea of a chance combination of sound signal and object action and the subsequent fixation of this connection.

The new type of communication represents a synthesis of two previous methods of communication, sound signal and gesture. The need to free the hand may have been one of the driving forces for this peculiar interiorization of gesture, in which emerged the different "poses" of tongue, lips, soft palate, and so on, required for utterance of different sounds. Constant input of kinesthetic information from muscles of the speech organs paves the way for the formation of an internal topological scheme, articulatory images classified by sound type. As a result, a new type of praxis is formed—articulatory praxis. If praxis is action with the object, what, then, is the object? We think it is *sound*. Functional signs—those aspects of the object that determine how to act with it—in speech sounds are those acoustic features which are formed in specific "articulatory poses." It is the articulatory origin of speech sounds that distinguishes them from all other acoustic signals.

Speech sounds are characterized by their own articulatory patterns, by the place and type of their formation. The Russian consonant *m*, for example, is formed with participation of the lower lip and teeth; the English *n*, *l*, and *r* are apical consonants, formed with participation of the anterior part of the tongue (place of formation). In Russian *m* and *n* are formed with shape change of the oral cavity resonator; *l* is formed with the speech organs closing up and leaving a unilateral passage for the air stream (type of formation).

The emergence of articulatory praxis means that a particular sound changes to a fixed utterance; further development of articulatory praxis results in more precise definition of articulations and mastering new articulations. The emergence of articulatorily fixed sounds, the widening repertoire of articulations, and their combinations into syllables is considered by some authors as the initial stage of sound articulate speech (Bunak, 1966). At the corresponding stage of brain phylogenesis, there is as yet, we believe, no definite "separation" of left and right cognitive mechanisms to within a single hemisphere. Nevertheless, the existence of functional signs of sound and their articulatory features at this stage indicates the emergence of the left hemispheric mode of information processing. On the other hand, articulated sound may be a sound symbol of action (right hemispheric reduction of situation series). The early development stage of the signified (as we speculated in chapter 1) is also characterized by duality: topological scheme in the visual modality (left hemispheric type of information processing) coexists with a visual symbol of action (right hemispheric reduction of situation series).

The complex of articulations that had served at the early stage of sound speech development to implement communication is not the signifier in the sense we imply when defining language phonological code. The sound of the word at that stage was not yet separated from word meaning (sound symbol of the situation). Moreover, sound and visual image–situations likely composed the whole right hemispheric representation. Authors have indicated that, at the early stage of its historical development, speech probably had elements of singing: change in intonation, rhythm, frequency, and other individual characteristics of sound changed word meaning (Blonsky, 1935). We assume that these rhythmic and melodic parameters were included into the whole image, gestalt, visual-acoustic representation, which was at the same time both word and message. Speech at these early stages was not, we think, articulate speech as it is today. Articulatory complexes—"words"—were not successive sequences of separate units, a left hemispheric cognitive process, but the result of *agglutination* of sound symbols formed in accordance with right hemispheric cognitive rules. As Blonsky (1935) stated, "Initial words were very short; however, through agglutination of these very general words, highly image-laden speech strings could be produced" (p. 97).

In language, Jakobson wrote, there is neither signifier without signified nor signified without signifier. If we paraphrase this thesis, we may say that at the early stages of human articulate speech there was neither signified nor signifier. At these stages there did exist, however, the rudiments of the empirical component of word meaning (signified), as well as fixed speech sounds (articulatory praxis), which is the basis for the further use of sounds for language (signifier).

We conclude that the relatively autonomous line in the development of articulate speech discussed here reflects formation of the gnostic-praxic level of the emerging speech functional system. The cortical zones providing for articulatory praxis are secondary cytoarchitectural fields in the inferior part of the postcentral region connected with kinesthetic praxis. Auditory speech gnosis, which serves as afferentation for articulatory praxis, is provided by secondary cytoarchitectural fields 22 and 42 in the superior-posterior temporal region. Blinkov, who examined in detail the cytoarchitectonics of the temporal cortex in different animal species and in the human brain, has shown that fields 22 and 42 are present in anthropoid apes and undergo only slight differentiation in man. Blinkov found field 22 characterized by mainly the same signs not only in the orangutang and chimpanzee but in nonanthropoid, lower primates as well (Blinkov, 1955). Note that we are speaking here about Wernicke's zone, the area of the cortex that, when damaged, has been connected with sensory aphasia!

The second line we trace in the development of human articulate speech is the symbolic (language) level in the emerging speech functional system. We suppose that the development of the symbolic functional level was possible and connected with the fundamental reorganization of the brain beginning with the Neanderthal man. This included (1) the emergence of hemispheric asymmetry, and (2) the division of the initial focus of growth in the temporal-parietal-occipital region into two epicenters: anterior, corresponding to field 40, and posterior, corresponding to fields 39 and 37 (Kochetkova, 1973). The posterior epicenter, according to Kochetkova, expands considerably faster than the anterior, with field 37 in particular undergoing even further differentiation (recall that we have connected field 37 of the left hemisphere, in modern man, with categorical signs, the world of meanings). Paleoneurologic data from the Neanderthal also indicate that the posterior epicenter extended along the temporal lobe in the anterior direction, occupying the cortical surface that corresponds, in addition to field 37, to field 21 in modern man (Kochetkova, 1973). Blinkov describes marked differences in the structural organization of field 21 and secondary temporal fields 22 and 42 in modern man: "Field 21 is highly differentiated. It represents a complex system of cortical structures, in which one distinguishes both new, specific for this field features, as well as transitional formations with bordering areas. In apes, it is a transitional structure, gradually turning into bordering areas" (Blinkov, 1955). Blinkov also emphasized that in modern man there is great individual variability within this field, indicating its young phylogenetic age. In modern man, field 21 is a tertiary, specifically human field that is formed as a result of differentiation of the temporal (auditory) cortex; its symbolic function is built upon the auditory, modality-specific system. Is this field that morphologic substratum which enabled the emergence in phylogenesis of the ability to make use of sounds for the formation of language code, or phonological system? There is some clinical evidence that field 21 is concerned with the symbolic, phonological level. Vinarskaya (1971) observed an impairment in the phonologic level with intact gnostic-praxic function within the sound speech functional system in patients with lesions in the left temporal area corresponding to field 21.

Examination of endocrane molds of Neanderthal man has revealed a new focus of intensive growth not only in the area corresponding to field 21 in modern man but also in the region corresponding to field 44 (Broca's zone) (Kochetkova, 1973). Field 44 is situated in the inferior part of the premotor region and is a secondary field implementing speech praxis. It is unusual that a secondary, not tertiary, field developed at the final stage of anthropogenesis. In contrast to the postcentral zones of the cortex that are responsible in modern man for inner topological schemes of separate articulations (kinesthetic-articulatory patterns of speech sounds), the role of premotor field 44 is the inner programming of temporal, successive sequence of separate articulations (sequence that forms syllables and words). Broca's zone is only relevant or useful in the presence of sound articulate speech, where meaning is transmitted through linear sequence of separate units (articulation). At present, it is hardly possible to explain distinctly the mechanism of phonological system formation, that qualitative leap from the whole sound–rhythmical structures to sound code of the word. Available data allow some aspects of this complicated problem to be distinguished. First, it is interesting to note that three events may be attributed to the same stage of brain phylogenesis: the further differentiation of field 37, the formation of field 21, and interhemispheric separation of two polar types of cognition (left brain–right brain). This brain development might be the structure–function base for both the signified and signifier in language. We may suppose that, in its phylogenesis, the signifier was formed together with the signified: analysis of object images with distinguishing of categorical signs (signified) and separation of word sound from visual image. If, previously, not well-differentiated sound-visual complexes made up each individual word message, now the function of distinguishing one word from another will be fulfilled by sound complexes. In other words, the sound complex separated from the meaning of the word becomes its designation (code). In this process, properties of sounds *sui generis* become relevant (in connection with their ability as the coding system); that is, contrasting distinctions of sounds from one another.

Distinctive features of sound, which may be considered as a result of left hemispheric analysis, will consequently constitute the system—the phonological system of the given language. Structuring of distinctive features onto the phoneme may be considered a result of left hemispheric synthesis. Jakobson characterized the phoneme as a differential sign devoid of any meaning (Jakobson, 1976). Formation of the phoneme must go in parallel with the restructuring of articulatory complexes. The phoneme sequence forming the word, the linear sequence of units, is realized through kinetic articulatory praxis, the linear sequence of articulations subserved by Broca's zone.

Thus, in phylogenesis of sound articulate speech, we consider two relatively independent lines: (1) articulatory praxis (the gnostic-praxic level), and (2) sound code (symbolic, language level). They are connected with the development of the different cortical areas, which are formed at different stages of human brain phylogenesis. The need for communication played a significant role in the development of the first line. The second line, the phonological code of language, is intimately connected with thinking—it is the code to translate thought into sound. The code itself is the set of sound "ideas" based on the logical principle of binary oppositions. The symbolic (phonological) level, although developing relatively independently and probably later in phylogenesis, is nevertheless functionally connected with the gnostic-praxic level (sound "ideas" need sound "matter"). Selective impairment of the gnostic-praxic or symbolic levels of the sound part of the speech functional system (see preceding) indicates the necessity of developing new rehabilitative techniques specifically targeting the involved level. This methodology should consider both the relative autonomy of the two levels and their functional connections.

3.2. Left Temporal Region and Sound Code of the Word

The phonological system includes a few distinctive features, approximately the same for different languages. For example, in the Russian language, there are 11 distinctive features whose grouping in various combinations gives 42 phonemes

(Jakobson, Halle, & Cherry, 1962; Jakobson & Halle, 1962). The quantity and character of distinctive features that form the phonological system present the base for the statistic regularities of language; that is, those rules according to which higher-order phonological units—phonemes (combination of certain distinctive features) and words (combination of certain phonemes in a certain sequence)—are formed. These sound sequences are not all equally probable. The potentially possible number of combinations is restricted by the task of the phonological code—distinction of these combinations from each other. Those combinations that contain the danger of confusion are rejected both at the phonemic and word levels. For example, in the Russian language, the vowel is a necessary component of the syllable and cannot be repeated or omitted in the borders of one syllable; syllables are formed based on the contrasts vowel/consonant, consonant/vowel, consonant/vowel/consonant (Jakobson & Halle, 1962; Jakobson et al., 1962). Jakobson indicated that in the phoneme one distinctive feature "appears in combination with certain other concurrent features, and the repertory of combinations of these features into phonemes such as *p*, *b*, *t*, *d*, *k*, *g*, etc., is limited by the code of the given language. The code sets limitations on the possible combinations of the phoneme *p* with other following and or preceding phonemes; and only part of the permissible phoneme–sequences are actually utilized in the lexical stock of a given language" (Jakobson, 1971d, p. 242). The phoneme is the minimal sense-distinguishing unit of the language sound code. Words similar in all other aspects can be distinguished from one another by a single different phoneme. For example, in Russian, the words "*d*otchka," "*t*otchka," "*p*otchka," different in only one phoneme, have completely different meanings (daughter, point, bud). In English, a similar example is "*b*ig," "*p*ig," "*f*ig," "*d*ig." Phonemes themselves may be different from one another in only a single distinctive feature, such as in the words "big" and "pig," in which *b* and *p* differ only in the feature "voicing": one is voiced and the other voiceless (oppositional phonemes). The sense-distinguishing function is also performed by the sequence of phonemes in the word: for example, in Russian "*sde*lat" and "*sed*lat" (to make and to saddle a horse).

Thus, the phoneme does not have its own meaning: its semantic value in language derives from its role in differentiating the meaningful units of language—words and morphemes. According to Jakobson (1976), the phoneme is not a sound but brought-together signs of sound. If we consider the phoneme in terms of its representation in the brain, we may speculate that its distinctive features are units of the supramodal symbolic functional level. Originally extracted from sound, they are reorganized at this higher symbolic level to construct a phoneme, a "sound idea," a left hemispheric representation, not whole but constructed from discrete signs. We think that this process parallels Bernstein's basic conception of the reorganization of space (afferentation) at each successive function level, proceeding from the objective to the schematic, with each level having a different representation based on the original modality, progressing from the "real" toward the symbolic. We have earlier drawn an extensive parallel with the visual modality, proceeding from the visual image to topological scheme to categorical thought (see chapter 2). Here we deal with the auditory modality, with representation of sound as physical features at the gnostic-praxic level and sound ideas, an abstraction, at the symbolic level. This implies that information is processed in parallel at different levels, and that there are bidirectional interconnections between them. We recall Zeki's new work on connections in the visual cortex and his conclusion that backward connections (from higher to lower) exist and function in the same manner throughout the cortex, and that an area with specialized higher functioning will draw on any source of useful information (Zeki & Lamb, 1994). Zeki and Lamb further concluded that the brain does not just analyze the visual world but *constructs* it, and we hypothesized that the brain "constructs" the visual world at every function level, each with its own interpretation. This, we think, is what the symbolic level of the temporal region is doing with sounds: it uses information from lower levels for its own needs, to make a phonological system to distinguish meaningful units of language.

Phonological signs are units that correspond to the supramodal symbolic function level, but they are described in terms of their physical equivalents, acoustic and articulatory signs of speech sound. In terms of cerebral organization, these physical features correspond to the gnostic-praxic level: auditory speech gnosis and kinesthetic (articulatory) gnosis and praxis.

The physical equivalents of distinctive phonological features, acoustic and articulatory, make communication with the outside world possible. The listener identifies the distinctive features in speech sounds; the speaker produces them with the articulatory organs (Jakobson, 1976). According to Jakobson, "The phonemes of a language are not sounds but merely sound features lumped together 'which the speakers have been trained to produce and recognize in the current of speech sounds—just as motorists are trained to stop before a red signal, be it an electric signal light, a lamp, a flag, or what not, although there is no disembodied redness apart from these actual signals'[here Jakobson cites linguist L. Bloomenfield, 1933]. The speaker has learned to make sound-producing movements in such a way that the distinctive features are present in the sound waves, and the listener has learned to extract them from these waves" (Jakobson & Halle, 1962, p. 468).

Language pathology as a result of damage to the left temporal area (sensory aphasia) gives an illustration of inability to distinguish phonological features of the given language. For example, in languages in which the contrast of long and short vowels has phonemic significance, a patient with damage to the left temporal area (sensory aphasia) may lose the ability to distinguish long or short vowels in hearing or his own speech (Jakobson, 1971c). Jakobson pointed out: "There is no question of inability to hear or articulate vowels of longer or shorter duration; what is lost is the distinctive semantic value of the difference between long and short signals in the phonemic code" (Jakobson, 1971b, p. 312).

3.3. Cerebral Organization of Single Word Processing

Figure 14 represents our summary of the processing of a single word, presented auditorily or visually, at the gnostic-praxic and symbolic levels. As shown, auditory speech gnosis consists of two links: auditory-gnostic proper (*a*) and auditory-spatial (*b*). The first link includes analysis of acoustic, "useful-for-speech" signs in the sound sequence of speech flow; the second includes the establishment of the spatial relations between speech sounds or syllables. Articulatory praxis also includes two links: kinesthetic (*c*) and kinetic (*d*) (Luria, 1947/1970; Luria, 1966/1980). Perception of the acoustic parameters of words serves as an afferentation ("object") for articulatory praxis, inducing the kinesthetic image, the topological scheme of articulation specific for producing this sound (*c*). The spatial interrelations of the speech sounds within the word constitute that specific afferentation which is necessary for unfolding in time of articulation sequences; in other words, the acoustic-spatial link of speech gnosis is connected with the kinetic link of articulatory praxis. Thus, while auditory parameters of speech sound are the "object" (afferentation) for the kinesthetic link of articulatory praxis, the "object" for the kinetic link of articulatory praxis is acoustic-spatial parameters of the word. At the gnostic-praxic level, there can be mechanical repetition of a word without any understanding, due to the direct connection between recognition and articulation; if the symbolic level is involved, which is usually the case, repetition includes recognition—comprehension—articulation (see Figure 14).

Processing of auditory parameters of a word is the basis for phonological qualification of sound sequences at the symbolic (language) level (Figure 14, *e*). According to Jakobson and Halle (1962), as the listener receives a message in the language he knows, he compares it to the code he possesses and, by means of the code, interprets the message. We think that these operations are connected with tertiary field 21 in the temporal region of the left hemisphere. Operations may include: 1) recognition of phonologic signs; 2) recognition of phonologic series as phonemes; and 3) recognition of phoneme sequences as the sound code of the word.

The first two steps represent the ability to use the rules (code) of language, which is connected, in our opinion, with symbolic function of field 21. This is realized during the first years of life, when the child is immersed in the speech flow of the given language. The realization of the third step, however, depends not only on the ability to use language rules but also on the ability for categorical and visual-symbolic thinking. This step includes the connection: signifier (field 21, left hemisphere) → signified (field 37, left hemisphere) → right hemispheric equivalent of word meaning (field 37, right hemisphere). All these components taken together form the vocabulary of an individual. Thus, the complex ability to accumulate sound (phoneme) sequences as meaningful units of the given language depends on the relative development, and efficient functioning, of several regions

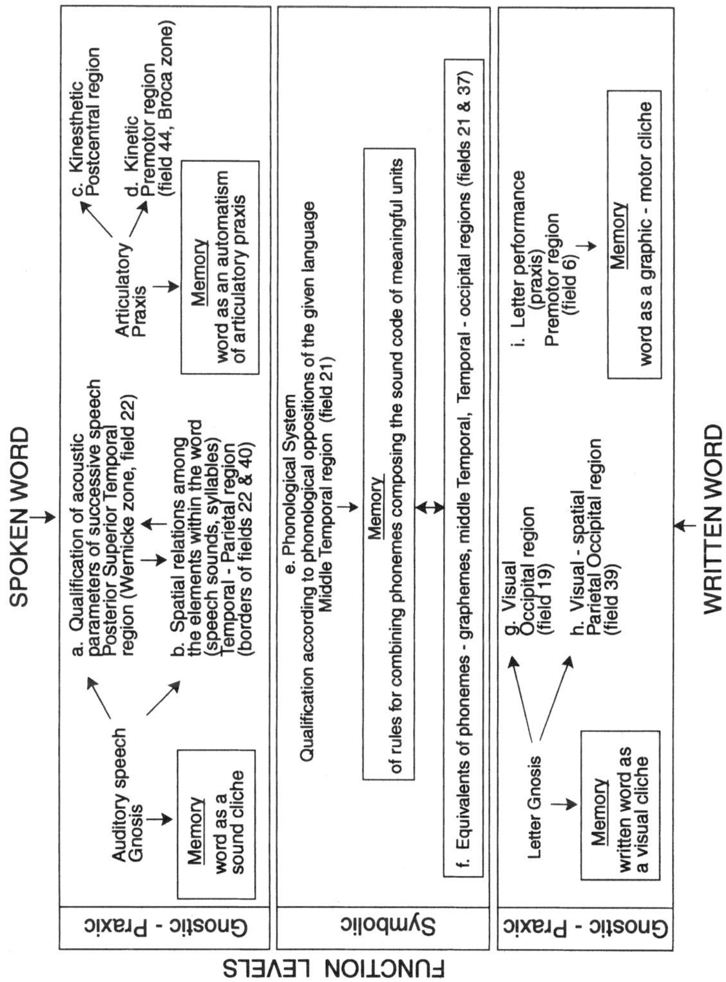

FIGURE 14. Cerebral organization of single word processing.

of the left and right hemispheres. Vocabulary is acquired throughout an individual's lifetime and is to a significant extent dependent upon the quality and volume of speech information received; that is, on the character of the language environment, education, profession, and so on. Possession of a code means the possibility of its decoding: recognition of the sound word code (signifier) activates all connections behind it, resulting in the comprehension of the word. As these components depend on different regions and hemispheres and their interconnections, and considering the relative independence of individual variability of different cortical formations in the norm (Blinkov & Glezer, 1964), the abilities that form one's vocabulary may also vary independently from one another within the individual, producing unique patterns of vocabulary.

If an individual is presented with a meaningful word of his language that he does not know, the phonological processing is stopped at the second step; that is, possessing the statistical rules of his language, the individual is able to recognize that the word belongs to the given language but cannot decode the sound code. Thus, the sound code of a word represents the hierarchical formation that includes combination of distinctive (phonological) features, phoneme, and phoneme sequence (word sound). In written speech, a word is a one level formation, representing a sequence of signs (graphemes), which are equivalents to phonemes. The recoding of a phoneme into a grapheme is an operation of the symbolic (phonological) level (Figure 14, *f*). It is based upon human ability to master the graphic code of the phoneme, established in the given language and acquired in the process of learning to read and write. It is interesting to note that the mean ages at which complete tertiary field formation, hemispheric specialization, and development of the phonological system take place coincide with one another and coincide with the optimal age of learning to read and write. In the grapheme, the phoneme has become a single unit (not a combination of signs); it acquired "wholeness" within the limits of the left hemispheric cognitive mechanism.

The equivalent of the grapheme at the gnostic-praxic level is the letter with its physical parameters: visual and visual-spatial (Figure 14, *g*, *h*). The visual image of the letter at the gnostic-praxic level is a left hemispheric generalized image—topological scheme. Bernstein emphasized that topological scheme is a combination of qualitative characteristics independent from metric and geometric ones, such as size, form, curvature of its outline, and so on. He distinguished each letter as a separate topological class (scheme). For example, *A*'s of small, large, bold, italic types; *A*'s from different handwriting; and so on—all belong to that topological class of letter *A* (Bernstein, 1947). "Regarding the letter as an object, not only its semantic essence depends exclusively on mutual relations of its elements ... but also movements connected with writing a letter are topological as well" (Bernstein, 1947, p. 124). As we discussed before, unfolding of the topological scheme of an object in time (object action) is connected with the function of the left premotor region. Thus, the topological scheme of a letter (visual pattern) is converted into its dynamic scheme (motor pattern) in the left premotor region (see Figure 14, *i*).

Analysis of visual signs of the letter is connected with the secondary (gnostic) fields of the occipital regions, whereas the analysis of visual-spatial signs is connected with the neighboring field 39 of parietal-occipital region. Whereas spontaneous writing requires comprehension of the word, both copying and mechanical repetition may be realized at the gnostic-praxic level, although, as a rule, they are realized with participation of the symbolic level when performed by a person of normal intellect.

The existence of separate cerebral regions for single word processing in the norm has been demonstrated in positron emission tomographic studies. Different and specific cortical areas were activated for modality-specific word form: superior temporal for the auditory word image, secondary occipital fields and temporal-occipital boundaries for the visual word image. The premotor structures, including Broca's area, were activated for the articulatory word image (the subject was asked to repeat the spoken or written word) (Peterson, Fox, Posner, Mintin, & Raichle, 1988).

Figure 14 shows that each link of the word sound processing system includes in itself not only the ability to perceive, produce and understand words but also a corresponding memory connected with the separate gnostic, praxic, or symbolic function. The notion of memory is an extremely broad one, and related disciplines—psychology, neuropsychology, cognitive psychology, neurophysiology—have often concerned themselves with different aspects. Because of this, it is necessary to define more clearly what memory parameters are considered in the given context. As is known, memory is implemented by the functioning of many different cerebral regions. The limbic-hippocampal system is the primary mnestic system, responsible for the so-called general memory factor—the ability to register any new information, which underlies mnestic processes. Among the different classifications of memory are short-term and long-term, which may be disordered selectively depending on the area of damage to the brain. Long-term memory includes the ability to consolidate newly formed "traces" (newly learned information) and to store it (memory archives). Lesions in the subcortical areas involving the reticular activating system result in disorder of short-term memory and the ability to consolidate newly learned information, whereas damage to the cortex results in disorder of short-term memory and both aspects of long-term memory. Traugott (1973) concluded that the ability to store information is connected with the cortex; however, mnestic deficits observed in patients with focal cortical lesions are not global memory disorders but impairments in remembering information that is normally processed in the area damaged. These selective deficits will be in both short- and long-term memory. Traugott thought that there was a common deficit in gnostic (for example, disorder of visual gnosis) and mnestic disorders (memory of visual object images), whereas Luria believed that mnestic deficits generally accompany gnostic deficits but reflect disorder of higher function levels. Based on the observation that mnestic and gnostic deficits may occur independently, we suppose that mnestic and gnostic deficits might have a different localization within the modality-specific area (Glezerman, 1983). In connection with this, it is worth noting that distinguishing only two higher function levels (gnostic-praxic and symbolic) within the cortex is a methodologically justifiable simplification. Bernstein (1947) spoke of the likely existence of transitional levels. In any case, we think that there is a separate memory (short- and long-term) corresponding to the gnostic-praxic and symbolic levels, respectively (see Figure 14).

We will at this point attempt to substantiate our model of spoken and written word processing with examples of syndromes involving selective disorder of the different links in single word processing, presented in Figure 14. Disorder of auditory speech gnosis proper, link *a* in Figure 14, may be illustrated by the syndrome known in the literature as pure word deafness, or verbal auditory agnosia (Auerbach, Allard, Vaeser, Alexander, & Albert, 1982). This is a rare syndrome first described by Kussmaul (1877) and characterized by inability to comprehend spoken words with otherwise intact spontaneous speech, writing, and understanding of printed words. For example, Albert and Bear (1957) described a 52-year-old right-handed male who suffered a stroke in the left frontal temporal area. He could not understand or repeat spoken words, but at the same time his spontaneous speech was fluent, his reading aloud and reading comprehension were well preserved, and his spontaneous writing was excellent. His writing to dictation (auditory presentation) was severely impaired. This patient's disorder was selective for the gnostic-praxic level: pure tone audimetry showed that his hearing in the speech range was nearly normal and was symmetrical, indicating that the lower, sensory motor level was intact. In contrast to his impaired perception of auditory speech stimuli, the patient showed no difficulty in recognizing and naming nonverbal, meaningful sounds such as a telephone ringing, a train, a horse trotting. It is of interest that, in magnetic imaging studies in normal subjects, involvement of superior temporal gyrus (Wernicke's zone, secondary fields 22 and 42—link *a* in our model) in the perception of acoustic-phonetic features of speech, rather than in the processing of semantic features, has been documented (Binder, Rao, Hemmeke, and colleagues, 1994).

Classically described conduction aphasia, we believe, is comparable in our model to disorder of auditory speech gnosis involving spatial relations among the elements within the word, link *b* in Figure 14. Conduction aphasia is characterized by a profound deficit in verbal repetition on a background of normal comprehension of spoken language and fluent and meaningful spontaneous and conversational speech, although with literal paraphasias. Reading aloud in some patients may be contaminated by literal paralexias, although reading comprehension is preserved. We see here a selective disorder of spoken language (repetition) with intact comprehension of both spoken and written language, suggesting dysfunction at the gnostic-praxic level in this disorder, rather than the symbolic level involvement that, in our view, is characteristic of aphasia. Patients with this disorder are aware of their errors and will recognize the correct variant, which patients with symbolic level disorders will not. Analysis of the specific types of paraphasia present in conduction aphasia, extensively detailed by Blinkov, allows us to identify that the gnostic-praxic deficit involves spatial relations among word elements, link *b* in our model. Blinkov studied a right-handed 32-year-old male Russian-speaking patient with a gunshot wound to the left hemisphere, in the region surrounding the junction between the superior temporal gyrus (field 22 and 42) and supramarginal gyrus (inferior parietal field 40). This patient manifested severely impaired repetition, with intact fluent spontaneous speech and intact speech comprehension. He could understand words, but when asked to repeat them, severely distorted their sound structure: the word *vol* (bull) was repeated as "lov" (meaningless, a literal paraphasia), and then "lom" (crow, a verbal paraphasia); *pot*olok (ceiling)—"po*lot*ok" (meaningless, literal paraphasia), and then "platok" (kerchief, verbal paraphasia); "*rif*ma" (rhyme)—"*fir*ma" (firm, verbal paraphasia); "e*kip*azh" (carriage)—"e*pik*azh" (meaningless, literal paraphasia); "*dob*ro" (goodness)—"*bod*ro" (briskly, verbal paraphasia) (communicated by Blinkov to Glezerman, 1983). In these examples, there is a mirror reversal of syllables by the patient—an auditory-spatial deficit—resulting most often in a meaningless word. This is followed at times by a compensatory attempt to find a meaningful word using an intact phonological code: the patient conducts an "internal" phonological search for meaningful words using the rules for combining phonemes—link *e* in our model.

Conduction aphasia has been considered to be the result of disconnection between the left hemispheric temporal auditory comprehension area (Wernicke's area) and the left frontal motor speech area (Broca's area), based on damage to the white matter pathways between these zones. However, the existence of lesions in these tracts does not prove that they are responsible for the disorder, because accompanying lesions in gray matter may be the cause (Mendez & Benson, 1985). Modern imaging data have shown consistently damage to the temporal-parietal junction on the left, involving bordering parts of 22, 42, and 40, as well as the white matter directly underlying them, in patients with symptoms of conduction aphasia (Damasio & Damasio, 1980; Mendez & Benson, 1985). In our view, it is the damage to the cortical areas themselves that is responsible for these symptoms. Secondary temporal fields (22, 42) implement auditory analysis of speech sounds; inferior parietal field 40 implements multimodal and supramodal spatial analysis. We think that the junction area between these two regions might perform spatial analysis in auditory speech perception, link *b* in our model, which is afferentation for articulatory praxis, link *d*. Analysis of patients' mistakes supports this hypothesis: patients produce mirror reversals of syllables with in an intact word contour. In any case, we believe that the disconnection theory is incorrect because gnostic-praxic and symbolic levels work in parallel; one doesn't need to understand a word in order to repeat it. Other modern models of the cortical anatomy of single word processing support multiple parallel, relatively independently functioning links of the speech functional system (Peterson et al., 1988).

A pure syndrome of postcentral (kinesthetic) articulatory apraxia (link *c*) is described by Vinarskaya (1973), which she called postcentral apraxic dysarthria. In contrast to an aphasic disturbance, patients with this syndrome have no difficulties in comprehension of spoken and written language, and their writing is normal. They manifest disorder of fluency and literal paraphasia in spontaneous speech, repetition, and reading aloud. In this type of literal paraphasia, the key to the syndrome, the general configuration or contour of the word is intact, but there is replacement of separate articulations within the word by articulations similar in place or type of formation. There are mainly difficulties in pronunciation of consonants, not vowels. Vinarskaya considered the disorder of speech fluency in patients with this syndrome as secondary, the result of spontaneous "compensation": the patients perceive auditorily their defects in articulation (indicating intact links *a* and *b* in our model) and seek for the correct articulation by kinesthetic trial (compensation for disorder of cortical topological kinesthetic schema of articulation). This results in pauses within the rhythmical structure of the word, speech slowness, and nonfluency. Active search for the correct kinesthetic pose (correct articulation) is very characteristic of this syndrome. In mild cases, disorder of speech fluency, behind which the "apraxic search" is concealed, stands in the foreground. Vinarskaya differentiated disorder of postcentral (kinesthetic) articulatory praxis from dysarthrias (both cortical and subcortical, bulbar and pseudobulbar) by neurophonetic analysis.

Vinarskaya also described a pure syndrome of premotor (kinetic) articulatory apraxia (link *d* in our model), which she called *premotor apraxic dysarthria* (Vinarskaya, 1973). Although occurring as a rule together with anterior aphasia, in the rare cases of the pure syndrome only speech articulatory performance disorder is present, whereas other aspects of speech production, such as grammar in sentence construction, vocabulary, and written language, are intact (Vinarskaya, 1973). The core of this syndrome is the breakdown of the unfolding in time of articulations (kinetic programming). This underlies the characteristic features of the clinical picture: difficulties shifting from one articulation to another, speech sound omissions resulting in simplification of the articulatory form of the word, and perseverations of the previous articulations. Articulatory movements are tense, slow, and inert.

Disorder of letter gnosis is illustrated by the syndrome of optical dyslexia, described by several authors (Koch, 1967). We observed this syndrome in children with dyslexia (Glezerman, 1983; Glezerman & Dmitrova, 1989). For example, a 9-year-old right-handed male with an above-average IQ of 115 manifested no difficulties on verbal, including vocabulary, and nonverbal tasks in the Wechsler Intelligence Scale for Children (WISC). He had a selective disorder of reading and writing, replacing letters similar in their visual form (literal paragraphia). There was also a selective, modality-specific short term verbal memory deficit, with a significant impairment in memory for words presented visually and a more intact memory for words presented auditorily.

Disorder at the level of single letter performance (link *e* in our model) may be illustrated by the syndrome of apraxic agraphia. It is "pure" agraphia with otherwise intact verbal and nonverbal functions, including spoken language and reading. The syndrome was first described by Exner in 1881 and was localized in the superior part of the left premotor area above Broca's area (fields 44, 45), known as "Exner's writing center." The clinical manifestation of apraxic agraphia is different from the much more commonly found agraphia that occurs as part of aphasia. Agraphia secondary to aphasia will be due to disorder of letter selection, with letter omissions and substitutions—literal paragraphia—which will be qualitatively the same as mistakes in spoken language (literal paraphasia). "Pure" apraxic agraphia represents disorder of letter formation. Writing is slow and laborious, with poorly formed letters and distorted drawing pattern of letters. Patients not only cannot recode from uppercase to lowercase letter or from print to handwriting, but they cannot "slavishly" copy printed letters (Roeltgen, 1985). This very rare syndrome was recently observed by Anderson, Damasio, and Damasio (1990) in a patient following a stroke; CT scan of the brain revealed that the lesion was localized in Exner's center.

Selective disorders at the gnostic-praxic level would result in speech agnosia, speech apraxia, and visual dyslexia and dysgraphia. The mechanism that underlies these disorders is blocking of input (agnosia) or output (apraxia) in the corresponding modality, acoustic or visual. We will call "aphasia" only that disorder which involves the phonological level of the single word processing system (link *e* in our model), presumably field 21. Considering that the main deficit in this syndrome is a disorder of the "signifier," or sound code of the word, we will call this form of aphasia *lexical (phonological) aphasia*. The basic mechanism of this form of aphasia is disorder of the usage of language rules, losing the regularities according to which the combinations of distinctive features and phoneme sequences are formed, resulting in the breakdown of the word sound code. The spoken or written word is processed successfully at the gnostic level—it can be repeated—but is not analyzed at the phonological level. In other words, the sound sequences that have been perceived are not compared with the patient's own code, because the latter has been lost; hence, they cannot be interpreted. The native language is perceived as a foreign one, as when receiving a message and not knowing its underlying code beforehand. Thus, disorder of word comprehension will be central in the clinical picture of phonological aphasia. Disorder in recognition of sound code of the word will block the activation of the connections behind it necessary for word comprehension: topological scheme of the object, categorical component of word meaning, object image, and individual symbol.

As we discussed before, it is only in the word, its "sound," that categorical signs are incorporated into the hierarchical structure of the categorical component of word meaning (signified). Word sound code is, for the categorical component, that integral formation which implies and activates structuring of the definite sequence of the categorical signs characteristic for the given word. If word sound is impaired, categorical signs lose that form through which they assume the appearance of the concept. This disorder in concept formation in these patients will result in a specific and selective thought disorder.

Historically, various aphasic syndromes have been described in the literature that are analogous to disorders at the symbolic level and fit our concept of phonological aphasia. We review some of these syndromes in the following paragraphs.

Luria's *acoustic-mnestic aphasia* is connected with lesions in the middle and inferior left temporal area (field 21) (Luria, 1966/1980; Luria, 1947/1970). The basic deficit, according to Luria, is disorder of higher level integration of acoustic processes, in contradistinction to acoustic gnostic deficit, which Luria called *acoustic-gnostic aphasia*, caused by a lesion in Wernicke's zone and equivalent to the classical *sensory aphasia*. Description of the clinical picture of the acoustic-mnestic aphasia, however, is very similar to our understanding of disorder at the symbolic (phonological) level: "The word, while preserving its clear phonetic composition, loses its meaning and is not recognized anymore, sounding like a new word ... the word starts to seem foreign" (Luria, 1947, p. 142).

Another comparable example is transcortical sensory aphasia, one of the classical aphasic syndromes in clinical neurology, which was attributed to interruption of connections between Wernicke's zone and the so-called center of concepts. Characteristics of this syndrome include disorder of speech comprehension, intact repetition, and preserved ability to read aloud and write from dictation without understanding what is read or written. Imaging studies have shown involvement of the inferior-middle temporal region, close to field 21, of the left hemisphere (Alexander, Hiltbrunner, & Fischer, 1989; Kertesz, Sheppard, & MacKenzie, 1982).

Vinarskaya distinguished speech disorders at the gnostic-praxic and symbolic levels, and her conception of what she called *phonological aphasia*, involving disorder of both spoken and written language comprehension of phonological origin in patients with selective damage to field 21, is closest to our own (Vinarskaya, 1971).

The classical sensory aphasia—which includes disorder of speech comprehension and disorder of repetition, with massive phonemic paraphasias—although usually linked to Wernicke's zone (secondary temporal fields 22 and 42, which we think are responsible for auditory word gnosis) probably most often includes a combination of gnostic-praxic and symbolic disorders.

3.4. Lexical Phonological Aphasia

Phonological aphasia is based on selective disorder of the phonological (symbolic) level in the sound side of the speech functional system (Figure 14, *e*); speech sounds perceived are not recognized as language symbols and thus sound sequences cannot be interpreted as meaningful language units. Categorical signs characteristic for a word (that which is signified) cannot be actualized beyond the sound form, which results in disorder in word comprehension as well as a specific, peculiar disorder of conceptual thought.

The speech status of the patient with phonological aphasia is characterized by massive paraphasias, literal and verbal. Literal paraphasias are a direct consequence of the basic deficit, disorder of the phonological code: they are due to confusion of phonemes similar in their distinctive features. The most vulnerable and first to be affected are the so-called oppositional phonemes, which differ from each other by one distinctive feature. For example: the Russian word *frukti* (fruits) becomes *vrugti** (meaningless). (When the disorder is more severe, even phonemes that are not similar can be replaced—the Russian word *chemodan* [suitcase] becomes *naliman** [meaningless]). This last example illustrates another feature of literal paraphasia in phonological aphasia: while the phonemic "filling" of the word sound structure is impaired, the rhythmic contour and general sound form is intact (right hemispheric gestalt). In the most extreme form, replacement of phonemes becomes so extensive that speech is completely incomprehensible, presenting as neologistic jargon, as in the example from Kertesz and Benson (1970) of a conversation with a 71-year-old right-handed patient:

> *Question*: What is this? (A pen is shown)
> *Answer*: Kind of ateuna is emessage, card.
> *Question*: What do you use it for?
> *Answer*: This is a tape of brouse to make buke deproed in the auria.
> *Question*: What do you call it?
> *Answer*: That's a moista groise.

Phoneme replacements may result in verbal paraphasias, in which meaningful words similar in sound replace the target word: for example, Russian "kod"–"god" (code–year), "tochka"–"dochka" (point–daughter). In verbal paraphasias, the replacing word is usually a more frequently used word in the given language than the target word (Vinarskaya, 1971). Replacing the phoneme with a meaningful word is a more complex phenomenon: it is the result of both the deficit and a spontaneous attempt at compensation. In pursuit of the elusive, slipping-away sound of the word, the patient finds support in a similar-sounding, more frequently used word, whose sound code has a greater chance of being spared. Also, frequent words may become automatisms, "descending" to the gnostic-praxic level, becoming sound clichés and automatisms of articulatory praxis (see our model of cortical organization of single word processing, Figure 14). Vinarskaya emphasized the frequency of the word as an important factor in the origin of verbal paraphasia in patients with selective disorder of the phonological level. Vinarskaya understands frequency as the comparative frequency of word usage in the given language and in the individual language (vocabulary) of the patient.

Our model of the stepwise hierarchy of operations within the phonological level—that is, recognition of phonological signs, recognition of phonemes, recognition of phoneme sequence as the sound code of the word, memory of the probable (in accordance with the rules of the given language) phoneme sequences composing the sound code of the word—allows, at least in theory, selective disorder of each of these steps. If disorder of the last step (memory) prevails, impoverishment of vocabulary acquired during past language experience will stand at the foreground of the clinical picture instead of phonemic paraphasias.

Thus far, we have described symptoms of phonological aphasia that are due to the deficit in the phonological link in single word processing (damage to the temporal region of the left hemisphere) and attempts at spontaneous compensation, which in this type of paraphasia utilize islands of spared function of the damaged area and the gnostic-praxic level of the left hemisphere. In general, however, the clinical picture of phonological aphasia is broader, and, we believe, is the product of complex interactions between the deficit and intact functions. The deficit is a deficit in word sound (signifier), whereas the intact function is word meaning (signified), which cannot be decoded. Cortical organization of word meaning in our model consists of discrete left hemispheric components and their right hemispheric equivalents (see Figure 12 in chapter 2). We see the interaction between the deficit and the intact players as the basic and necessary mechanism for pathological speech patterns in phonological (sensory) aphasias; the actual, heterogeneous clinical picture is a result of the variability in the relative prominence of the intact components—empirical component and categorical component of word meaning in the left hemisphere and the visual situation, object image and individual symbol in the right hemisphere. The varying clinical picture will depend on the relative development and previously established patterns of usage of the intact components, reflecting the premorbid profile of the individual. Thus, although the clinical presentation of sensory aphasia is quite variable and even chaotic, this basic mechanism of interaction of deficit and intact components in our model can explain and encompass the range of variations, both in degree of severity and the qualitative characteristics of speech patterns, which, in severe cases, would constitute the picture of what has been described in the literature as the fluent incomprehensible speech of sensory aphasics. Figure 15 is a modification

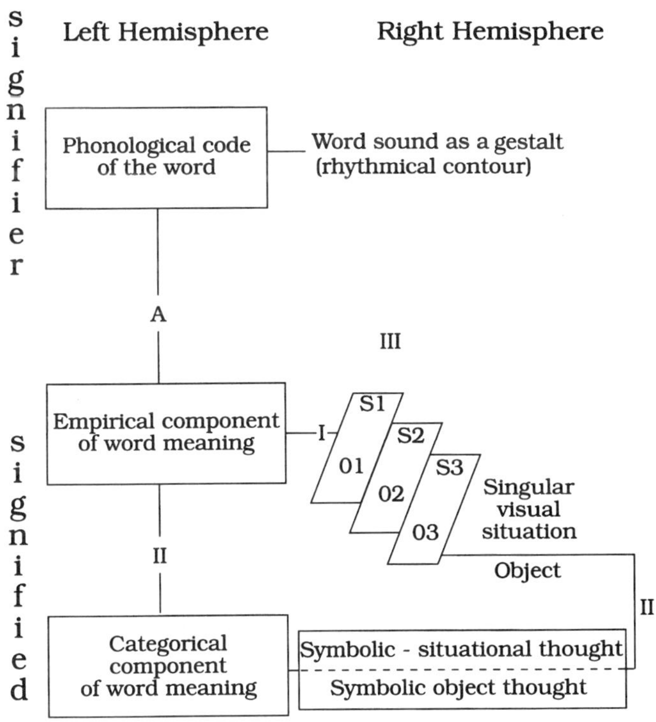

FIGURE 15. Word sound and word meaning.

of Figure 12, presented here to emphasize the different possible interactions between disordered phonological code and the intact components of word meaning. We will describe how these interactions may be involved in the mechanism of verbal paraphasia. We start with the interaction between disordered phonological code and object image; link AI in Figure 15. Unstable and diffuse, word sound can no longer, as it would in the norm, give definite direction through the empirical component of word meaning (left hemisphere) toward the object image (right hemisphere). Instead, right hemispheric associations become dominant; now they "direct" the selection of word sound (the sound shell of the word). Floating right hemispheric gestalts emerge on the surface, and visual images may underlie verbal paraphasias. Examples of such paraphasias* include ditch—"plate," wall—"sheet," suitcase—"well." We see here typical right hemispheric associations by resemblance of holistic forms. Glimpses of visual images and unsuccessful searches of word sounds can result in cases of severe phonological aphasia, at times to the degree known as "word salad." Sometimes, however, the image helps to find a necessary word. Thus, deficit and compensation go hand in hand, and verbal paraphasia represents the manifestation of unsuccessful attempts at spontaneous compensation.

Although we have been speaking of noun replacements, adjective replacements also occur in sensory aphasia. Adjectives do not designate the object; instead they represent a feature or sign of the object that is actual in the particular situational context, in which content and emotional attitude are inseparable. Although the inability to "fix" constantly glimpsed images by word sound in phonological aphasia secondarily results in the "depletion" of image content, its subjective emotional attitude remains, which is the basis for adjective replacements. The literature gives the examples of these replacements in patients with sensory aphasia*: lean—"shortish," poor—"low," lazy—"weak," hot weather—"big weather," beautiful—"strong." As can be seen, replacement here is based on the subjective emotional evaluation of the sign constituting the meaning of the adjective, suggesting the right hemispheric origin of these paraphasias.

To summarize this process: as we see in Figure 15, phonological code is directly connected with the empirical component of word meaning and through it with the right hemispheric object image (link AI). Using linguistic terms, we may call the empirical component of word meaning *object reference*, and it is exactly that which is mostly impaired in patients with sensory aphasia. Even though the word sound has quickly slipped away, the instant it was heard it evoked an object image in the right hemisphere. Now the image, without the support of word sound, is subject to right hemispheric rules and brings to the fore the particular group of associated holistic forms, the revolving roundabout of images similar in appearance but different in content, any one of which may come forth and push out a word sound. Thus, there is a vicious circle: the object image is secondarily impaired, diffuse, but at the same time it is the image which is the basic source and support in search of word sound.

We will now discuss the interaction between disordered phonological code and categorical component of word meaning, link AII of Figure 15. This attempt at compensation uses resources from the left hemisphere. As was discussed earlier, when the sound of the word is impaired, the categorical component of word meaning loses that form in which it is embodied into a concept. In the norm, the complex sound code of the word/object image directs categorical signs into a certain channel of hierarchical succession, from the general categorical sign to the more specific categorical sign that is the marker of a particular category. When this complex is weakened due to a primary disorder of word sound code and secondary diffuseness of the object image, categorical classification is not complete even though categorical attitude and categorical ability themselves are intact. In this instance, it is primarily the most specific categorical sign, which marks the particular category, that is impaired. The equivalent of the specific categorical sign at the gnostic-praxic level is the functional sign, which corresponds to Bernstein's concept of topological scheme of the object. Topological scheme corresponds to the object image in the right hemisphere (see Figure 15). For example, the hierarchical sequence of the categorical component of the word "knife" is: objectness → nonanimation → thingness → artifactness → instrumentality (toolness). The functional sign of the empirical component of the word "knife" will be the definite operation with it as a tool—to cut—which corresponds to the object image of the knife and the visual action situation behind it in the right hemisphere. The verbal paraphasia that results from loss of the more specific categorical signs of word meaning is the replacement of concrete words by more abstract words within one categorical group. For example,* the more specific word "economics" might be replaced by the more general "science," and "notebook" by "stationary." Uncovering of the more general categorical sign creates the impression that abstract thinking is intact in patients with phonological (in the literature, sensory) aphasia; this is, in fact, not infrequently asserted in the literature. However, in special tasks on verbal abstraction, these patients cannot find the word sound for the abstract concept because the word with abstract meaning has no corresponding object image in the right hemisphere.

Interaction between disordered phonological code and right hemispheric symbolic associations represents another avenue for attempts to compensate. Word sound code is indirectly (through the empirical component of word meaning and object image) connected with right hemispheric symbolic associations that form the individual sense of the word. This, in turn, through interhemispheric connections, interacts with the categorical component of word meaning. "The living word does not designate the object but freely chooses, as though for lodging, this or that object meaning, the dear body. And around this thing, the word wanders freely, as the soul around the cast off but not forgotten body" (our translation from Mandelshtam, 1921). In patients with phonological aphasia, the "body" of the word (sound code–object image) is not only "cast off" but lost. Again, the separate constituents of word meaning—categorical signs—are intact, but not having joined with word sound, do not turn into signified; thus, word meaning is impaired. The individual sense of the word and the right hemispheric symbolic associations connected with it, far away from the phonological code, remain most intact and become an important source for compensation.

The individual sense of the word is its out-of-language content and, in contrast to the categorical component of word meaning, is relatively autonomous from word sound. Diffuseness of the object image may even strengthen the symbolic component, "emancipate" it. The intact categorical thinking and disordered conceptual thinking that characterize phonological aphasia are also favorable for the individual sense of the word, as the lack of framing concept may liberate right hemispheric associations. On the other hand, the absence of control of right hemispheric associations by the left hemispheric cognitive mechanism presents a danger of losing altogether the connection with the word. To paraphrase Kretchmer (quoted earlier, see Chapter 2), one may formulate that in phonological aphasia the hard nucleus of the uttered word gets weaker, exposing therewith the surrounding vapors from merged-together images and strong affects.

At this point, we will briefly review symbolic thought of the right hemisphere, whose different aspects may come into play as the compensatory mechanisms that make up a part of the pathological picture in phonological aphasia. Right hemispheric symbolic thought is heterogeneous; it includes "layers" that have arisen at various stages of phylogenetic development. Situational and symbolic situational thought emerged earlier than separate object images and abstract concepts, the latter based on reduction of situation series. It corresponds to the functional signs of the object, and further, with the empirical component of word meaning on the left. The later layer of symbolic object thought, based on reduction of image-series, was possible because of the "developing" of the object image in the right hemisphere, in which the object image is identified within the situation through interaction with topological scheme of the object on the left. Symbolic object thought was further developed through interaction with categorical thought of the left hemisphere. However, it is important to emphasize that the mutual interactions within the symbolic levels of both hemispheres was a constant factor in their further development. This dynamic—which holds true both in phylogenesis and ontogenesis—should be kept in mind as we discuss compensatory mechanisms in aphasia.

This constant developmental interaction of right and left in truth explains why language is not a pure, logical left cognitive mechanism code system. In fact, even within the left hemisphere, the different components of left hemispheric thinking each contribute to a word's meaning, with categorical and functional signs coexisting and defining its history. Mandelshtam (1921) was able to state this perhaps more clearly in a metaphoric sense: "A word has become not a seven-barreled but a thousand-barreled reed, brought to life by the breath of all centuries at once."

The type of spontaneous compensation that may predominate in patients with phonological aphasia often reflects that type of right hemispheric thought which was prominent in the particular individual premorbidly. Again, word sound is unstable; before it evaporates it may evoke situational, situational-symbolic, or symbolic-object associations, any of these becoming the source of the peculiar narrowings of word meaning seen in different patients. The following is an example of situational associations that emerge in a patient with phonological aphasia as he attempts to explain the word *sonorousness*: "It is when it feels quickly and easily ... wait a minute ... some telephones ... just a moment ... sonorousness ... sonorousness ... flight away ... but it was so close ... and word sonorousness ... it is a bell ... not music ... such cracking of a bell ... to ring ... not music rings ... at summer sonorousness happens also in the forest. This is Pushkin

... the sonorousness of a bird, when one may hear their sonorousness ... they ring at spring, for example, when birds ... and beautiful sonorousness ... however, in the beginning it appeared as something else ... but when they showed, I just understood ... skylark."*

The next examples (our observations) illustrate various individual symbolic associations. Asked to explain the word *sentence*, a patient responded: "A novel by Stendhal, *The Red and the Black*, the main personage hangs himself." For the word *sentence*, the much more common meaning is a part of speech, yet this meaning bears no emotional content. The patient responded to the less common meaning as verdict, which has a powerful emotional connotation that he could convey through symbolic associations, even though he actually could not define that meaning either (the "hard nucleus" of the word has disappeared, while the "vapors" from associated images and affects remains). In this example, we see cultural symbols in the patient's unique associations (see other examples of this patient's cultural symbols in chapter 2). In the peculiarly narrowed word meaning seen in sensory aphasia, the object reference and abstract meaning of the word may be lost, whereas aspects of figurative meaning from the powerful emotionally loaded individual symbols remain. To illustrate this phenomenon, we repeat some of the examples given in chapter 2: *pipe*—"peace pipe"; *dwarf*—"dwarf's thought ... diminutive, pygmean soul"; *sharp*—"well, sharp question, it means complex, unpleasant, and also tongue, sharp tongue, everybody is afraid of it."* The patient's attempts to convey the figurative meaning of the word may lead to so-called idiomatic paraphasias: for example, *head*—"document of preciseness"; *history*—"depth shadow of centuries enlivening."* The examples show that it is possible to catch the meaning that the patient tries to express through defective language usage. However, the major factor that shapes the message is the gestalt evoked by the word in the patient at the moment, and his desire to convey all this. Sometimes patients use nonroot morphemes, which are intact in phonological aphasia and which help push across the meaning they wish to convey. In addition, the intensity of the patient's emotional response to the word seems heightened. For example, a Russian patient said "*Vilupila*, ja nakonetc statju" ("I finally *hatched out* the article").* Here we see an idiomatic paraphasia with nonroot morpheme *vi* expressing the emotional intensity of extruding something with effort; the equivalent in the English translation would be *out*. Bein noted that these idioms are almost never repeated by the patient. The distinctive character of the idioms results not only from the individuality of the symbols but their defective sound expression.

The clinical presentation in phonological aphasia in an ambidextrous patient gives another illustration of the interaction between defective and intact links in the

*Examples of literal and verbal paraphasias are taken from Bein (1961). The material is used here to illustrate our interpretation of the mechanism of verbal paraphasia and the importance of the right hemisphere in this process. Bein herself presents a different understanding of the mechanism underlying paraphasia.

circumstances of less lateralized hemispheric distribution of speech function and more intimately interconnected verbal and nonverbal functions premorbidly. A 40-year-old ambidextrous Russian-speaking patient, a professor of history, with a diagnosis of phonological aphasia in residual phase was observed by one of us (Glezerman, 1986). What stood in the foreground at that point was a disorder or delay in word comprehension. It was interesting that the patient appeared to use the visual modality at the gnostic-praxic level to help him understand a difficult word. The patient reported a visual image of a blackboard emerging in his mind upon which he mentally "wrote" the word that had been presented to him auditorily. The visual image of the written word evoked a visual situation and visual images connected with that word. The patient was asked to explain the meaning of the word *plagiarism*. He averted his face to write the word mentally and then reported the association process that followed (translated to English except for the key Russian words): "OK, I wrote *plagi-at*, then together—*plagiat* (plagiarism). Immediately Roman, for some reason, *lati* (armor), for some reason so beautiful, to make the Roman *gla* ... *gladiat* ... *gladiator*." The patient explained that he first had a visual image of a gladiator and then the word "gladiator" emerged. Here again, the image pushed out a word that was a verbal paraphasia to the target word plagiarism. The word stimulus "plagiarism" is an abstract concept that does not have a visual situational equivalent in the right hemisphere. The word stimulus led to a visual image of the written word, which evoked visual images of objects whose words sounded like the word stimulus. He tried to make the word more stable by "writing" it mentally, and then by repeating it verbally, but he also attempted to make it more decipherable by making it visual, by "looking" at its written form. Despite all these strategies, he could not decode the word meaning. At the same time, a search was under way for similar sounding words; some of these popped to the surface with their accompanying images, and these resulted in the verbal paraphasia. The resultant verbal paraphasia is typical of the phonological aphasia patient, based on similarity of word sound: plagiat–lati–gla–gladiat–gladiator. The mechanism that has produced it, however, may reflect peculiarities of cortical speech representation in this ambidextrous patient: the intimate connection between the written word and visual object image.

This example demonstrates utilization of visual images; we give another example of word explanation in this same patient that illustrates interaction with a different right hemispheric "layer"—visual situational associations (AIII, Figure 15). It shows that different right hemispheric strategies may be utilized in the same patient and reflect the premorbid profile of relative abilities. The patient was asked to explain the meaning of the Russian word *bedstvie* (disaster); he responded "bezshatsja" (meaningless, similar in sound to "run away"). This was a literal paraphasia based on similarity in sound but at same time situational association: when you have a disaster, you run away. He continued (our English translation except for key Russian words): "Ah! *Bedstvie*! Understand! I see! To the habitual is destroyed. Habitual ... established ... permanent ... is destroyed. This we remember ... about war, in family, about death, death, gravely, disabled, without eyes, without arms, horrible. The third ... when happens ... water, typhoon in Japan, something like that. About water and fire. Briefly, this is correct. Houses are burning, horrible fire. Also, water, good, in Leningrad ... they waited for water." Question from examiner: "How do you call this type of disaster?" Patient: "There is a *ure* ... *urovenj* (level), that is, I think, all life higher than man, and children are running to the deck." Examiner: "What is the word?" Patient: "*Beda*" (misfortune). Examiner: "No, this is not the correct word. How do you say the word for 'water rising?' " Patient: "*Utop–top–potop*" (the first two are literal paraphasias to the word *sink* and the third is the word *deluge*). "The TV says! They will show word!" (He was distressed, gesticulating, almost agitated.) "*Potop*. They say in the past, *potop*, but now *na-vod-nenije*!" (flood). The patient conveyed that the process that led him to arrive at the word "flood" was a series of visual scenes that came to him that he had seen previously on the television about a flood, and that the word "flood" emerged at the moment he had a visual image of the announcer uttering this word.

We will outline our interpretation of the compensation process used by the patient. When he was asked to explain the word *bedstvie*, he was unable to decode the phonological code of the word and he was searching for the word sound code, resulting in a literal paraphasia, *bezshatsja* (similar in sound to "run away"). This is in fact not just a literal paraphasia but a "double" paraphasia: literal in that it is similar in sound to the target word, a left hemispheric mechanism, but at the same time based on situational associations to the target word, a right hemispheric mechanism (one runs away when there is a disaster). In Chapter 2 we gave another example of double paraphasia from this same patient, which occurred when he was asked to explain the word *tirade* and came up with *terrace*, a replacement by sound resemblance as well as by right hemispheric symbolic associations. Again, the pattern of attempts at compensation in this patient reflects his unusual premorbid cortical speech organization and a more intimate connection between speech and visual perception. It is important to note that the patient, in the literal paraphasia *bezshatsja* (run away), turns a noun *bedstvie* (disaster) into a verb. Next the patient repeated the target word correctly, and extracting the categorical sign of destructive change implicit in the word disaster, he then produced an awkward metaphor, a linguistically agrammatic expression—"to the habitual is destroyed." This idiomatic paraphasia, we think, is a result of the interaction of left (categorical signs) and right (symbolic associations) hemispheric mechanisms. Then numerous visual scene-situations symbolizing various types of disaster emerged. Although they are flashing with kaleidoscopic speed, each conveys basic elements—destruction and negative emotional effect—and each helps to "push out" a word that designates an event—war, typhoon, fire, deluge, flood. The biblical symbol was easier for him than the more common, literal word *flood*. In the final step we see the use of the gnostic-praxic level—word as a sound cliché—which he remembered and could identify with phonological code and repeat.

We will present a detailed description of a patient with phonological aphasia (Glezerman, 1986). Patient S was a 30-year-old, right-handed engineer who presented for rehabilitation one year after suffering a traumatic head injury and fracture with resultant hematoma of the left temporal region. At the time of presentation, his full scale IQ (FS IQ) was 83, with verbal IQ (V-IQ) of 72 and performance IQ (P-IQ) of 101. The main deficit observed was a disorder of speech comprehension. He was slow at understanding instructions and it was necessary for him to repeat key words aloud in order to understand what he was supposed to do; even then, he was not always successful. He could repeat words but to understand the meaning of words he needed to repeat them out loud; again, this did not always improve comprehension. Words with complex sound structure he could not repeat at all, but he could read these words, with occasional improved understanding. For example, when asked to explain the word "enormous," he replied: "Enormously ... it says nothing to me ... I need to read it, then something will start. (Reads word correctly.) It's similar to something, but not understandable ... something familiar but I don't understand what."

In general, the patient reacted to his deficit in the following fashion when presented with a word: "It's something familiar but I don't understand what. I cannot remember. I understand that I spoke it sometimes ... as though it's mine, but what it means flies away."

The speech of patient S was characterized by searching for word sound and paraphasias. This speech pattern was observed in repetition, naming, spontaneous speech, responses to tasks of word definition, and so on. Most often there were literal paraphasias secondary to oppositional phoneme replacement: "pechka" (stove)—"bechka" (repetition); "pozdno" (late)—"*b*ozno" (spontaneous speech). Both represent *b—p* replacement, resulting in literal paraphasia—a meaningless word. In a word-definition task, when presented with *grivennik* (dime), S responded: "Is this money? You said so, but this word is not mine, I don't understand. Krivennik?" Note that in the very beginning the word is heard and the semantic sphere of word meaning is evoked, but it is quickly lost, probably due to instability of word sound. As we see, *g—k* replacement leads to a meaningless word. There were also verbal paraphasias, meaningful words similar in sound to the target word. For example, chram (temple)—gramm (gram). The following is an example of verbal paraphasia of right hemispheric origin. In response to the task, "Draw the cross under the circle," S stated, "Cross? It says nothing to me. What is a cross?" Then the patient said the word for under, *pod*, but associated it to the Russian word for sweat, *pot*, saying, "Is it ... water dropping? I don't understand." The word *pod* (under) in spoken Russian is pronounced as *pot*. So the words *pod* and *pot* are homonyms. The word *pod* has only grammatical meaning, expressing spatial relations of two objects. The word *pot*, thus, is the only one of these two words that has a right hemispheric equivalent, a visual situation. So here it is the visual situational context—not the content of the sentence—that determines word comprehension. Patient S had low scores on WAIS** subtests that assess verbal-logical thinking: Vocabulary (scaled score 6 against average of 10) and Similarities (4 against average of 10). In the vocabulary subtest, the low score was due to disorder of word comprehension with literal and verbal paraphasias. We illustrate this with a few examples. Asked to define the word *speshitj* (to hurry), S replied *bezshkom* which is not a word but a literal paraphasia (with *b–p* and *zsh–sh* replacements for the word *peshkom* [an adverb] which means "on foot," as in traveling by foot). *Bezshkom* is more than a literal paraphasia: the root morpheme "bezsh" is from the verb *bezshatj* (to run). Thus, this fascinating neologism represents what we have called an *idiomatic paraphasia*, an overdetermined, idiosyncratic invention reflecting the patient's response to the word. *Speshitj* (hurry) evoked the visual situation of hurrying, but he could not produce the correct definition; instead, he made up a new word by aggrammatically combining two pieces associated emotionally with the visual situation behind the word. When asked to define *stojkostj* (fortitude), S responded "*stojka*? (counter) ... stojkostj ... something familiar ... but what?... It is to stand (*stojat*), somewhere and to do something." Here we have a verbal paraphasia, a meaningful word of similar sound to the target word, which is the result of the search for the meaning as the unstable word sound is slipping away. The patient produced words that are similar in sound but, unlike the target word, do have a clear right hemispheric equivalent, a visual image or situation. Sometimes patient S adequately explained words by their visual-situational context. For example, the word "brave": "I understand brave ... during the war ... how to say ... so to speak ... not running away but attacking."

On the similarities subtest—a test of verbal abstraction in which S was asked how two words are alike—S's score was severely impaired, at the level of moderate mental retardation. His answers were all situational; that is, he united the two words according to their situational context but there was no one abstract word in his responses.

To explore further his ability to abstract, patient S was given the task to group similar picture cards. He correctly united objects belonging to the same categories, but he could not name the categories. For example, he united domestic animals into one group, and when asked to give the group a name, he responded: "How to say, they live ... not people but live." In this clumsy, aggrammatic attempt, he did not give us an abstract name of category, but he did show that he extracted the

**The Wechsler Adult Intelligence Scale (WAIS) is the most widely used method of intelligence measurement. It is a standardized test of general intelligence composed of 11 subtests (six verbal and five nonverbal or performance subtests) measuring many different abilities and summarized in a single number, the intelligence quotient (IQ) (Rapaport, 1946).

categorical sign of animal. He did not go to the specific categorical sign of domestic animal, which will be the specific categorical sign for the category (objectness → animated object → animal → domestic animal). In another test of abstraction, he was presented with a series of cards, each with one geometric figure of a different color and size, and asked to classify them. He classified according to any one of categorical signs (shape, color, size) correctly, and when asked to change his strategy, he easily switched and grouped cards by another sign.

His scores on subtests assessing nonverbal logical thinking were above average: he scored 13, with the average being 10, on both Block Design (spatial analysis, constructional thinking) and Picture Arrangement (ability to organize situations, planning and logical sequential relationship). On the Verbal subtest Arithmetic (verbal logical thinking that is similar to Vocabulary and Similarities but also contains elements of sequential logical operations similar to Picture Arrangement and spatial ability or spatial analysis, which is similar to Block Design), he received an average score when he was presented the tasks visually instead of auditorily. This again indicates that his sequential logical thinking and spatial ability were still adequate.

In summary, the patient's speech deficit might be characterized as a disorder of phonological symbolic code (Figure 14, *e*). The disorder was selective, with the gnostic-praxic level of single word processing basically preserved. The patient was unable to understand a specific word but could repeat it; moreover, his repetition of the word was a necessary part of his attempt to comprehend it and represented his spontaneously acquired partial compensation. With words that were very complicated in their sound structure, difficulties with repetition could arise in this patient. On these occasions, the patient again spontaneously referred to the gnostic-praxic level of the visual modality, trying to understand the word through reading and writing. In both reading and writing, the same type of phonemic paraphasia as in spoken language was observed, indicating that it was word sound disorder that caused errors in written language. As we mentioned earlier, in the norm, repetition, reading, and writing are never purely mechanical, with the symbolic level influencing performance that may require only the gnostic-praxic level. Similarly, in pathology, disorder of the symbolic level may influence the intact gnostic-praxic level.

There is further evidence that the gnostic-praxic level was intact in patient S. It was not by chance that the patient constantly emphasized both the familiarity of word sound and the difficulty understanding it. Here we see traces of this individual's speech experience at the gnostic-praxic level—word as a sound cliché (Figure 14, *a*, *b*) and word as an automatism of articulatory praxis (Figure 14, *c*, *d*)—in other words, acoustic and articulatory characteristics of words beyond their role as meaningful units of language. Finally, dissociation between gnostic-praxic and symbolic (language) level is suggested by the significant latent period between perception of the spoken word and its comprehension in cases in which the patient was able to understand the word.

Regarding manifestation of the syndrome of phonological deficit in this particular patient, his speech production was not increased in amount and word salad was not observed. He presented primarily with literal, rather than verbal, paraphasias due to replacement of oppositional phonemes, those generally the first to be affected with disruption of phonological code. When verbal paraphasias did appear, they were the result of spontaneous compensation using right hemisphere visual image and visual situation. When his ability to abstract was examined, a significant discrepancy was observed between his performance on the similarities task and his good ability to categorize objects. This discrepancy was due to his inability to find the abstract word concept. We attribute this to at least two factors, which we will examine. There was a discrepancy in patient S's performance in the classification of objects, in which he could not arrive at the most specific category, as opposed to his ability to classify according to a single categorical sign (shape, color, size), which he did well. This discrepancy reflects the fact that the combination of categorical signs involved in object classification (concepts) requires word sound, which is needed to channel the particular combination of categorical signs fixed in language history (categorical component of word meaning). This constitutes the specific disorder in conceptual thought. This factor is common to any patient who has phonological aphasia. The second factor concerns an individual peculiarity of patient S.

In patient S, right hemispheric object images and right hemisphere situation played a leading role in finding a word, even if the word was a verbal paraphasia. However, an abstract word does not have a visual object image and visual situation as a direct equivalent in the right hemisphere, although it may have a symbol as a right hemispheric equivalent. In the case of patient S, individual symbols were not used as a means for compensation, so an intact right hemisphere did not help him to "push out" the sound of the abstract word. But his use of object images and situations helped him to push out the sound of a concrete word similar in sound to the target abstract word (verbal paraphasia).

Again, patient S's FS IQ was 83, with V IQ 72 and P IQ 101, which we understand as a decrease from his premorbid level. There was a very large difference between his V IQ (72) and P IQ (101). Qualitative analysis of his performance on the verbal subtests clarified that his low performance was due to his specific neuropsychological deficits. The pattern of S's spontaneous compensation reveals that in his premorbid neuropsychological profile, right hemispheric visual situational thought, but not right hemispheric symbolic thought, prevailed. This should be considered in developing an individualized program of rehabilitation in this patient.

4

Parietal-Occipital Region: Spatial Perception and Word Form

4.1. Delineation of Anatomical Region

The parietal-occipital region includes the most phylogenetically young part of the parietal cortex. This consists of cytoarchitectonic fields 39 and 40, which are situated between parietal, temporal, and occipital fields and are consequently denominated as a junction of tactile-kinesthetic, auditory-vestibular and visual cortical centers. The major function of these fields is orientation in extrapersonal and intrapersonal space. From the point of view of cytoarchitectural hierarchy, fields 39 and 40 are considered intermediate, secondary-tertiary fields (Stankevich & Shevchenko, 1935). This transition is reflected in some contradiction in the definition of spatial function, which, on the one hand, is closely connected with modality specific functions, and on the other hand, is not reducible to them. Spatial function is considered a supramodal function, and it is suggested that "a supramodal spatial framework could be constructed out of converging inputs to posterior parietal cortex from all exteroceptive sensory modalities, with a significant contribution from vision" (Ungerleider & Mishkin, 1982). Studying the primate visual system, Mishkin et al. (1983) distinguished two cortical pathways: the ventral visual pathway, which interconnects the primary and secondary occipital fields with inferior temporal areas, and the dorsal visual pathway, which interconnects the primary and secondary occipital fields with inferior parietal areas. The ventral visual pathway is concerned with object vision (the "what" system) and the dorsal visual pathway is concerned with spatial vision (the "where" system) (Figure 16). Object vision is crucial for the visual identification of objects. Spatial

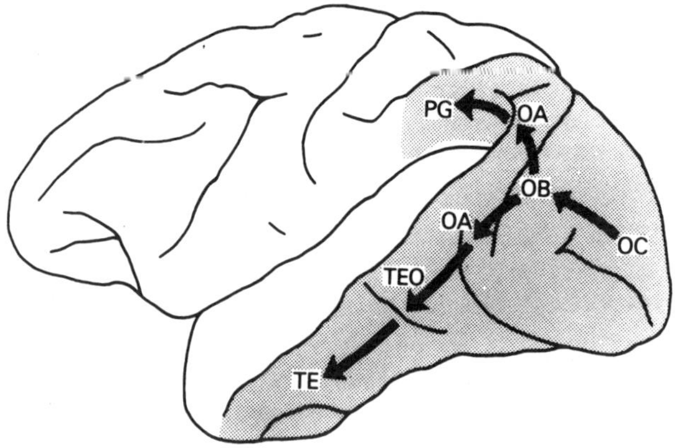

FIGURE 16. Two visual cortical pathways. From "Object Vision and Spatial Vision," by M. Mishkin, L. C. Ungerleider, and K. A. Macko, 1983, *Trends Neurosci.*, *6*, pp. 414–417.

vision is critical for the visual location of objects (Ungerleider & Mishkin, 1982). In contrast to the ventral pathway, which remains modality-specific throughout its course, the dorsal pathway appears to receive convergent input from other modalities (Mishkin et al., 1983).

Function divergence within the primate visual system starts at early levels and evolves from specialized cells distributed within area VI (corresponding to primary visual field 17 in the human). The color and form selective cells of the ventral, object vision, pathway receive their input from the subcortical Parvo system, and the form and movement selective cells of the dorsal, spatial vision, pathway receive input from the Magno system (Livingstone & Hubel, 1988). The cortical endpoint of the ventral object vision pathway is the inferior temporal area (tertiary cytoarchitectural field 37 in the human). In the inferior temporal region, form is processed for the purpose of identifying the visual stimulus and assigning it some meaning (Mishkin et al., 1983). The dorsal spatial visual pathway culminates in the parietal-occipital region (secondary-tertiary cytoarchitectural fields 39 and 40). Spatial vision does not require object recognition but does require efficient analysis of shape properties and spatial relations (Ullman, 1995).

4.2. Neurobehavioral Correlates: Visual-Spatial Ability Connected with the Parietal-Occipital Region

It is well established that damage to the parietal-occipital region (fields 39, 40) results in visual-spatial deficits, with right and left hemispheric lesions producing qualitatively different syndromes. Right parietal-occipital syndrome is manifested by: (1) neglect of the left part of visual space; (2) visuospatial agnosia; (3) constructional apraxia; and (4) dressing apraxia (Critchley, 1953; Brain, 1941). Patients with lesions in the right parietal-occipital area cannot accurately describe or visualize familiar routes, losing the ability to find their way in a previously familiar environment: their neighborhood, their street, their room in the hospital, and so on (Landis, Cummings, Benson, & Palmer, 1986; Heilman, Watson, & Valenstein, 1985). The mechanism underlying the neuropsychological deficit in this syndrome is the breakdown of the whole visual picture, or gestalt, which is expressed in fragmentation of the spatial situation. In the drawings of these patients there are serious distortions in perspective and proportions, spatial displacements, and lack of important components, breaking the object as a spatial image into separate fragments (Kock, 1967; Lezak, 1983).

Patients with lesions in the left parietal-occipital region may orient well in the environment but experience difficulties in performing tasks requiring schematic concepts of space: drawing maps, spatial planning, and any other schemes of spatial relations among objects. Typical are mistakes in line directions (left–right, up–down, forward–backward). For example, in performing the "clock test," in which the patient is asked to draw a clock designating a certain time, patients with left parietal-occipital lesions, in contrast to patients with right parietal-occipital lesion, preserve the whole spatial image of the clock but generally reverse the directions of the clock hands 180 degrees (mirror reversal). This deficit, called by Potzl (cited by Kock, 1967) *geometric-optic agnosia*, underlies the constructional apraxia observed in these patients. Forgetting and estrangement of words that designate spatial relations and directions are also characteristic of these patients (Kock, 1967). Luria (1966/1980) described the left parietal-occipital syndrome as an entity that included: (1) visual-spatial agnosia and spatial apraxia (spatial deficit itself); (2) semantic aphasia; and (3) dyscalculia. The term *semantic aphasia* was introduced by Head (1926/1963) to describe the language disorder in which patients could understand isolated words yet were unable to grasp the meaning of a word combination (syntagma). Patients with semantic aphasia speak fluently and effortlessly, without evidence of agrammatism, dysprosody, or literal paraphasia. They understand the meaning of isolated words, both concrete and abstract, including complex concepts. Yet they are unable to comprehend a short sentence or even specific two-word combinations whose meaning goes beyond the meaning of the component words to imply relations between the words (for example, the combination "father's brother"). Luria proposed that the difficulty for these patients derives from the fact that they are unable to understand certain grammatical constructions that express the relationship. Luria distinguished several grammatical constructions that present comprehension difficulties for patients with semantic aphasia, and designed corresponding diagnostic task sets (Luria, 1966/1980; Goodglass & Kaplan, 1972). Examples of such constructions include these:

- Constructions with prepositions and adverbs of location ("Draw a cross under a circle," "Draw a cross to the right of a circle," and so on)
- Instrumental constructions expressed by the preposition *with* in English; in Russian, this is expressed by the inflectional ending of the instrumental case *om* ("Touch the pencil *with* the key," "Pokazshi karandash kluch*om*")
- Constructions with prepositions of temporal relations ("Which sentence is correct: 'Spring comes before summer' or 'Summer comes before spring'?")
- Possessive constructions; in Russian, this relationship will be expressed by the substantive genitive case inflectional ending ("father's brother," "otets brat*a*")
- Passive constructions to distinguish subject and object ("The lion was killed by the tiger" versus "The tiger was killed by the lion")
- Comparative constructions ("Which boy is shorter if Tom is taller than Steve?")

According to Luria (1966/1980), spatial relations underlie, directly or indirectly, the grammatical constructions listed here. In other words, the relation between words expressed by grammatical structures is a linguistic analog of the relations between objects in external space. Indeed, the history of language shows that adverbs expressing temporal relations derived from adverbs expressing location; those expressing quantitative concepts derived from spatial concepts (Levy-Bruhl, 1930). In its early stages, language described first of all and with great exactness positions and arrangements of objects, the distance between them, and other spatial relations. Spatial relations were included in the content word's meaning. For example, in the Yagham language of Terra del Fuego, examined in the 19th century, the pronouns *he* and *she* had many different words that expressed the location of the man or woman: whether he or she was located inside the wigwam or to the right or left, and so on (Levy-Bruhl, 1930). Only with further language development were spatial relations abstracted from content words and embedded in grammatical structures and grammatical words (e.g., prepositions).

So-called primary or spatial dyscalculia is the next symptom component of left parietal-occipital syndrome, according to Luria. It is due to disruption of that "inner spatial schema" which is the framework for the concept of a number (as a combination of order and quantity) and calculating operations. Luria compared the meaning contained in the structure of numbers with meaning in a syntagma: the separate numerals in a number have their own meaning, as do the separate words in a syntagma; the general meaning of a number is determined by both the numerals and their relations to each other, as meaning in a syntagma derives from both the meaning of the separate words and their relations. In numbers, however, spatial relations are more literal, with meaning derived from the relative positions of numerals from left to right in the linear structure of a number.

Patients with dyscalculia typically make specific mistakes that reveal the spatial character of the disorder. For example, if asked to read the number 729 the patient may say "7, 2, 9"; if asked to write 1,001 they may write two numbers (1,000 and 1) or 101. They may also evaluate numbers by their numerals' value, ignoring their positional rank: thus, 1,897 may be evaluated as larger than 3,001 (Luria, 1966/1980). Typical mistakes in calculations include mirror reversals in direction.

According to Luria, constructional apraxia, semantic aphasia, and dyscalculia constitute a distinct syndrome, because all three are due to a "falling out" of a common factor: spatial function of the left parietal-occipital region. Gerstmann (1940) had earlier described another symptom complex, partially overlapping Luria's, in patients with left parietal-occipital disorder. Gerstmann's syndrome includes dyscalculia, left–right disorientation, finger agnosia, and agraphia. Gerstmann attributed this tetrad of symptoms to a discrete lesion within the left parietal-occipital region, the angular gyrus in its transition to the second occipital convolution (corresponding to cytoarchitectural field 39). Gerstmann assumed that a single basic deficit underlies the disorder. He regarded finger agnosia as the principle item of the four features, and for him the disorder of finger localization represented a fragment of autotopagnosia (body schema disorder). Gerstman said, "It is as though the body schema were affected in one sphere only ... the sphere concerned with the individual fingers—as though the optic-tactile-kinesthetic image pertaining to the fingers were split off from the total body image, the finger schema from the total body schema" (cited by Critchley, 1966, p. 190).

As we will discuss later in this chapter, body schema is mostly connected with the right parietal-occipital region. It is of interest that only the hand as a skilled manipulative tool is additionally represented in the left parietal-occipital region. The inner connection between finger agnosia and dyscalculia might be explained by the common historical roots of the corresponding cortical functions. Anthropology and historical linguistics provide rich material regarding the correspondence between the use of the fingers and the act of calculation (Levy-Bruhl, 1930; Critchley, 1966). Since Gerstmann's syndrome was initially described, numerous examples followed in the literature. Cases of the incomplete syndrome were described in which one or more of the four features were not present (Critchley, 1966). In fact, pure Gerstmann's syndrome is a rare event, occurring most often accompanied by other symptoms such as constructional apraxia, ideomotor apraxia, dysphasia (Critchley, 1966; Benton, 1992). Authors also indicated that individual items of the syndrome may occur as parts of other symptom complexes; for example, dysgraphia may be of dysphasia origin.

A renewed interest in Gerstmann's syndrome took place when it became possible to identify relatively small lesions within the left parietal-occipital territory using neuroimaging techniques. Authors have pointed out that the rarity of Gerstmann's syndrome without other accompanying deficits is probably explained by the fact that brain lesions are very likely to compromise a more extensive cerebral region and thus produce a larger symptom complex, of which Gerstmann's syndrome is only a part (Benton, 1992).

Gerstmann's syndrome has also been described in children without apparent brain injury—developmental Gerstmann's syndrome (Kinsbourne & Warrington, 1963; Kinsbourne, 1968; Benson & Geschwind, 1970; Glezerman & Novinskaja, 1983; Novinskaja & Glezerman, 1986). Developmental Gerstmann's syndrome was described in a boy with Fragile X syndrome, a genetic disorder connected with anomaly of the X chromosome (Grigsby, Kemper, & Hagerman, 1987).

In our observation, left parietal-occipital syndromes, including Gerstmann's syndrome, were observed upon neuropsychological examination of a group of children with learning disability and normal intellect; in some instances, these syndromes were observed not only in probands but in their relatives (familial neuropsychological syndromes) (Glezerman, 1983). From among the broad spectrum of symptoms that can be attributed to dysfunction of the parietal-occipital region, only some were observed in individual children, and different symptom combinations were present in the different children. Interestingly, different variants of symptom complexes of left parietal-occipital syndrome observed could be family-specific. We observed both complete Luria's left parietal-occipital syndrome (constructional apraxia, semantic aphasia, and dyscalculia) and an incomplete syndrome, in which semantic aphasia and dyscalculia were found not accompanied by spatial deficit itself (Glezerman, 1983). We also observed striking dyscalculia with elements of semantic aphasia in two siblings who nonetheless had high nonverbal IQs; although they both had a mild manifestation of left–right disorientation, their visual-spatial ability and constructional thinking were in the high range (Glezerman, 1983).

As mentioned earlier, Luria considered that visual-spatial-constructional disorder, semantic aphasia, and dyscalculia constitute one syndrome, and that its symptoms are connected pathogenetically by a single underlying basic spatial deficit. It seems, however, that these symptoms are at least relatively independent: we observed semantic aphasia and dyscalculia without visual-spatial deficit; semantic aphasia without dyscalculia; dyscalculia without semantic aphasia; and even dissociation within semantic aphasias (Glezerman, 1983). As above, when we observed parietal-occipital syndrome in relatives, the particular combination of symptoms was family-specific. This lent support to the suggestion that the separate parietal-occipital symptoms might be inherited independently from each other.

The dissociability of symptoms within these focal brain syndromes and the possibility of their genetic transmission may be evidence that the regions accounting for these symptoms are topographically different. Indeed, direct stimulation of different loci in the post perisylvian region in a 17-year-old epileptic male elicited discrete left parietal-occipital symptoms: acalculia, agraphia, alexia, anomia, constructional apraxia, finger agnosia, and right–left disorientation (Morris, Luders, Lesser, Dinner, & Hahn, 1984).

On the other hand, although symptoms associated with left parietal-occipital dysfunction are independent, we think that Luria's basic insight that visual-spatial construction disorder, semantic aphasia, and dyscalculia have a common root in spatial disorder remains valid. For us this means that spatial function is the basic function of the left parietal-occipital region, and that it has evolved into quite anatomically and functionally discrete subsets over the long course of human brain development. In other words, we see this as a manifestation of differentiation of this region along horizontal and vertical dimensions. Horizontal means that spatial function is expressed in various domains: physical relations between objects, in speech in formation of grammatical constructions expressing directly or indirectly spatial relations, in formation of concept of number. By vertical we understand that these functions are hierarchically different. Spatial relations in physical space belong to the gnostic-praxic level; spatial relations in semantic space (speech, calculation) belong to the symbolic level.

Comparative cytoarchitectural investigations have allowed the suggestion that the phylogenetically new, specifically human cortical formations came into existence at the junction of different fields as a result of differentiation of so-called transitional structures (Blinkov, 1955). Interestingly, cortical organization along the hemispheric surface was described by Blinkov as changing continuously and gradually, although he noted "critical points" at which the structure changes more or less sharply, involving all cortical layers (Blinkov, 1938). Typical characteristics of the new structures are most distinctly manifested in the central part of the formation, whereas in the peripheral part there are transitional structures that combine features specific for the given formation with the cytoarchitectural parameters specific for the neighboring regions (Blinkov, 1955). Blinkov determined the degree of development of a cytoarchitectural field by the ratio of specific to transitional type formation, with higher development reflected in a greater ratio of specific to transitional type. He emphasized that there is a great deal of individual variability in this ratio.

The functional significance of these transitional areas and the extent of interindividual variability may be illustrated by clinical examples. In our cases of developmental Gerstmann's syndrome in three children, we observed additional symptoms: disorder of phoneme discrimination, visual object agnosia, anomia, color agnosia, letter agnosia with disorder of memorizing visually presented words (Glezerman, 1983). Although these additional symptoms were expressed to a very mild degree, they were specific, corresponding to dysfunction of those regions in the temporal and occipital lobes bordering the parietal-occipital region. Dysgraphia in these cases was caused by disorder of phoneme discrimination and visual verbal memory disorder. It was of interest that two siblings in whom we observed

Gerstmann's syndrome had also the identical "set" of accompanying symptoms. In these cases, the symptoms of Gerstmann's syndrome and the accompanying "echo" of dysfunction of bordering areas might reflect the degree of development, in Blinkov's terms, of field 39. Thus, we may hypothesize that the central, specific region was less well differentiated, and the peripheral, transitional areas were more prominent, resulting in a low degree of development of field 39 associated with increased expression of poorly differentiated nuclear and transitional formations and blurred boundaries with neighboring fields 22, 42, 37, 19 of the temporal and occipital regions. In general, learning disabilities may reflect relatively lower end development of particular brain formations, and can occur on a continuum ranging from severe to very mild and almost indistinguishable without specific neuropsychological examination or environmental "challenge."

4.3. Left Parietal-Occipital Region: Hierarchy in Spatial Processing and Morphological Language Code

As we discussed previously, the left hemisphere operates not with whole objects but with their signs. Regarding spatial function proper, it may be assumed that the left hemisphere implements analysis of the spatial situation by distinguishing signs of spatial relations. "Meaningful" analysis of space (the "what" system, discussed in chapter 2) results in the distinguishing of functional and then categorical signs. These signs characterize the object as such and reveal the inner, logical relations between objects. "Geometrical" analysis of space (the "where" system) results in the distinguishing of spatial-temporal signs, which serve to characterize the outer relations between objects—that is, the relations of contiguity in space and time.

Bernstein (1947) considered the inferior-posterior parietal region (fields 39 and 40) the basic anatomical substratum for afferentation of the D level (gnostic-praxic level in our terms). What is the specific contribution of this area to afferentation for praxis? We discussed earlier the specific contribution of the left temporal-occipital area (chapter 2) and mentioned the contribution of the left postcentral parietal field to afferentation for praxis. It is difficult and artificial to divide afferentation of the D level into component parts corresponding to the function of definite cortical fields, yet we need to understand the contribution of fields 39 and 40 to the D level in order to understand the connection of the spatial function of these same fields with higher symbolic levels. We emphasize the functional heterogeneity of fields 39 and 40, which subserve both gnostic-praxic and symbolic levels.

To review, the left temporal-occipital region (field 37) is responsible for the topological scheme of an object in the visual modality, which, as we understand it, is a combination of the functional signs of the object, providing comprehension of the object as a tool. The left postcentral parietal region is responsible for topological scheme in the kinesthetic modality, the kinesthetic engram of object action; that is, the plan for those hand poses and positions necessary to manipulate the object. We propose that the specific contribution to praxis afferentation of the left parietal-occipital region will be to provide spatial coordinates for hand movement relative to extrapersonal, visual space. In this connection, it is of interest to note that there is a group of neurons found close to the intraparietal sulcus that discharge in relation to manual reaching toward visual stimuli in extrapersonal space. They accelerate their discharge prior to the movement. It was found that the activity of these hand-projecting neurons was neither related to proprioceptive input nor to the details of the movement. None of the hand-projecting neurons could be activated solely by mechanical, auditory, or visual stimuli (Mountcastle et al., 1975). Another study of movement-related cortical potentials has shown that human parietal association cortex provides modulatory input to the sensory motor cortex beginning at least 400 msec prior to movement (Knight, Singh, & Woods, 1989). Roland, Skinhoj, Lassen, and Larsen (1980) found that voluntary movements in extrapersonal space are associated with activation of the parietal regions. These areas were assumed to provide information about the demanded direction of motion in extrapersonal space in relation to the proprioceptive reference system (in our terms, topological scheme in the kinesthetic modality).

Bernstein emphasized that apraxia due to lesions in the inferior parietal region is characterized by the breakup of the general plan of action. "Movements of such patients are not discoordinated or amorphous, they are confused and not adequate to the meaning or purpose of the action. Patients can imitate actions which they are unable to perform on command. Such dissociation gives support to the assumption that the basic plan for the movement, but not its motor composition, is impaired" (Bernstein, 1947, p. 137) Bernstein also indicated that the actions most often affected with lesions in the inferior parietal region are complex, goal-directed, successive chains of movement. We will cite here an example of *ideational apraxia*, a term often used in the literature for parietal apraxia, in a patient with left parietal lesion, described by Heilman (1973). The patient "was able to describe what objects were used for and could always select the correct object on a multiple choice type question. The patient's comprehension was intact.... On apraxia testing, when asked to pantomime certain motions with either her right or left arm (i.e., 'Show me how you would use a key') the patient looked at her outstretched hand, then asked the examiner to repeat the question. When the question was repeated, the patient would look down at her hand and say, 'I can't do it.' When asked if she understood the question, the patient would verbally demonstrate comprehension, i.e., 'Keys are used for opening locks.' When the correct movement was shown to this patient in a multiple choice fashion, she was always able to select the correct movement. She was able to imitate in a flawless manner even with minimal cueing and she performed extremely well with the actual object. She had no difficulty with multiple object sequencing (i.e., taking a cigarette out of a pack, putting it in her mouth, lighting it, and then smoking)" (Heilman, 1973, p. 862).

As we see, topological scheme in the visual modality was intact: the patient had a comprehension of the object as a tool. The engrams for motor sequences were also intact: the patient could imitate the action and could manipulate with the actual object. Heilman described the same apraxic syndrome in two more patients with lesions in the left parietal area. All three patients had difficulty with isolated movements (pantomimed object use to command) but not with a series of movements. In other studies, patients with left parietal lesions could not perform a series of complex separate movements to commands (Pick, 1905; Marcuse, 1904; Bernstein, 1947). The dissociation between pantomime of object use and complex goal-directed chains of movements allows the suggestion that these two types of movements might have separate anatomical substrata within the left parietal region.

Within the left parietal-occipital region, field 39 is one that borders upon the occipital, visual region. Its basic function concerns visual-spatial perception. This field is probably related to complex, multistepped object actions, series of movements with several objects. The contribution of left field 39 to afferentation for praxis (keeping in mind Bernstein's thesis, "Afferentation determines not only what to do with objects but also in which consecutive order") is the analysis of the spatial scene where the action is to be realized. By this we mean the distinguishing of spatial relations between the topological schemes of objects participating in the action. It is in these spatial relations that the potential temporal sequence of the separate links of the action is implicitly embedded. The sequence is realized by the effector counterpart of the D level (motor composition engrams) connected with the left premotor area.

We think that field 39 is involved in distinguishing the following spatial signs: the signs of direction, of spatial coordinates, of quantity, and also the spatial-temporal signs of approaching–receding.

Field 40 is closely connected with the postcentral cortical fields and topographically presents the continuation of that territory of the parietal region which is related to the hand (Chlenov, 1934). Hand poses and positions in object action and stereognosis (object recognition by signs determined by touch, primarily shape) are among its known functions. It may be supposed that the most basic underlying function of field 40 is spatial-kinesthetic, to determine the spatial relations among the parts within one object, whereas visual-spatial analysis is the basic function of field 39, necessary for distinguishing spatial relations between separate objects.

Thus, we think that field 40 of the left hemisphere, like left field 37 (see chapter 1) implements recognition of the single object. The difference in their functions, however, lies in their modalities: field 37 is linked with the visual modality and field 40 with spatial-kinesthetic. Visual object perception implemented by field 37 of the left hemisphere is related to distinguishing of functional and categorical signs of the object and the subsequent formation of the corresponding image (topological scheme of object) and concept (the "what" system). Field 40 relies on distinguishing of "spatial-geometric" signs of the object. signs of shape, size, dimensionality, proportionality, volumetric characteristics, and so on (the "where" system, even though it deals with one object). These two fields work in parallel and independently of one another, interpreting the same object according to their different strategies (the "what" and "where" systems).

We have presented here our conception of the basic spatial functions of fields 39 and 40 at the gnostic-praxic level. Following our model of vertical differentiation of these fields, we will now discuss the symbolic (language) level, showing how signs of spatial relations evolved into the semantic signs of language. An analogous hypothesis was suggested in our discussion of field 37 functions in the previous chapter; namely, that categorical signs resulting from the analysis of the object world form the categorical component of word meaning and serve as the base for so-called covert grammar. In this chapter, we will show that signs of spatial, temporal and quantitative relations connected with fields 39 and 40 serve as an evolutionary root of word elements that contain signs of relations, overt grammar. It should be kept in mind that these basic functions of fields 39 and 40, although they served as the source for evolutionary development of higher symbolic functions, did not "disappear" but remain important in the vertical functional hierarchy of these fields. At the same time, the higher symbolic functions became relatively independent, presumably subserved by different areas within these fields.

In order to present our theory regarding the relation of fields 39 and 40 to morphological language code, we need to introduce some linguistic terms and concepts. The linguist Roman Jakobson proposed the theory that language has two axes, paradigmatic and syntagmatic, both operative in any speech event (Jakobson, 1971b, 1971d). By paradigmatic, Jakobson referred to the selection of words, and by syntagmatic, the combination of words. "If, for instance," wrote Jakobson, "I intend to tell something about my father, I have to make a conscious or subconscious choice of one of the possible terms—father, parent, papa, dada, daddy; then, if I want to say that he is in bad shape, again I select one of the suitable words: ill, sick, indisposed, not healthy, ailing. Selections are one aspect of the twofold event and the combination of the two selected verbal entities 'Father is sick' is its other aspect" (Jakobson, 1971b, p. 308).

The entities among which we make our selection are mutually connected by various forms and degrees of similarity: likeness, equivalence, analogy, contrast (paradigmatic relations). Contrary to selection, combination involves the external relation of words by contiguity (syntagmatic relations) (Jakobson, 1971b). Jakobson indicated that selection and combination may be impaired separately, distinguishing two types of aphasia: selection disorders (paradigmatic relations), connected with lesions in posterior cortical regions, and combination disorders (syntagmatic relations) connected with lesions in anterior cortical regions.

Linguistic entities mutually connected by similarity are united into one of several types of paradigmatic series. Among these, there are morphological paradigmatic series, or the morphological language code. The units of the morphological language code are nonroot morphemes: inflectional endings (in English, these are the regular forms of plural, the possessive form of the noun, the simple past of the verb, the third person singular present indicative of the verb, the comparative and superlative of the adjective, the progressive "ing" form of the verb), suffixes (relation*ship*, work*er*), prefixes and prepositions (*over*shoe, a sign *over* the entrance). Jakobson stated that morphological paradigms represent the system of correlative series in which the grammatical meaning of morphemes is expressed by their mutual oppositions (Jakobson, 1971b). For example, in the word *dogs*, the inflectional ending *s* is the indicator of the plural, contrary to the word *dog*, which does not contain this sign. Thus, we have opposition, with the presence of the sign: *plural*, and absence of the sign: *singular*.

When a nonroot morpheme is added to a root morpheme, we deal with word form: book*s*, deliver*er*, editor*ship*, walk*s*, play*ed*.

The grammatical meaning of nonroot morphemes presents those additional signs that the ground lexical meaning acquires as the corresponding word forms are being built. According to Katznelson (1972), lexical meaning gets in word form its formal mark, which gives reference to specific meaning within morphologic paradigmatic series, just as a book's code indicates its proper place on the shelf in the library. On the other hand, word forms also play a role in the connections of the words in a sentence (syntactical role of nonroot morpheme). For example, the nonroot morpheme *s* signifying the plural of a noun will combine with the word form of the verb that also signifies plural: the boys play, the boy plays.

Thus, nonroot morphemes have a dual nature: their grammatical meaning (referring to morphology) and their function in the arrangement of words into sentences (referring to syntax). There is evidence that these two major components of grammatical function are subserved by two different parts of the brain: (1) morphology is connected with the posterior cortex, encompassing the selection of inflectional endings according to the rules of grammar; and (2) syntax is connected with the frontal lobes, encompassing construction of the overall structure of a sentence (Nadeau & Rothi, 1992).

Nadeau and Rothi presented a stroke patient with impaired morphology but, unlike Broca's aphasics, relative sparing of syntax. Goodglass and Berko (1960) also emphasized that impairment in the capacity to use grammatical form to build a sentence (syntax) may vary independently of impairment in the capacity to discriminate their grammatical meaning (morphology); thus, the morphological aspect of agrammatism might be studied separately from the syntactical. The syntactic function of nonroot morphemes will be discussed in chapter 5 on frontal lobe function. Here we will attempt to demonstrate that the signs of spatial relations, connected with function of the left parietal-occipital region, are the cerebral base for grammatical meaning of nonroot morphemes.

The paradigmatic series of prefixes and prepositions is organized by the type of simple oppositions: (1) quantitative, that is, increase–decrease, negation–affirmation (e.g., *under*value—*over*value; desirable—*un*desirable); (2) quantitative-spatial (e.g., spread *out*, lay *out*, build *over*); (3) spatial (e.g., *under* the table—*above* the table, *under*water, slip *into*, *in*set); (4) direction of movement in space, approaching–withdrawing (e.g., go *in*–go *out*, come *to*—come *away*, come *toward*—come *from*). The role of prefixes and prepositions is to mark simple discrete oppositions; they are relatively autonomous from the root part of the word. The prefix or preposition in relation to the root morpheme (lexical meaning) acts as that additional sign which reveals the potential of the categorical sign of the word in the given context. For example, the word "come" contains the categorical sign of movement; the addition of a preposition "to" or "from" will actualize the direction of movement within the more general categorical sign of movement.

To illustrate the spatial radical in the grammatical meaning of inflectional endings, we will look at the declension paradigm, best studied in a highly inflectional language such as Russian. According to Jakobson (1971c), the six primary cases of the Russian declension are grouped into classes, each of which is characterized by the presence or absence of a particular semantic mark: (1) quantifiers (genitive, locative) versus nonquantifiers; (2) directional cases (accusative, dative) versus nondirectional; (3) marginal cases (instrumental, dative, locative) versus nonmarginal. The nominative is opposed to all other cases as markless versus marked. The semantic mark is understood as a minimal discrete unit of grammatical information, so that the general grammatical meaning or morphological invariant of any case within the given declension system is composed of semantic marks. When we consider the grammatical meaning of the cases in more detail, we will see that semantic marks correspond to the signs of "outer" relations, spatial relations.

The accusative case, according to Jakobson (1971c), is the carrier of one semantic mark—"relatedness to action." In relation to the subject–object of the sentence, the accusative indicates that action (of an acting subject) is directed at the object. For example, in the sentence in Russian "On chetal knig<u>u</u>" ("He was reading a book"), the underlined *u* is an inflexional ending of the accusative. In English, the meaning is expressed by word order, with the direct object following the predicate. Jakobson indicates that the accusative signals the subordinate character of the object in regard to the subject.

The genitive indicates that the extent to which the object takes part in the message is less than the full volume of the object. The degree is indicated by the context: (1) the object may be partially presented in the message—"Poel hleb<u>a</u>" ("Eat some bread"). The underlined is the inflexional ending of the genitive; in

English, the object's volume limits are expressed by the pronoun "some." (2) The object may be placed outside of the message. In this case, depending on the context, the object may be on the margin of the message ("Odnoj nogoj kasajas' pol*a*"—"Touching the floor with her one foot"); the message may move away from the object ("uzbezshat' smert*i*"—"escape death"); the message excludes the object ("Mi ne nashli kvartir*i*"—"We did not find an apartment") (Jakobson, 1971c). Finally, the genitive may be subordinated directly by the noun on which it is dependent, either limiting the volume of the object directly ("stakan vod*i*"—"glass of water"; "chast' dom*a*"—"part of the house") or abstracting from the object a feature ("krasota devushk*i*"—"beauty of the girl"), some of its belongings ("snarjazshenije rabocheg*o*"—"equipment of the worker"), something from its surroundings ("sosed rabocheg*o*"—"worker's neighbor") (Jakobson, 1971c).

Like the accusative, the dative is a case of relation; it carries the semantic mark "relatedness to action" ("On dal shofer*u* adres"—"He gave an address to the driver"). The underlined is the inflexional ending in Russian; in English, this relation is expressed by the indirect object with the preposition "to." In contrast to the accusative, the dative has a second semantic mark: the sign of marginal position. It imparts to the object the position at the periphery of the message. The general meaning of the dative includes direction of action without coming into contact with the object (Jakobson, 1971c).

Analysis of the Russian declension paradigm shows an astonishing correspondence between the linguistic organization of the declension system and the function of field 39 of the left hemisphere. Indeed, semantic marks of the cases represent the signs of spatial relations; the structuring of the general meaning of the case (morphological invariant) corresponds to the left hemispheric synthesis of discrete units (signs). Considering possible function level heterogeneity of field 39, we will try to speculate about the operations connected with it at both the gnostic-praxic and symbolic levels.

The role of field 39 in praxis (object action) consists of distinguishing spatial relations of topological schemes of the objects taking part in the action. At the symbolic level, the signs of spatial relations between objects are implemented already in the declension system (grammatical relations of words in the sentence). In essence, the grammatical meaning of the case is represented by signs indicating the spatial relations of objects taking part in the message. The event that is reported in the message occurs in the visual-spatial situation; spatial analysis of situational context, in our view, is connected with field 39. This analysis is realized through the selection of the corresponding morpheme from the declension paradigm and its projection upon the plane of the message. As we discussed earlier, at the gnostic-praxic level, the "unfolding" of the spatial pattern of object action into successive structure is connected with the frontal region of the left hemisphere. At the symbolic level, the unfolding of the space of the message into a linear structure (sentence) in which the grammatical meaning of the case obtains sound expression (grammatical form) is also connected with this region. Grammatical meaning (morphological invariant, according to Jakobson) is an abstraction, distinguishing signs and combining them according to the opposition principle into a paradigmatic system. In a sentence, however, it is the combinatory variant that is realized by acquiring the sound form (Jakobson, 1971c). In general, the grammatical meaning of the case inflexional ending indicates the relation between subject and object in the sentence; its form, however, performs the connection—"articulation" of words in the linear structure of a sentence (syntactic function).

Another group of nonroot morphemes is suffixes. They, in contrast to other nonroot morphemes, do not just mark but are closely interwoven into lexical meaning. The suffix's grammatical meaning reflects spatial relationship of the parts within a single object. Nouns suffixes may express the general concept of form. For example, in Russian the suffix *nits* is the sign of containment, a receptacle. In the Russian word "sachar*nits*a" (sugar pot), "sachar" (sugar) is the root morpheme and *nits* is the suffix, the sign of containment; in the word "bol*nits*a" (hospital), "bol" (pain) is the root morpheme and *nits* is the suffix, sign of containment. In English, an example of a similar suffix of containment may be *ium*, as in aquarium or sanitarium.

Suffixes may reflect representation of a great number of single objects "molded" into one form, giving to this quantity wholeness or collectiveness; for example, man*kind*, brother*hood*. The sign of collectiveness may be applied to both concrete and abstract words (for example, relation*ship*). Suffixes may have the grammatical meaning of a person's profession or occupation; for example, act*or*, work*er*, jewel*er*, doct*or*. We suggest that these suffixes reflect a general concept, form of "I" existence, which is an equivalent to the right hemispheric I-space–non-I-space integration. There are suffixes whose grammatical meaning designates the extent of the object in the outer space: the Russian *dom* (house)—"dom*ishk*o" (small house)—"dom*isch*e" (big house); in English, pig*let*, ham*let*. Suffixes with quantitative meaning sometimes acquire a metaphorical sense. For example, Russian diminutive suffixes *ik*, *enk*, *onk* may also denote endearment or gentleness: "kot" (cat)—"kot*ik*" (small cat). Augmentative suffixes like the Russian *ish* and *esh* often have an additional meaning of crudeness: "ruka"—"ruch*ish*a" (hand—big hand). In other cases, the augmentative suffixes acquire a positive connotation; for example, the Russian "chelovech*ish*e" (man of great value), while a diminutive suffix will add a negative evaluative sign, with a hint of despising, "chelovech*ishk*o" (little man). Polysemantics and emotional saturation of the secondary nuance meanings of suffixes indicate the role of the right hemisphere in their origin. The grammatical meaning of suffixes does not depend on context. In producing word form, the suffix closely interacts with root morpheme lexical meaning. In this case, the parts of the word are connected with each other to give new meaning. This gives an additional argument for the possible different topographical cerebral basis of prefixes and inflectional endings (field 39) and suffixes (field 40).

The genitive case in the declension paradigm gives us, however, an example of the intermediate category between the two groups of nonroot morphemes mentioned. The grammatical meaning of genitive, according to Jakobson, includes only one sign: volumeness. The sign of volumeness in itself corresponds to spatial relations within one object (part–whole); that is, the function of field 40. However, the degree to which volume of the object will be limited depends on the situational context of the message. In other words, the limits of object participation in the message are determined by relations of the object with the other object acting in the message (which corresponds to the visual-spatial signs, function of field 39), and not relations of parts within the object. There is one variation of the genitive, the genitive subordinated directly by the noun, which does not depend on situational context. Depending on the categorical signs of the noun lexical meaning, this subordinated-to-noun genitive may reflect the relation part–whole literally (the hand of the brother) or indirectly (brother's shirt, sign of possessiveness or belonging, or brother's father, sign of relationship).

Let us summarize our concept of morphological language code in its relation to cortical fields 39 and 40. Nonroot morphemes have grammatical meaning and formal syntactic functions in sentence arrangement. The grammatical meaning of nonroot morphemes reflects or comes from the outer relations between objects (relations in space and time). The signs of outer relations, in contrast to categorical signs, have formal expression in language—"overt" grammar (sound form of nonroot morphemes). Neurolinguistically, we suppose nonroot morphemes might be divided into two groups: prefixes (prepositions) and inflectional endings, whose grammatical meaning originates in visual spatial analysis (spatial relations between objects, field 39) and suffixes, whose grammatical meaning has its origin in kinesthetic spatial analysis (spatial relations within the object, relations between parts and whole, field 40). In light of what we have said in this chapter, that which has been called semantic aphasia requires a neurolinguistic reinterpretation. We propose that the basic mechanism of this disorder is the impairment in the comprehension of the grammatical meaning of the nonroot morphemes, or a selective disorder of morphological language code. It is more appropriate, therefore, to call this disorder *lexical morphological aphasia*. Patients with morphological aphasia will experience special difficulties in comprehension of the grammatical relation between words in circumstances when grammatical redundancy is absent; that is, when they cannot compensate their deficit of morphological code by means of formal syntactic relations, and in particular, word order (remember that syntactic code is intact in these patients). In English, syntactic code prescribes word order in the sentence as follows: subject (actor)–predicate (action)–object (acted upon). A patient suffering from morphological aphasia will understand the sentence "Steve likes Nicole" even if he has problems with the inflectional ending of the third person singular present indicative of the verb "like." However, he will have great difficulty understanding reversible, passive constructions such as "The tiger was killed by the lion" versus "The lion was killed by the tiger." As a compensatory mechanism, the order subject–object ("prescribed" by the syntactic code in English and intact in patients with morphological aphasia) becomes irreversible in patients with morphological aphasia. This is true even in languages with freer word order than English. For example, in Russian "Luk*a* pomnit Ol'g*u*" and "Ol'g*u* pomnit Luk*a*" both mean "Luka remembers Olga" (*a*, Russian nominative declensional ending; *u*, accusative declensional ending). As Jakobson stated, "For a Russian with semantic aphasia, any noun which precedes the verb becomes a subject, and any postverbal noun is comprehended as an object notwithstanding the inflectional ending" (Jakobson, 1971b, p. 315). It is especially difficult for patients with morphological aphasia to comprehend the subordinate in groups of two nouns (for example, father's brother and brother's father) that do not depend on the syntactic environment and whose meaning may be extracted only from nonroot morphemes. Note that tests designed for diagnosis of semantic aphasia (Luria, 1966/1980; Goodglass & Kaplan, 1972) use nonredundant grammatical constructions. Finally, it should be considered that if prefixes (prepositions) and inflectional endings are related to field 39, and suffixes to field 40, as we proposed, dissociation within the syndrome of morphological aphasia might be expected depending on relative involvement of these cytoarchitectural fields. The question arises whether specific thinking disorder is associated with morphological aphasia. The signs that underlie the morphological paradigmatic series reflect not inner connections of the objects (logical classification) but rather their connections by contiguity in space and time. It is not by chance that they always have an outer marking (overt grammar). In contrast to "covert" grammar, overt grammar is not universal but different for each given language. Covert grammar, in creating the logical frame of language, is inseparably linked with thinking. The impairment in the outer markings of categorical signs of word meaning that occurs in morphological aphasia theoretically may result in some secondary thinking disorder.

4.4. Right Parietal-Occipital Region and Its Contribution to Self

4.4.1. Cerebral Organization of "Body Scheme"

Visual spatial disorders connected with lesions in the right parietal-occipital regions involve not only extrapersonal space but also cause disorientation in intrapersonal space (body scheme). There is some evidence that extrapersonal and intrapersonal space might have discrete anatomical representation within the right parietal occipital region. Guariolia and Antonucci (1992) reported a case of severe personal neglect in the absence of a deficit in extrapersonal space. An extensive neuropsychological assessment demonstrated a severe representational deficit of the left side of the body (intrapersonal space) in the absence of cognitive impairments in visuospatial processing.

Body scheme is defined as a tridimensional inner diagram that includes form, relative size, and spatial relations of separate body parts in an autonomous whole. It is not part of conscious self-awareness; indeed, we only have learned about body scheme from instances of pathology that have resulted in body scheme disorders. Various syndromes of partial (regarding separate body parts) and general body scheme disorders following lesions of the right parietal region are described in the literature (Smirnov, 1976; Critchley, 1950, 1953). Authors identify the thalamoparietal system as the cerebral base of body scheme (Chlenoff, 1934; Schmaryan, 1949; Smirnov, 1976). We will consider the contributions to the formation of body scheme of the different regions of the thalamoparietal system, applying our three-dimensional framework of vertical levels and horizontal differentiation within levels.

4.4.1.1. Thalamic Level and Body Scheme.
Bernstein's Level of Synergetic Movements and I-space. We will start with Bernstein's definition of B level afferentation-space, expanding this concept to the right hemispheric cognitive mechanism. Then, we will show how modern data regarding thalamic structure-function organization may complement Bernstein's insights, and how they may be further interpreted using our three dimensional framework. According to Bernstein, the contribution of the thalamus to the B level is provision of afferentation for the highly coordinated simultaneous contraction of the numerous muscle groups, the "synergetic chorus." Afferentation is composed of mostly proprioceptive sense from one's own body. Bernstein emphasized that it is from this constant inflow of proprioceptive information about the ever-changing position of the body and its parts that one invariable entity, one's own body space, is built. One's body space is independent of the particular body position in the extrapersonal space at the moment and independent of the extrapersonal space itself.

Bernstein indicated that proprioceptive pathways terminate at several "phylogenetic stories" of the brain corresponding to the functional levels A, B, C; however, at each level, the "meaning" of the proprioceptive sense is different depending on the conceptual framework of the level (space, time) and the composition of sensory modalities integrated into the afferentation of each level. At level A, afferentation includes proprioceptive and vestibular sense needed to maintain muscle tone and vertical body position in the gravitational field. At this level there is no division into outer and inner space (or body space).

At the "extroverted" C level, afferentation is composed of distant modalities—visual and auditory—as well as vestibular, proprioceptive, and tactile. The proprioceptive modality together with other modalities included into afferentation of the C level will participate in evaluation of external objects.

Bernstein calls the B level "introverted," because for the first time, and only at this level, proprioception exists in pure form, without admixture of other modalities. At the B level, Bernstein writes, it is proprioception par excellence, aimed, directed, projected inward at one's body. It is afferentation of one's own body without any connection with the external world (if we theoretically disconnect level B from the other levels). Proprioceptive sensations integrated along one's body spatial coordinate system, according to Bernstein, provide afferentation for B-level movements, a function of the left hemispheric cognitive mechanism. At the same time, we think, proprioception "forms" the body space, defining its borders or spatial contour and giving "sensational filling," composing an indivisible whole that we call I-space. It is this basic formation, I-space, that the right hemispheric cognitive mechanism uses in construction of the self, incorporating it into higher function levels. I-space at the thalamic B level, we believe, has already incorporated in itself a subjective sense of internal body mass and visceral sensation of the lower A level (see Chapter 5). In this context of a unique indivisible whole of one's own body space with its sensational content, it is worth recalling that in classical neurology and psychiatry, authors mentioned various sensations from the body: undefined body feelings (Seshenov, 1952; cited by Anufriyev, 1979); "protopathic sensations" (Shmaryan, 1949); somatic feeling (Schneider, 1959); one's own body sensational feelings (Gruhle, 1915). They were considered subconscious, unrealized, and unapproachable for discriminative sensational analysis in the norm; but of great importance for the internal subjective world and fundamental psychical condition or living tone, vigor vitalis, psychical tone.

The role of separate thalamic nuclei in the formation of "body scheme." Studies utilizing electrical stimulation of the thalamus have been helpful in elucidating the different functions of the various thalamic nucleii and further differentiation within nucleii. We will consider those thalamic nuclei that, in our opinion, are key players in the formation of body scheme. First is the ventral posterior nucleii group (VP), which is the principle relay station for somatic sensation (Kandel & Schwartz, 1985). It is a complex formation including numerous relatively autonomous units differentiated along somatotopic and modality-specific dimensions.

Stimulation of any unit within the VP causes so-called artificial sensations experienced in that particular part of the body (arm, leg, hand, and so on), which is represented in the stimulated VP unit (Smirnov, 1976). These sensations can be tactile, proprioceptive (sensation of moving, shifting, deep pressure), or pain or temperature sensation, depending on the modal specificity of the stimulated area. In general, sensations induced by stimulation of VP had neutral emotional tone. Patients called them a "shadow of sensations," "incomplete, indifferent sensations" (Smirnov, 1976). Different results were obtained with electrical stimulation of areas within the VP by low- versus high-frequency current. Low-frequency stimulation caused various local sensations, whereas stimulation of the same areas by high-frequency current produced instead distorted perception of spatial configuration (form, size, location) of that particular part of the body. It was assumed that stimulation by low-frequency current induces functional activation, whereas stimulation of the same region with high-frequency current may cause damage and corresponding dysfunction of the nucleus. Thus, we may say that low-frequency stimulation of the VP produces "actualization" of the corresponding body part's space expressed by the sensations that "form" this space, the "contours" of this space; whereas high-frequency stimulation causes distortion of that body part's spatial contours, producing disturbance that mimics the clinical syndromes qualified as partial body scheme disorders. In essence, then, the VP integrates sensations according to spatial coordinates, producing knowledge of one's own body parts and the rudiments of subjective experience. This is highlighted by phantom phenomena, in which false sensations of absent body parts persist after they are lost as a result of amputation, denervation, or trauma. This syndrome is particularly troubling clinically when significant persistent pain is experienced as originating from the absent part. The very existence of the phantom phenomenon indicates an extraordinary vitality of body scheme, which, once formed, is not destroyed despite the absence of incoming sensory information. However, ablation of the VP results in the disappearance of the phantom sensations, indicating the importance of this nucleus in the formation of scheme of separate parts of one's body.

The VP is directly projected on somatosensory cortex, cytoarchitectonic fields 1, 2, and 3 of the postcentral area and fields 5 and 7 of the superior-parietal area. The VP and its corresponding cortical counterparts form the VP—projective parietal cortex functional system. The ways in which the VP and the corresponding cortical areas interpret sensory input is quite different. In contrast to diffuse thalamic sensations, sensations provided by somatosensory cortex are discriminative and gnostic (directed at the object stimulus). Considering the difference between the thalamic and cortical components of this subsystem from the position of function level hierarchy, we may say the following. Using Bernstein's terminology, somatosensory cortex is the anatomical substratum for afferentation of function level C. In contrast to the fully introverted (thalamic) B level, the C level is completely extroverted. Afferentation of this level provided by projective sensory cortical fields gives information about the external spatial field, i.e., non-I-space. The distant sensations, visual and auditory, play the major role in afferentation of this level. Tactile-proprioceptive sensations give additional information about features and qualities of stimuli (objects) from the external world. Tactile-proprioceptive sensation is refracted in the opposite direction in thalamic and cortical levels: thalamic sensation forms space of one's own body parts and is very important in formation of body scheme during ontogenesis; cortical sensation, in contrast, is turned to the object. "The body is represented in somatosensory cortex with grossly distorted proportions which parallel the importance of a particular part of the body for tactile sensibility to evaluate the external stimuli. In humans, in whom language and the handling of tools are so well developed, the tongue and hand predominate, both with large representations" (Kandel & Schwartz, 1985, p. 322). In most cases, lesions in the postcentral cortex give simple sensory disorders without body scheme impairment.

Within the thalamus, the VP projects upon the integrative or associative nucleii of the lateral group, including the lateral posterior (LP). The LP receives multimodal afferent input and functions as an integrative nucleus in regard to both modality and somatotopic dimensions. It is the LP which is the major thalamic part of the thalamoparietal body scheme system, connecting with the parietal-occipital region (Kandel & Schwartz, 1985). We speculate that the LP is concerned with integration of sensations according to body spatial coordinates, which corresponds in Bernstein's terms to afferentation of one's own body. Integration of sensations that make up afferentation of one's own body forms the whole and singular spatial image of the body, which we will term I-space, the indivisible wholeness in which both spatial form and content (sensational filling) are implicitly incorporated.

Oliver Sacks (1987), in his essay "Disembodied Lady," described a patient whose clinical picture may be an eloquent illustration of the importance of sensational filling for the formation of I-space. Patient C was a young woman with high premorbid intellectual level who suffered selective damage to proprioceptive fibers throughout the neuroaxis, an extremely rare occurrence, as a consequence of an acute sensory polyneuritis. In other words, she sustained proprioceptive deafferentation of the whole body. The patient felt her body was dead, not real, not hers. She could not identify herself when she was shown home movies of herself with her children taken just a few weeks before her polyneuritis, stating: "She is gone, I can't remember her, I can't even imagine her. It's like something's been scooped out of me, right in the center" (Sacks, 1987, p. 51). Patient C could rationally understand and even invent herself, and employed compensatory strategies by using other sensory modalities (hearing, vision). It is she who found this word—*disembodied*—to describe her experience. Using her extraordinary premorbid abilities and her great insight, C was very successful at her rehabilitation work. Yet, as Sacks put it, "She had succeeded in operating, but not in being." We see in this case that even a peripheral disorder, if it causes global absence of proprioceptive inflow (which in the norm is being integrated according to body spatial coordinates at the thalamic level and reinforces body scheme each moment of our lives) may lead to some sort of depersonalization. We speculate that the associative LP nucleus plays important role in this process.

Another example of clinical pathology supports the existence of sensational filling in the formation of the physical sense of self in the norm, which is not part of conscious awareness. An autoscopic experience is a perception of one's own body image, usually visual, projected into the external space. Some patients, however, "experience the presence of their own body image projected outside their actual physical bodies by means of *senses other than vision*" (Lukianowicz, 1967, p. 34). Lukianowicz reported on a 34-year-old patient who described that during migraine headaches he would frequently momentarily feel "that I have two separate bodies. They are both 'me' or 'I'.... Yet, I have never seen him with my eyes, though I feel his presence very intensely" (p. 34). Another patient, a 23-year-old schizophrenic, described feeling "that my 'other self' got up, walked to the window, and looked out, when my 'real self' was sitting in the chair, like a shadow" (p. 35).

The mediodorsal (MD) is an integrative nucleus that, within the thalamus, is mainly connected with nonspecific nuclei and associative nuclei, including the LP. Because the LP is an integrative, multisensory formation we may suppose that information which comes from the LP to the MD is presented with several modalities (mainly kinesthetic and tactile) in an integrated form, irreducible into separate modalities again. Yet although sensory information is integrated into a general bodily sense, it is somatotopographically organized according to body parts, as opposed to one's body scheme as a whole. Actually, the MD is not part of the thalamoparietal system of body scheme; its main connection is not with tertiary parietal fields but with prefrontal cortex. It is the most important thalamic connection to the prefrontal cortex (Fuster, 1985). In fact, Fuster emphasized that the MD is "a nucleus so heavily and distinctly projecting to the prefrontal cortex that the latter is conventionally defined as the cortical territory of MD projection" (p. 151). Given the extensive anatomical connections of the MD to the most important human-specific part of the brain, its input would seem to be of crucial importance in higher cortical functions. Yet little is known about the MD's role. We will attempt to offer some speculations about this role, drawing general inferences based on patterns connected with thalamic pathology.

Direct stimulation of the MD may cause unusual and undeterminable, incomprehensible, unexpressable sensations accompanied by emotional tone, usually negative (Smirnov, 1976). Pathological sensations due to stimulation of the separate sections of the MD will involve the corresponding body parts rather than the body as a whole. Literature from both neurology and psychiatry gives information about unusual, pathological bodily feelings similar to those found with stimulation of the MD. Classical neurologists identified the syndrome of thalamic hyperpathy, characterized by extremely distressing, agonizing, poignant sensations that are diffuse and obscure at the same time. This syndrome is found in patients with damage to the thalamus. Another syndrome, so-called senesthopathy,* has been described in the psychiatric literature. Senesthopathy is somewhat similar to hyperpathy but less intense and more narrowly localized in different parts of the body. Senesthopathy is characterized by distressing, restraining sensations difficult to attribute to any one modality. These sensations are qualitatively "new," different from what may be felt in the norm, and described as unusual, queer,

*The term "senesthopathy" is equivalent to "coenesthesia" (or cenesthesia, cenesthopatic symptoms) used elsewhere to mean disagreeable or unusual bodily sensations.

incomprehensible, vague, and unexpressable sensations accompanied by negative (much more rarely, positive) emotional tone, projected on the separate parts of the body. In its quality, senesthopathy cannot be compared with any other somatic sensation. It often has a sense of something impeding or moving from place to place, and the word *tension* is frequently applicable to characterize senesthopathic sensations and feelings. Trying to describe their sensations, patients relate that certain parts of the body "got narrow" or "became thicker" or "swell," or are "displaced" or "fixed," or are "squeezed, compressed with some bonds," or "shrunk." "Alien objects are struck into the body," "gases are infiltrated," "current is circulated." Patients feel "boiling," "crackling," "crepitation," "swelling." As a rule, although these are unpleasant sensations, they are not characterized as pain.

Senesthopathy was occasionally described in patients with known focal damage to the thalamus as well. In these patient populations, senesthopathy might be observed both in combination with hyperpathy and separately (Shmaryian, 1949; Dobrochotova, 1974). Anufriyev (1979) suggested that senesthopathy is not just a disturbance of sensation but a disorder of subjective experience of the body, or a disorder of bodily feelings, and is of thalamic origin. Dobrochotova also believed that senesthopathy is not just a disturbance of bodily sensation but, as a subjective experience, involves affect (Dobochotova, 1974).

The affect or emotion associated with senesthopathy involves the subjective attitude to the sensation, its specific sensual tone and timbre. This "emotion" is qualitatively different from the usual positive or negative colored feelings, such as joy or sorrow, in which somatic sensations play no role. In contrast, the senesthopathic "emotion" is localized in the separate body parts; further, it is important to emphasize that in senesthopathy, the sensational component is not separate from the emotional component. The classic literature in neurology contains examples of patients with focal unilateral thalamic lesions and localized disturbances in somatic sensation accompanied by equally circumscribed emotional changes. For example, Head described patients with focal thalamic lesions and peculiar half sided changes of emotional tone. One patient reported that he could not go to church because horrible feelings emerged on his affected side when the chorus started singing. Another patient reported that the right side of his body became more gentle: "I have extreme desire to put my right hand on the soft skin of a woman. My right hand needs consolation. It feels that my right side is seeking consolation. My right hand feels more artistic." A third patient reported that his soul on the left side of his body was different from the soul on his right. Head concluded that the emotional tone of somatic or visceral sensation is a product of thalamic activity (cited by Kretschmer, 1927, p. 27).

Senesthopathy as a symptom is included into the structure of many psychopathological syndromes as one of the "notes of the pathological orchestra," yet it may also be found in a pure form against a background of intact emotion, intellect, and personality. Such occurrences are usually considered due to an inborn thalamus anomaly. Cases of senesthopathy in a pure form allowed description of this condition by itself. According to Djupre (cited by Anufriyev, 1979), senesthopathy in such instances is characterized by strange sensations, often defying description, which are explained by patients with use of various image-bearing expressions. The capability of conveying one's own experience by means of figurative comparisons, the preciseness and picturesqueness of this capacity, corresponds to the patient's intellectual level and wealth of imagination. Djupre stated that patients often understood the incompleteness, the imperfection of their descriptive tales. He suggested that language cannot express these absolutely new, unusual sensations, which are unique and idiosyncratic for each patient and which are not related to past experience.

We are also of the opinion that these sensations are virtually impossible to express directly. In our theoretical framework, these sensations belong to the introverted B level; as such, they are not subject by themselves to external influences and are thus unique to the individual. Further, the B level has no direct input from the visual modality. As we have shown in Chapter 2, the development of language in phylogenesis is closely associated with the visual modality and is a function of higher levels, although incorporating information from the B level. Thus, the sensations of level B, without direct access to language for expression, in some way are translated into visual images that may then be expressed in language, though generally imperfectly. Metaphors seem to capture the feelings more accurately than direct efforts at expression. The translation of the sensations of level B into visual images and then language is a universal phenomenon, and we think a major source of creativity in art forms as well as all kinds of creative thinking. We conceive of the bodily feelings of level B as the fundamental basis of the space of the self, yet their inability to be directly accessed creates a constant drive for their translation, their decoding, into knowable forms, the drive to understand or express them through their projection onto external space, where they may be seen or touched or heard. We emphasize that this is not a literal translation of bodily sensations but relies on metaphor, behind which is visual symbolic thinking. Indeed, artists and great thinkers must have higher development of level B and a concomitant increased drive to decode this space of the self, creating infinite numbers of metaphors, symbols of I-space, using their particular media—art, music, literature, and so on.

Another extreme of this sensation coming to the surface occurs in psychiatric disorders, as we will show in subsequent chapters, in which pathological bodily sensations (senesthopathy) may represent the fundamental defect that is the nidus of subsequent complex pathological processes. They will be "translated" into visual or auditory modalities of non-I-space, with further delusional interpretation.

4.4.1.2. Cortical Level and Body Scheme.

Partial body scheme disorders. Right parietal lesions may cause both partial and general body scheme disorders. Partial disorders of body scheme include the experience of an increase or decrease in the size of body parts, distortion of their shape, reduplication of parts of the body, illusions concerning the position of the limb in space, and so on. For example, patients with right parietal lesions or epilepsy with right parietal or temporal focus have reported distortions in experience, such as the left hand grew larger, the left leg lengthened, "my fingers looked very long, my feet seemed so big they wouldn't go under the seat in the bus" (examples from Arseni et al., 1966, and Hecaen & Ajuriaguerra, 1952, both cited by Cutting, 1990). An illustration of disorder of shape is given by Dobrotochova and Bragina (1977) from a patient with a focus of epileptogenic activity in the right parietal-occipital region, who alleged that her head had turned into a cone and the frontal part of her head was absent. An example of body part duplication was presented by Weinstein, Kahn, Malitz, and Rozanaski (1954), reporting on a patient with extensive lesions in the right hemisphere who claimed that she had two left hands; similarly, a patient described by Chlenoff (1934) claimed that he had two left ears, and so on.

The fact that parietal cortical cytoarchitectural fields 39 and 40 are junction points for several modalities leaves an imprint on body scheme. As Chlenoff (1934) put it, the borders of body scheme coincide with the borders of the inferior parietal region. However, the closer to the central convolution, the more proprioceptive-tactile components appear; the closer to the occipital lobe, the more visual components of body scheme arise. Passing over to the temporal lobe, there are more vestibular elements to body scheme. Patients with lesions in the right temporal-parietal region may manifest body scheme disorders that clearly reflect vestibular influences, such as feelings of weightlessness which might be accompanied by a change in body position relative to other objects. One such patient, described by Dobrochotova and Bragina (1977), felt as though objects were moving away, losing their volume, and becoming small, while she herself was flying up to the ceiling and looking down. Another of their patients with an epileptogenic focus in the same area experienced a paroxysmal feeling that her left arm was flying away. Such symptoms as sudden perceived increase in the size of limbs or the whole body or cutting off the extremities also reflect vestibular influences (Shmaryan, 1949). In general, right cortical parietal disorders of body scheme are characterized by disturbance in spatial contours of the body and its parts, whereas disorder of body scheme that are a consequence of thalamic level dysfunction produce more of a disturbance of sensational filling, which forms body space in general and space of body parts. The disturbance in spatial contours seen with right parietal disorders is accompanied by subjective experience that is completely different from that seen with thalamic lesions. The subjective phenomenon that is regularly observed with this spatial distortion of body scheme can be characterized as a sense of estrangement or loss of belonging, which includes physical sense (often of detachment) and psychic feeling (alienation) pertaining to a body part. Different patients with right-sided lesions made the following statements: "I feel as if the left arm is cut off from the shoulder, as if it just left me."

"I feel as if the left leg is leaving me and that my left eye seems to be leaving its socket." "I felt as if instead of left leg, there was something that did not belong to me, a piece of meat, as if I'd no leg" (cases reported by Hecaen & Ajuriaguerra, 1952, cited by Cutting, 1990). Patients described by Critchley (1950) felt their arm as strange or as not belonging to them. Summarizing extensive literature as well as his own observations, Cutting (1990) concluded that alienation or detachment was the single common theme apparent in the anomalous experiences reported with right-sided lesions. Patients with right parietal lesions and body scheme disorders may develop confabulation, yet we note that their themes remain in the realm of "not belonging." For example, a patient described by Roth (1949) complained bitterly that there was another man's arm in bed with him. Another patient described by the same author was unable to find her hand and was seen searching for it; noticing that she was observed, she said, "It feels as if someone had stolen it." We note the overall similarities of these symptoms with the well-known right parietal syndrome—hemiasomatognosia, neglect of one side of the body. The symptoms of unawareness, unconcern, and anosognosia (denial of deficit, usually of right-sided hemiplegia) that are observed may be seen in body scheme disorders.

General body scheme disorders and "psychic I." Partial body scheme disorders, body scheme disorder involving the left half of the body, and general body scheme disorders may all be found in patients with lesions in the right parietal area. Patients who have body scheme disorder involving the left half of the body may relate that part to another person or have a sense of a "double." The term *general body scheme disorder* is sometimes used interchangeably with *physical depersonalization*, in which patients feel that their bodies are dead, and they often experience doubt in their own existence. Dobrochotova and Bragina (1977) described patients with lesions in the right parietal-occipital region who experienced a feeling of disappearance of their body or who had a sense of estrangement of the left half of their body. Patients described by Gertsberg (1948) and Gurevitch (1948) (cited by Dobrochotova and Bragina, 1977) felt themselves to be a "casing," a "cover," believing their "I" to have been separated and located outside their body, close to it, more to the left. Thus, disorder involving half of the body and general body scheme disorders are accompanied by feeling of estrangement, a continuum from estrangement of one's own body to estrangement of one's own self.

The experience of estrangement from one's own self represents the psychopathological symptom *depersonalization*. This experience may be accompanied by a sense of change in or foreignness of one's own self, disconnection from the world, loss of feeling of belonging to the world. Patients described feeling like insensible automatons, puppets, spectators, devoid of feelings and cut off from contact with the surrounding world (Dobrochotova and Bragina, 1977; see also "The Disembodied Lady," described by Sachs, 1987). It is interesting to note that in patients with partial body scheme disorders, there is an estrangement of that body part whose spatial image is distorted. Patients usually deny ownership of that part (see preceding examples), yet the sense of self as a whole is usually intact. Along the continuum from partial to general body scheme disorders, it becomes increasingly difficult to distinguish between traditionally neurological (body scheme disorders) and psychiatric (disorders in the sphere of the psychic "I," depersonalization) symptoms. Shmaryan (1949) described a patient who suffered for many years from a slowly growing brain tumor located in the posterior parietal region parasagitally (and probably came to involve thalamic areas as well). The patient complained that he felt himself to be fossilized; his heart and body became wooden; he was just a piece of flesh; he was somehow separated from the world; he was split in two; he had double thoughts; some force outside of him made him think; somebody gave commands to his brain. Here we see general body scheme disorder and depersonalization as part of a psychopathological picture that includes schizophrenic-like delusions of control.

Some patients with right parietal lesions seem to have a specific type of depersonalization, and describe passing "into another space," feeling as though they are leaving the space in which other people and the world remain and can be observed by the patient, and going to another space containing only themselves (Dobrochotova and Bragina, 1977). The authors emphasized that these patients struggle to convey to the examiner their experience with astonishingly uniform descriptions, using the word "space" regardless of their intellectual level, age, education, occupation, and so on. Patients used the expression "I go to another space" not to designate that they had a sensation of movement but in an attempt to express the complex subjective experience of their altered existence in regard to the outside world.

Table 4 shows our summary of the contributions of the different formations of the thalamoparietal system to cerebral organization of body scheme. The VP nucleus of the thalamus provides a unity of sensations and that spatial form in which they are embedded. "Form" here is represented by shape of separate body parts. Sensations can be proprioceptive, tactile, temperature, and pain, each having its own form (body part representation). The LP nucleus integrates sensory information so that it is no longer modality-specific or divided into body parts, instead producing a generalized image of one's own body space filled with sensational content (I-space). We think that LP function is close to what Bernstein described as afferentation-space of the B level, one's own body kinesthetic space with its own propriomotor rhythm.

Although the MD is not part of the thalamoparietal system, we include it because it is an important part of the thalamic contribution to self. In the MD we again have, instead of a whole physical sense of self, feelings of separate body parts. Here the feelings are integrated, unlike modality-specific sensations in the VP, and acquire a new subjective or emotional tone. This is the information that the

TABLE 4. Cerebral Organization of Body Scheme: Contribution of the Components of the Thalamo-Parietal System

	Space/ form Space of the whole body	Space/ form Space of boyd parts	Sensation/ content Modality-specific	Sensation/ content Integrated	Subjective tone
Thalamus					
VP nuclei		+	+		
LP nuclei	+			+	
MD nuclei		+		+	+
Cortex					
Postcentral and superior parietal regions (fields 1, 2, 3, 5, 7)			+		
Parietal-occipital regions (fields 39, 40)	+				

MD provides to the prefrontal lobe: a transformation of I-space into a new subjective sense of bodily feelings, "thalamic emotion."

In the primary parietal cortex (fields 1, 2, 3, 5, 7), sensation is modality-specific but it is split from its spatial form. Sensation is important for evaluation of the object, not the body (subject). In secondary-tertiary parietal occipital cortex (fields 39, 40), form (coordinate system of one's own body) is split from its sensational content.

4.4.2. Right Parietal-Occipital Region and Integration of I-Space and Non-I-Space

As we discussed earlier, the right parietal-occipital region plays a leading role in orientation in real space. Patients with damage to this region experience difficulties in visual recognition of landmarks, with their unique orienting value (Landis et al., 1986). They lose a sense of locality: they cannot find their way in their own neighborhoods or their wards in a hospital, in contrast to patients with left parietal-occipital damage, who can orient in real spatial situations but cannot draw a map or plans. Other terms that have been used for spatial disorder connected with right parietal lesions include *environmental agnosia, disordered sense of familiarity, topographical amnesia*. Cutting views the role of the right hemisphere in the norm as "setting the scene in a matrix of spatial coordinates in the same way as we would perceive the scene if it were actually happening" (Cutting, 1990, p. 38). We have discussed in chapter 2 our understanding that the visual scene is stored in the right hemisphere as a single whole, as it was perceived. However, we emphasize now that the same situation will be interpreted and stored differently by the parietal and temporal lobes, based on the fundamental differences in functioning of ventral and dorsal visual pathways (object vision versus spatial vision). The visual scene situations in the right inferior temporal lobe are represented as a single whole within which objects are included; the objects are recognized within the situation and forever linked with it, and this situation-with-object is a unit for operation in symbolic thinking. Within the right parietal region, the visual situation is interpreted as a purely spatial gestalt. It is the space of the visual scene, its dimensions and contours, and the objects within it as spatial forms in the unique locations and positions in which they were perceived. Thus, we assume that in the right parietal region, singular spatial situations are represented. There are as many spatial situations stored as are perceived, and each situation is unique regarding its spatial configuration and the location of objects within it. We consider these multiple *situation-spaces* as non-I-spaces of the C level, meaning that they are a right hemispheric equivalent of Bernstein's C level external spatial field.

The structural heterogeneity within fields 39 and 40 suggests that they contain regions with different functions of varying degrees of complexity and of different phylogenetic age. Specific spatial functions we have discussed include spatial forms: the coordinate system of one's own body and non-I-space. Although we are discussing cortical levels, it should be recalled that in the right hemisphere, information processing reflects extensive integration of cortical and subcortical activity. The right parietal region (fields 39, 40) is intimately integrated with the thalamus (the thalamoparietal system). The thalamic contribution to this system will be one's own space filled with proprioceptive sense: I-space. We suggest that a key function of the parietal-occipital region is integration of spatial forms of the cortical level and I-space of thalamic level (see Figure 17).

We think that integration of I-space and non-I-space plays a major role in the formation of the self (space for psychic I). Depersonalization may illustrate selective disorder of integration of these components, with different types of depersonalization reflecting disintegration of one or more components. Depersonalization includes two aspects that are always together: change in the self and estrangement from the world, or a loss of one's own self belonging to the world, the feeling of natural contact with the world lost. These are psychic metaphors and underneath them, literal physical spatial equivalents, disintegration of spaces. There are several types of depersonalization in which disintegration of one of the spatial forms prevails; for example, the spatial form–coordinate system of one's own body is separated, as illustrated by Gurevitch's patient noted earlier, who felt like a "casing," a "cover," feeling the self located outside of the body. In another type of depersonalization, there is splitting of I-space and non-I-space, as in the patient who reported "passing to another space," with two real spaces arising—

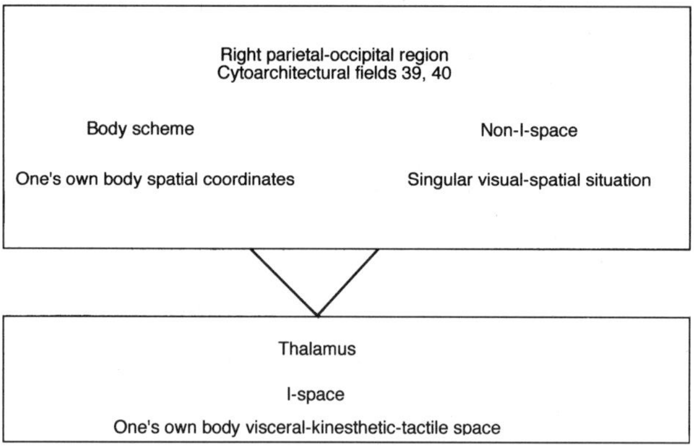

FIGURE 17. Thalamoparietal system and integration of I-space and non-I-space.

one in which the world remains and the other containing himself. Finally, in physical depersonalization sensational content is removed from spatial form, as in Sacks's case noted earlier of global proprioceptive deafferentation: the patient felt estranged, not the same, "like something has been scooped out of me, right at the center." By these selective disorders we infer that in the norm they are integrated, giving us a normal sense of self.

The degree of integration between I-space and non-I-space varies according to the historical layers of the psychic I. The first stage in the process of psychic I development was identification of I-space and non-I-space, the latter resulting from projection of I-space upon objects within spatial situations and simultaneous identification of these objects in situations belonging to the same symbolic system. We will discuss phylogenetic development of the psychic I related to cerebral hierarchical organization in Chapter 5, in which we discuss the frontal lobe. Here we will dwell only on spatial aspects of the psychic I provided by the right parietal region.

As a result of I-space and non-I-space identification, spatial situations in the right hemisphere became imbued with subjective sense. In this connection, it is of interest to mention the astonishing topographic memory of primitive peoples, their extraordinary sense of location and direction. Levy-Bruhl described the Australian aborigines' exhaustive knowledge of the territory they occupy and each unique landmark: "After hearing a shower they know where, in which rock, some water remains; they will know the location of that particular crevice where water will remain longer" (Levy-Bruhl, 1930, p. 74). Having once passed through a place, the aborigines remember it with fine detail. Analyzing these and other examples of the phenomenal memory of primitive peoples, we conclude that it is right hemispheric in origin. It is the single visual situations, subjectively felt space—the whole right hemispheric gestalt. The Australian aborigines recognized places with all their unique features as we recognize faces. Yet the same people who had this remarkable memory, Levy-Bruhl indicates, could only count up to three. These exposed right hemispheric abilities represent an extreme, on the opposite end of which we see patients with right parietal-occipital lesions, presenting with what was characterized by terms such as environmental agnosia, disorder of topographic memory, and disorder in sense of familiarity.

Because the right hemisphere spatial situations are singular, they are "separate spaces" that do not connect and do not cross with each other. Again, there are as many non-I-spaces as there are singular spatial situations represented in the right hemisphere but there is only one I-space. Considering that the contribution of the right parietal region to the psychic I will be the integration of I-space with separate non-I-spaces, resulting in multiple unconnected, subjectively felt spaces—multiple "selves"—we may assume that it is definitely not the right hemisphere which gives us our sense of our *one* individual I.

On the other hand, these "spaces" united in a symbolic system are identified with each other (reduction of situation series at the next level of right hemispheric cognition). To illustrate these last statements, we will use examples of so-called collective representations of primitive people described by Levy Bruhl in his book *Archaic Thought* (1922/1930). In collective representations, the right hemispheric symbolic systems, which, in modern man, are intimately incorporated into psychic content and displaced from conscious awareness, not only are exposed but are projected outside so that we may see deep layers of our psychic I in the structure and concepts of primitive societies.

Describing the Australian aborigines, Levy-Bruhl wrote, "For them, space is not something homogeneous and uniform, indifferent to what fills it, devoid of its own features and in all its parts identical to itself. Each social group is mystically connected with that part of the territory the group occupies and moves on. Between land and social group there are relations of mutual participation equal to some kind of mystical property which cannot be altered, taken away, or conquered. In addition, each place, with its characteristic landscape, is mystically connected with visible and invisible creatures, with personal 'spirits.' There is mutual participation between these creatures and place, neither place without creatures nor creatures without place would be the same as they are" (Levy-Bruhl, 1930).

Thus, within the symbolic system, which is expressed in collective representation, there is an identification with outside space that is manifested in the attribution of mystic forces to the external world. Self is not differentiated from the outside world. There is also no individual I—members of the tribe identified with each other. Self is not differentiated from "my people," the group of people that belong to the same symbolic systems, which in our terms will be a projection of I-space into objects within symbolic systems. Objects can be people or other animate and inanimate things within situations belonging to the same symbolic system.

As Levy-Bruhl emphasized about collective representations, there is no notion of one external space; the regions of external space are represented as qualitatively different, determined by their mystical participation with certain creatures and objects. "If you ask the aborigine what some drawings mean they may answer that these drawings are made for amusement and that they don't have any meaning.... However, the same drawings, if they are performed on some ritual objects or in a special place, have quite definite meaning" (Levy-Bruhl, 1930). One and the same drawing may mean different things if performed at different places; for example, a drawing performed on one ritual object may mean tree while on the other ritual object it may mean a frog. It should be emphasized that these people will name without mistake the object that is actually drawn and can act with it skillfully; however, this object has no meaning to them (at the symbolic level). Its role is the role of a messenger through which meaning can be carried.

Interpreting these data in terms of right hemispheric cognitive mechanisms at the stage of symbolic-situational thought, described in Chapter 2, we may say that if the same objects belong to those non-I-spaces (visual situations-spaces) which are not united into a symbolic system, or refer to the different symbolic systems, these same objects will be perceived correctly regarding their physical features but not identified with each other in consciousness (will have different meaning). Along these lines, different objects that belong to non-I-spaces united into one symbolic system, although recognized as physically different, will be identified with each other in consciousness according to the symbolic meaning. Two fascinating psychopathological syndromes, Capgras and Fregoli syndromes, seem to involve an analogous mechanism. Both syndromes include disorder of identification of a person with intact recognition of the person's appearance. Modern data suggest that these syndromes are associated with brain pathology with consistent localization in right frontal, temporal, and parietal regions, and their connections. In Capgras syndrome, the patient accurately perceives and recognizes physical features—the appearance—of a familiar person but does not correctly identify this person, believing that he is replaced by double, an imposter. We think that this disorder involves right hemispheric symbolic systems, those same systems that were apparent in archaic thought, in which one and the same object may have different meaning if it belongs to different symbolic systems. In Fregoli syndrome, an individual (usually a persecutor) is felt to change himself into the forms of various people that the patient encounters during his daily life, so that the same person (the persecutor) is identified in different people whose physical features (appearance) are recognized correctly but assigned with different meaning. Thus, in Fregoli syndrome we again see involvement of symbolic thought: different objects acquire the same meaning by coming to belong to the same symbolic system.

To review, right parietal-occipital (fields 39, 40) disorders of different levels replay the same motif of estrangement: estrangement from one's body parts; estrangement from one's own body; estrangement from the familiar (individual visual spatial situation); estrangement from the world (disintegration of I-space and non-I-space, disorder at the situational level); estrangement from symbolic meaning (not belonging to symbolic system, disorder at symbolic level). All these syndromes are due to dysfunction within the right parietal-occipital region with the exception of Capgras and Fregoli, which involve right parietal-temporal-frontal dysfunctions, and possibly disorder of connections of dorsal visual and ventral visual pathways to the frontal lobe, rather than just separate regions.

Situation-spaces are united into a symbolic system by affect, which is the contribution of pathways connecting the amygdala with the temporal and frontal lobes. In this section, we examine the contribution to the self of the parietal-occipital region, which is subjectively experienced kinesthetic I-space projected upon multiple non-I-spaces. If we artificially extract the contribution of right parietal-occipital region from the many systems that constitute the cerebral organization of self, we would have multiple selves, each representing I-space–non-I-space integration included into the context of symbolic systems (no subject–object division). Disorder of the separate aspect of the self contributed by the right parietal region in the norm allows us to understand the variants or unusual cases of Capgras syndrome and similar syndromes called in the literature *delusional misidentification syndromes* (DMS), which any other theories, organic or psychodynamic, have difficulty explaining.

In summary, the integration of I-spaces and non-I-spaces results in a right hemispheric self that is not divisible from the outside world. Projection of I-space upon the object within situations yields a right hemispheric self that is not divisible from people within the group or inanimate objects within the symbolic system. There are multiple non-I-spaces and there are multiple selves. This may explain the pathogenesis of variants of DMS such as multiple doubles, the syndrome of subjective doubles, inanimate object doubles, and other unusual variants. For example, a patient described by Anderson believed that over 300 objects had been removed from his home by his persecutor and been replaced by identical doubles (Anderson, 1988). In this case, the objects as aspects of the self, though recognized for what they were, had lost their association with symbolic systems and were not assigned meaning any longer. In another example, a patient stated that there were many impostor cities, each containing duplicates of himself and his wife and children. He believed that in his real hometown his real family was deceived by an impostor of himself who had replaced him (Thompson, Silk, & Haver, 1980). Here we think visual spatial situations with their content were out of context of symbolic systems and lost their meaning.

4.5. Lexical Morphological Aphasia

The basic deficit in lexical morphological aphasia is selective impairment of the morphological code of language. Patients with this type of aphasia have difficulties in comprehension and use of nonroot morphemes, especially in circumstances in which grammatical redundancy is absent or insignificant.

Patients with morphological aphasia will understand simple sentences but have difficulties understanding grammatical constructions expressed by nonroot morphemes. Repetition is intact. In naming objects and in spontaneous speech, patients may have difficulties using nonroot morphemes, replacing or omitting them.

We assume that within the deficit of morphological language code, there may be two main variants that depend on the location of the lesion. Patients with a prevalent deficiency in field 40 will experience difficulties understanding the meaning of word forms that contain suffixes. Patients with prevalent dysfunction of field 39 will have impairment in understanding of word forms that contain prefixes and inflectional endings, and grammatical words (prepositions). A detailed description of a patient with lexical morphological aphasia follows (Glezerman, 1986). Patient N was a 52-year-old Russian-speaking right-handed man who sustained a cerebrovascular accident in the region of the middle cerebral artery. N demonstrated a striking dissociation between his selective language deficit (an inability to understand the grammatical meaning of nonroot morphemes) and otherwise well-preserved intellectual and linguistic abilities. Patient had an average IQ of 99. His ability for verbal abstraction was above average range (scaled score on WAIS Similarities subtest was 12, against an average score of 10), his comprehension of separate words was excellent (scaled score on WAIS Vocabulary subtest was 16 against an average of 10). At the same time he was unable to comprehend relations expressed by inflectional ending of the genitive case subordinated directly by the noun (the equivalent in English would be inflectional ending of the possessive form of the noun). As we will see, his responses reveal the very specific, spatial origin of his deficit. When asked whether the expressions "chief's brother" and "brother's chief" were the same or different, N responded, "So, brother's chief and then chief's brother. In general it is the same." This indicated that the patient was unable to derive the relationship of the two words. When questioned whether the expressions "daughter's mother" and "mother's daughter" have the same meaning, N responded, "The daughter gave something to the mother and the mother gave something to the daughter." Again, we see that the patient was unable to extract the relationship of the two words, and we also see his unsuccessful attempt to compensate by including the words into a context, by building a sentence, which in turn demonstrates his intact syntactic ability. The sentence that the patient built is grammatically redundant: relations between the words *mother* and *daughter* are expressed both by word order (subject–object relations) and the preposition *to*. When asked to explain the meaning of the expressions "father's brother" and "brother's father," N replied, "These may be different people. For example, I have a brother Ivan ... brother to father ... one ... brothers may be different, this is quite different.... His father has a brother, uncle to him." The patient struggles to understand the relations expressed by the nonroot morpheme; again unable to do so, he constructs a grammatically redundant structure and at one point in this manner is able to extract the correct relation (uncle), although the correct answer is not stable. When the patient was given the direct question, "Father's brother, what relative is it?" the patient's response was "Grandfather, father's brother, it is my brother, and this is his father.... Just a moment, I need to think ... It is so ... it turns out ... they have two brothers. I have a father, but he also has a brother." Asked the question "Brother's father, what relative is this?" N responded, "It is their grandfather, and they have two brothers." In all these examples, the patient could not grasp the single, instantaneous meaning of the two-word expressions (spatial or simultaneous synthesis) and attempted to break it down in a sequential manner (temporal synthesis).

In addition to the difficulties patient N had in understanding the genitive case, he could not understand comparative grammatical constructions expressed by nonroot morphemes of comparative and superlative adjectives. For example, when asked to show the less light (in terms of color) of two strips, N showed the light strip, replying, "This is the light strip." When asked to show the less dark strip, he showed the dark strip, stating "This is the dark one."

Patient N also had a disorder of suffix selection. This was expressed by literal paraphasia that was due to replacement of the appropriate suffix for the particular word by an inappropriate one chosen from the paradigmatical series of suffixes of the Russian language. For example, instead of "iskluchit*eln*ij" (exceptional), N said, "iskluch*ivn*ij" (meaningless); instead of "rabot*nits*a" (female worker), "rabo*chinits*a" (meaningless); instead of "korabl*ik*" (small ship), it was said as "korab*nik*."

Interestingly, in this patient there was a quite striking dissociation within the syndrome of morphological aphasia itself: he had a severe deficit in understanding of the genitive case subordinated directly by the noun (possessive form of the noun in English) and difficulties with selection of suffixes, yet he did not experience any difficulties in understanding of prepositions and adverbs with direct temporal or spatial meaning (such as above, under, ahead, forward, and so on). He easily could perform triple constructions involving spatial relations of the object in regard to two other objects, such as drawing a cross to the right of a circle but the left of the triangle, a very difficult task for someone with semantic aphasia in Luria's term. His nonverbal spatial ability (visual-spatial analysis, constructional thinking) was intact (scaled score on WAIS Block design subtest was 10, the average range). Patient N did not have difficulties in visual recognition and categorization of objects, letters, and colors. His spatial concept of number and calculation abilities were intact.

Patient N had another selective speech deficit. As we noted earlier, his word comprehension was excellent. His reading ability was intact, but his repetition was impaired, with multiple literal paraphasias. Literal paraphasias were represented by mirror reversal of syllables and phonemes within the word with intactness of word contour. Instead of "schek*old*a" (latch), the patient said "schek*odl*a" (meaningless), instead of "zab*lu*ditsja" (lost), he said "zab*ul*ditsja" (meaningless); instead of "sa*rk*ofag" (sarcophagus), he said "*kr*askofag" (meaningless). This type of mistakes supports the diagnosis of auditory-spatial gnostic speech deficit (see Figure 14, *b*). There are several indications that it was a disorder of the gnostic-praxic but not symbolic level. First, word comprehension was intact. Then, patient distorted word sound with mirror reversal of phonemes within the word (acoustic spatial deficit), but he builds his even meaningless words (literal paraphasia) following the rules of phonological system of the Russian language (phoneme combinability). For example, when asked to repeat the word "no*gt*i" (nails), patient N said "no*tk*i," which we assume reflects the mirror reversal *gt* to *tg*, but because *tg* is not a permissible combination in the Russian phonological system, he made a voiceless *k* instead of voicing *g*. Similarly, the word "*pt*itsa" (bird) N repeated as "*sp*itsa" (spoke). Reversal of *pt* to *tp* called forth replacement of stop consonant *t* by fricative consonant *s* because the combination *tp* is not characteristic of the Russian language. Patient N was always aware of his mistakes in word pronunciation. His search for word sound was goal-directed and each subsequent attempt was closer to the correct variant. Intactness of the word phonological code was evident by the following reply while Patient N was searching for the word sound: "It's still far ... not near." The earlier described patient with phonological aphasia expressed his difficulties differently: "The word is not mine." Thus, we deal here with speech disorder of the gnostic-praxic and not the symbolic level and, in particular, acoustic spatial gnostic deficit (in the literature referred to as conduction aphasia).

The presentation of this particular patient N with morphological aphasia illustrates the dissociability of the syndrome of morphological aphasia, a thesis that we proposed from a theoretical perspective earlier in this chapter. Our theoretical assumption of dissociability was based on our understanding of the different functions of field 39 and 40 (differentiation within the left parietal-occipital region). Patient N's deficits indicate a greater relative involvement of field 40 than field 39. Interestingly, this patient had another deficit that is known to involve the area of the junction of fields 22, 42, and 40, lending further support to our proposal for the localization of this syndrome.

Patient N's FS IQ was 99, as were the V-IQ and P-IQ. Regarding the WAIS profile, as we mentioned earlier, very high scores were received on subtests of verbal logical thinking: Similarities (12) and Vocabulary (16). A very low scaled score of 2 was obtained in the Digit span subtest, which could be explained by his speech gnostic deficit, disorder of repetition. Very low results were achieved in the Comprehension subtest (scaled score of 6, in the range of mental retardation). Patient N frequently could not assess a simple situation and gave literal explanations for proverbs: "Strike while the iron is hot"—"Well, because a hot iron is easier to strike"; "One swallow does not make a spring"—"One should plant in the spring, or it will be late." When asked to elaborate, N stated, "All do the same and do it correctly, but only one speaks." It is of interest that there was also a very significant lowering (scaled score 7, in the mentally retarded range) in the nonverbal counterpart of the Comprehension subtest, Picture arrangement. The common factor in the Comprehension and Picture arrangement subtests is the ability for adequate emotional orientation to and evaluation of a situation. The Picture arrangement subtest also involves visual-figurative thinking, a right hemispheric ability. In turn, tasks of proverb explanation in the Comprehension subtest require right hemispheric symbolic associations.

Thus, in patient N a high level of verbal abstraction and precise and categorical word definition (left hemispheric strategy in explanation of word meaning) coexisted with significant deficiency in explanation of figurative (metaphorical) meaning of proverbs and inability to "catch" simultaneously all aspects of the situation. This suggests a right hemispheric deficiency, which we cannot connect with the morphological aphasia described in this patient. If his FS IQ of 99 does indeed represent a lowering from his premorbid level, we believe it is a result of his speech gnostic deficit, a disorder of speech repetition. There is no evident influence of his morphological aphasia on either WAIS neuropsychological profile or IQ. The right hemispheric deficiency we consider a peculiarity of his premorbid neuropsychological profile. This premorbid profile may have been a lifelong characteristic or the consequence of a previous, undetected cerebrovascular event. Therefore, for this particular patient, verbal-logical thinking rather than right hemispheric cognitive mechanism should be a key factor in his rehabilitation strategy.

5

Frontal Region: Thought and Sentence

> ... the principle and most characteristic function of the prefrontal cortex is the temporal organization of behavior.... (This) broad and parsimonious postulate ... may not be construed as implying the functional homogeneity of the prefrontal cortex at any level of analysis. In fact, the results of many studies point to the functional specialization of various areas of prefrontal cortex, although those various areas may be seen to subserve one way or another the supraordinate function of temporal organization.
>
> J. M. Fuster (1985). The prefrontal cortex and temporal integration. In A. Peters (ed.) *Cerebral Cortex, Vol. 4 Association and Auditory Cortices*. New York, Plenum Press, p. 152.

5.1. Delineation of Anatomical Region

The frontal region includes cerebral formations located anterior to the central sulcus of Rolando (the "anterior brain" system). The frontal cortex consists of 11 cytoarchitectural fields that occupy about one-third of the cerebral hemisphere surface. Several fields represent the central end of the motor system: primary field 4 (motor region) and secondary fields 6 and 8 (premotor region) with bordering speech articulatory field 44 (Broca's zone) and field 45 (a transitional field between premotor and prefrontal regions (Luria, 1966/1980)). The rest of the frontal cortical cytoarchitectural fields belong to the so-called prefrontal region and represent formations of the higher order of structural-functional differentiation—the specifically human tertiary fields (Kononova, 1962; Poljakov, 1966) (see Figure 1, *d*).

Modern anatomical data regarding cortical connectivity patterns substantiate the concept of both vertical (including subcortical function levels) and horizontal differentiation of the anterior brain system and the phylogenetic ties of these two dimensions of brain organization. Five parallel anatomically discrete fronto-subcortical circuits originate from the frontal regions: motor, oculomotor, dorsolateral prefrontal, lateral orbitofrontal, and anterior cingulate (Mega & Cummings, 1994). There is a sequential flow of connections in each circuit from the cortical site of origin through the striatum to globus pallidum, to substantia nigra. Feedback from the MD nucleus of the thalamus to the cortical point of origin closes the loop. Although the vertical organization of all circuits includes the same subcortical formations, each circuit occupies a specific place within them, so that the circuits remained anatomically segregated. Note that the striatum, globus pallidus and substantia nigra are identified by Bernstein as effector centers of function levels C, B, and A, respectively. Actually, the motor loop corresponds to what Bernstein described as the hierarchical organization of movement in the brain. New data about vertical and horizontal organization of the frontal lobe justify our attempt to extrapolate Bernstein's theory about cerebral organization of movement to cerebral organization of higher mental functions.

Phylogenetic development of the frontal lobe represents "sequential unfolding" of higher-order cortical zones in the direction from the central sulcus to the frontal pole, as transitions from the motor cortex to premotor to prefrontal (Poljakov, 1966). In this sense, the origin of the cortical frontal fields from the cortical center of the motor system determines the functional unity, in contrast to the functions of posterior brain formations.

Tertiary cortical fields of the posterior brain are supramodal, that is, they are built upon the definite modality-specific system. For example, connected with field 37, logical-grammatical categories are built over generalized visual object perception; signs of spatial outer relations underlying overt grammar are built over visual-spatial perception (fields 39, 40). In contrast, the whole frontal region is built upon one motor system. The axis of its function is "action," whose evolution and interiorization corresponds to sequential development of structural-functional organization of the frontal cytoarchitectural fields.

Thus, we will consider the common denominator of frontal lobe function from the point of view of general opposition in the intrahemispheric dimension: anterior brain system (temporal, successive synthesis) versus posterior brain system (spatial, simultaneous synthesis). There is a correspondence between evolution of information processing in the posterior brain system with units specific for each hemisphere (left hemisphere signs, right hemisphere symbols) and the evolution of operating with these units in the anterior brain system. Modern anatomical data support this notion. Pandya and Yeterian (1985) indicate that each sector of the sensory association areas is connected with a frontal lobe region that has basically similar architectonic features. "This would imply that each sensory association sector, from first-order through second-order to third-order, may have developed in parallel with a specific frontal lobe region, and with that region may constitute a functional subsystem within the cerebral cortex" (Pandya & Yeterian, 1985, p. 52).

Horizontal differentiation within the prefrontal cortex leaves its imprint upon different aspects of action. The function of circuits depends on the cortical site of origin (leading function level) and its connections with the posterior brain system. Speaking about the function of frontal-subcortical circuits as "effector mechanisms that allow the organism to act on the environment," Mega and Cummings specify that the dorsolateral prefrontal circuit is responsible for so-called executive functions (the ability to organize information to respond to a problem), the anterior cingulate subcortical circuit is responsible for motivated behavior, and the lateral orbitofrontal circuit integrates emotional information into a behavioral context (1994, p. 368). They indicate that "disorders observed with dysfunction of the frontal-subcortical circuits are disorders of action rather than of perception or of stimulus integration." Thus, disorders of these three prefrontal circuits are characterized by disorder of executive function, disorder of motivation, and disorder of socially appropriate behavior, respectively.

In subsequent sections, we will concentrate primarily on information processing in the dorsolateral prefrontal cortex.

5.2. Mechanism of Right Hemispheric Thought

In describing the cerebral organization of movement construction, Bernstein has, in fact, analyzed the cognitive mechanisms of the left hemisphere. It is the left hemispheric mode of information processing that enables the strict hierarchical order of function levels and the distinct definition of each level's contribution. In the left, any new level is built upon the preceding one, separate from it and becoming leading and principal as compared to the levels that have been formed earlier. In contrast, in our conception, the right brain's mode of information processing intermingles functioning of both cortical and subcortical structures, with each new level intimately interwoven into the preceding level.

We emphasize again that the function of the prefrontal cortex, in the most general sense, is the temporal organization of cognitive information coming from associative areas of the posterior cortex; that is, the prefrontal cortex creates a *representation of action* (Fuster, 1985). In his comprehensive analysis of the function of the prefrontal cortex based on patterns of connectivity and neurophysiological and neuropsychological data, Fuster concludes: "Conceivably, the critical role of prefrontal cell assemblies is to expand in the temporal dimension the cortical representation of stimuli inasmuch as they are associated with prospective action and, at the same time, the representation of actions before they are executed. Thus, by extending retention of stimuli and anticipation of actions, those assemblies would permit the bridging of temporal discontinuities between them and the integration of a consistent structure. Prefrontal neurons may accomplish that in intimate cooperation with the rest of the cortical network and by sustained excitation of local and corticocortical reverberating pathways" (Fuster, 1985, p. 171).

The left hemisphere plays the dominant role in the programming of actions, especially complex ones, and goal-directed behavior. The anterior left hemisphere's "representations of action," or internalized action, depends on the function level as well, ranging from a simple movement program to categorical thinking (operations with categorical signs). What is the representation of action or internalized action of the right hemisphere?

In the last several years there has been an increased interest in the differences between the right and left frontal lobes (Goldberg, Podell, & Lowell, 1994), yet there has been minimal investigation into the differences in their cognitive mechanisms. Although there is more interest and evidence in the literature recently about the importance of the right frontal lobe—which is, in fact, bigger than the left (Galaburda, LeMay, Kemper, & Geshwind, 1978)—little attention has been paid to the principles governing right hemispheric action. A review of the literature (see Cutting, 1990) indicates that there has been a sense that the right hemisphere gives a subjective view of the world incorporating time and space, providing a "framework for our *Weltbild* (world perspective)" (Cutting, 1990, citing Lange, 1936), or a "spatial-temporal background of one's world image" (Cutting, 1990, citing Cloning et al., 1968). In this chapter, we will present our working hypothesis about the action of the right frontal lobe.

As we have discussed, the unit for operation in the right frontal lobe is the *visual scene-situation*, provided by right posterior areas. An interesting illustration of the "exposed" visual scene-situation is presented by a patient who had undergone staged sectioning of the corpus callosum. The topology of callosal projections makes it possible to sever the posterior interhemispheric connections while sparing the connections between anterior areas. Figure 18 shows an example of a naming response to a visual stimulus (word or picture) presented to the left visual field (right hemisphere) before and after partial and complete callosal commissurotomy in patient JW, a 26-year-old right-handed male who underwent staged commissurotomy for intractable epilepsy (Sidtis, Volpe, Holtzman, Wilson, & Gazzaniga, 1981). After posterior commissurotomy, the patient could not immediately name the stimulus but described it in the context in which it might be found, a visual scene-situation, as in the patient's response to "knight" in Figure 18: "I have a picture in mind but can't say it ... Two fighters in a ring.... Ancient ... wearing uniforms and helmets ... on horses trying to knock each other off ... Knights?" (Sidtis et al., 1981). The patient described his thought processes in the following way: "It's like things are moving around constantly, and I'm trying to narrow it down to something that will just stop. I'm seeing a whole general picture but one thing is almost right in the middle." After the callosum was completely sectioned, the patient denied seeing anything following stimulus presentation to the left visual field. The authors concluded that after partial (posterior) commissurotomy the transfer of sensory information from right to left hemisphere was eliminated, although the left did have access to "stimulus-related semantic and episodic information from the right." From our perspective we see that following

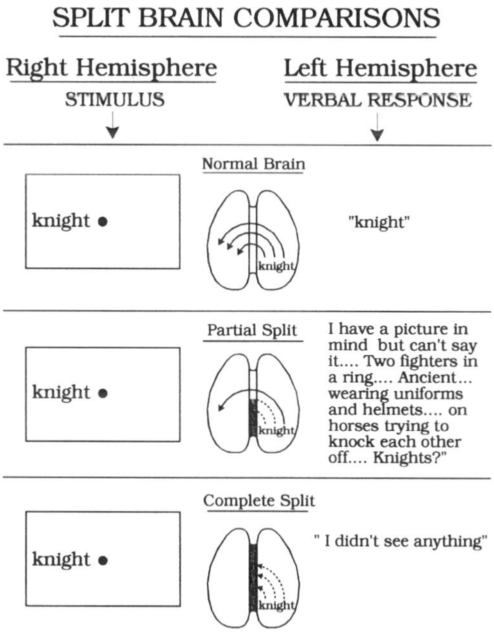

FIGURE 18. Schematic representation of naming ability in patient with staged callosal commissurotomy. From "Cognitive Interaction after Staged Callosal Section," by J. J. Sidtis, B. T. Volpe, J. D. Holtzman, D. H. Wilson, & M. S. Gazzaniga, 1981, *Science*, *212*, pp. 344–346. Reprinted by permission. Copyright 1981 American Association for the Advancement of Science.

partial (posterior) commissurotomy, the patient presents a description of a whole visual scene containing the object, giving a beautiful illustration of how, in our understanding, objects are represented and stored in the right hemisphere: *as included into situations*. It is in this form that information about objects from the right hemisphere is accessible to the left, which then interprets the information according to its own cognitive mechanism. As we discussed earlier, the posterior part of the left hemisphere analyzes the situation within the right and builds a topological scheme of an object as its own left hemispheric representation. Topological scheme is necessary to "develop" or distinguish the object within the scene, allowing naming of the separate object itself and not just a description of a scene containing the object. Thus the equivalent of the object image on the right is its topological scheme on the left. When the posterior inferior callosum is sectioned, right hemispheric representations are not immediately available to the left posterior hemisphere, and thus a separate object image is *not* developed and named right away. The left anterior hemisphere, however, still does have access to the visual scene containing the object; it can "read" the scene, describing the whole picture, and perhaps, we might say, make "conscious" right hemispheric content (compare to the patient's statement: "I don't see anything" after the callosum was cut completely, Figure 18). In the norm, we are usually not aware of the object's situational context.

Another example of right hemispheric representations' emergence may be found in patients with sensory aphasias (damage to the left posterior brain) in the recovery phase, who, being unable to find immediately the sound (phonological code) of a word may compensate and instead describe the visual scene situation connected with it. Damage to the left posterior brain in this case imitates disconnection between the left and right posterior parts when the right hemisphere representation is released from the analyzing censorship of the left. The flow of associations sometimes seen in certain situations (i.e., the free associations of psychoanalysis) may also represent right hemispheric contents. In both conditions, we suspect that the right hemisphere is released from the normally dominating influence of the left.

From these examples we can see that movement is an intimate part of the visual scene, that the flowing of the images creates action. It is our assumption that right frontal representation of action is, in fact, the continuous flowing of situations, a constantly moving picture of the world, of world movement. This reflects a high degree of integration between anterior (action) and posterior (situation) parts of the corticocortical pathways. In contrast, horizonal differentiation on the left is more distinct and characterized by strict division of the contribution of anterior and posterior regions in the corticocortical pathways; operations implemented by the left prefrontal region are distinguishable from the units with which it operates. This might explain why clinicians who had observed patients with right hemisphere damage came to the conclusion that the right hemisphere is considerably less locally differentiated than the left (Lebedinsky, 1941; Semmes, 1968). More recently, it has been found that patterns of activation in the two hemispheres are different: more focal in the left prefrontal region (Roland, 1985) and more diffuse in the right prefrontal region (Pardo et al., 1991).

Our goals in this chapter will be to characterize and describe operations with visual situations at different hierarchical levels, which constitute steps in right hemispheric cognition: situational thought, symbolic situational thought, and symbolic object thought.

5.2.1. Consciousness, Attention, and the Right Hemisphere

Although a discussion of consciousness is far beyond scope of this book, we nevertheless need to address it in the context of understanding the differences between right and left cognitive mechanisms. Up until the recent past, it was assumed that the left hemisphere directed consciousness. Currently, ideas about consciousness are more complex and perhaps can be best summarized in questions: Does each hemisphere have its own (different?) consciousness?; What is the contribution of each hemisphere to *one* consciousness?

It has been observed that left hemisphere damage or dysfunction leads to a disturbance in consciousness on the continuum from coma to wakefulness far more often than damage to the right hemisphere (Cutting, 1990). Patients who received left-sided unilateral ECT lost consciousness earlier and recovered more slowly than patients who received right-sided ECT (Balonov, Barkan, & Deglin, 1979). Some authors, trying to explain these phenomena, speculate that there might be closer functional connections between the left hemisphere and those brain stem and midbrain formations (reticular activating system) regulating "arousal" and providing general functional tone for the cortex (Dobrochotova & Bragina, 1977). Brain activation is subserved by the nonspecific system consisting of hierarchically organized formations which include midline structures (reticular formation) of several levels, starting with the brain stem and ending with the medial basal sections of frontal lobes. Lower levels of this activating system determine the generalized functional state of activation of the brain, creating a background against which information processing is possible. The state of brain activation in the continuum of wakefulness to coma depends mostly on the lower levels. The higher level of the system, its frontal lobe part, on the other hand, controls selective voluntary, directed attention. For example, performance of a task requiring voluntary attention is accompanied by increased spatial synchronization of areas of the frontal lobes on the EEG, which was interpreted as a reflection of the state of local functional activation. In patients with frontal lobe damage, there was a paradoxical decrease in frontal lobe activation during task performance compared to the activation before task performance. In patients with damage to the lower levels of the activating system, however, there were overall decreases and significant fluctuations in generalized synchronization on background EEG data, considered to represent an instability in the functional activation state of the brain (Slotinzeva, 1979). When these patients performed tasks, increased activation in the frontal lobes was observed, although there were still significant fluctuations in the level of local activation, explained by the instability of the activating influences maintaining the generalized functional activation state of the brain. There is an assumption in the literature that voluntary selective attention is connected with the left hemisphere (Cutting, 1990).

Regarding the right hemisphere, there is some consensus that its integrity is necessary for the ability to develop and maintain the general alert state (Cutting, 1990; Heilman et al., 1985; Harris, 1995). Posner and Peterson (1990) considered alertness as a subsystem of attention that acts on the posterior attention system. This system supports visual orienting and may also influence other attention systems. According to Posner and Peterson's hypothesis, this system is physiologically dependent on norepinephrine pathways that arise in the locus ceruleus and are lateralized to the right hemisphere. Interestingly, the notion of alertness includes both state of consciousness and attention (attentiveness). Evidence that the right hemisphere plays the dominant role in sustaining alertness came from analysis of a typical phenomenon following right hemisphere damage: left visual field (LVF) and/or left body side neglect (Cutting, 1990). Neglect after a right hemisphere lesion may occur not only in the visual arena but in the auditory, tactile, and olfactory modalities as well, and not only in perception but also in imagery and memory. The fact that left hemisphere damage does not cause right visual field neglect led to the hypothesis that the right hemisphere controls attention for both perceptory fields, whereas the left hemisphere controls only the contralateral field. If the right hemisphere controls both sides, unilateral left hemispheric damage will not result in right field neglect. Cutting suggests that the notion about right hemisphere dominance for attention even in the ipsilateral field is supported by attention deficits other than neglect occurring more often after right- than left-sided lesions. For example, reaction time was more prolonged after right-sided lesions than after left-sided (Coslett, Bowers, & Heilman, 1987). Cutting concluded that the right hemisphere plays the dominant role in maintenance of attention regardless of the site or type of cognitive activity (imagery, memory, perception).

Further information regarding the role of the right hemisphere in attentiveness comes from studies of memory in patients with right or left hemispheric damage and a control group. In these experiments, subjects were first given a list of 10 words to remember; those with left hemispheric damage performed significantly less well than controls. In the next two experiments, subjects were given different lists of 10 words and not asked to remember them but instead to perform some counting tasks involving the letters in the lists. They were then asked to remember the actual words that had been presented; subjects with right hemispheric damage showed a striking inability to recall words, with a much more modest deficit in those with left hemispheric damage compared with the control group. The authors concluded that the right hemisphere is responsible for this "involuntary memory" (Simernitskaya, 1978). In regard to attention, these experiments demonstrate that the right hemisphere is registering information based on a broader, more global attentiveness—in our view, not really involuntary because the right hemisphere is functioning according to its own rules of information processing. Goldberg et al. (1994, p. 375), citing several studies of lateralized activation of the prefrontal regions on PET and regional cerebral blood flow (rCBF) scans, concluded that "selective activation of the left or right prefrontal region depends solely on the nature of the task (following internalized instructions versus attending to external stimuli), and not on the side of stimulus delivery. The type of material (verbal or nonverbal) does not seem to matter, nor does the response hand." Examples of tasks leading to greater activation of the right prefrontal area include asking subjects to sustain their attention to a particular sensory stimulus, such as "just listen," or to perform a task requiring selective attention to external stimuli such as the Continuous Performance Task (CPT).

What, then, is the difference between right and left hemisphere cognitive mechanisms in regard to consciousness/attention? We surmise that the left hemisphere is more responsible for both wakefulness and focal attention, with wakefulness as a background against which focal attention may be established (focus of consciousness). In contrast, data suggest that the right hemisphere has the primary role in maintaining the alert state, a state of attentiveness in which the right hemisphere is constantly registering information—this is, indeed, the content of right hemisphere consciousness (flow of consciousness). We distinguish, then, between the focus of consciousness on the left and the flow of consciousness on the right, consistent with the general left–right difference in cognitive mechanisms: local versus global processing.

Right hemispheric representation is an indivisible unit including a visual picture (space filled by the visual modality), which may be accompanied by other perceptions (auditory, tactile, proprioceptive, olfactory), action, time, and affects. It is a global experience; for example, there are never affects without perceptual context. Thus, it is an experiential sense of events. Let us compare the paroxysmal state in temporal lobe epilepsy patients with right versus left focus. Patients with right focus appeared passive or stereotypically agitated, but later upon questioning reported experiencing intense multiple psychosensorial feelings (Dobrochotova & Bragina, 1977). The behavior of these patients was not informative; it did not reflect the content of subjective experiences. The authors stated that they had to describe these phenomena "from the point of view of the patient's subjective experience," that is, from "inside," not from the point of view of the objective observer. The authors concluded that behavior in this situation did not reflect, did not adequately subserve, the content of consciousness. In contrast, the clinical picture of paroxysmal states with a left hemispheric focus was of very complex and, at times, sophisticated psychomotor activity; when asked later what they had experienced, patients had no awareness of anything having happened. It seems clear to us, then, that patients' subjective experience is important in understanding the right hemisphere cognitive mechanism and offers an approach for examination of qualitative differences between right and left cognition.

One of the famous experiments performed by Gazzaniga and Le Doux (1978) on a split-brain patient gives further data regarding the right hemisphere's subjective experience. The patient, a young man who had undergone callosal section for intractable epilepsy, was briefly presented two different pictures simultaneously—a snow scene to his right hemisphere and a chicken claw to his left. He was then asked to select by pointing with both hands which among several different pictures related to what he had seen (see Figure 19). His right hand pointed to a picture of a chicken, while his left pointed to a shovel. The patient was asked to explain his choices and answered without hesitation: "I saw a claw and I picked the chicken, and you have to clean out the chicken shed with a shovel." Did patient experience the snow scene by his isolated right hemisphere? Yes. Do we have access to the right hemisphere's experience? No—what we have is the left hemisphere's expla-

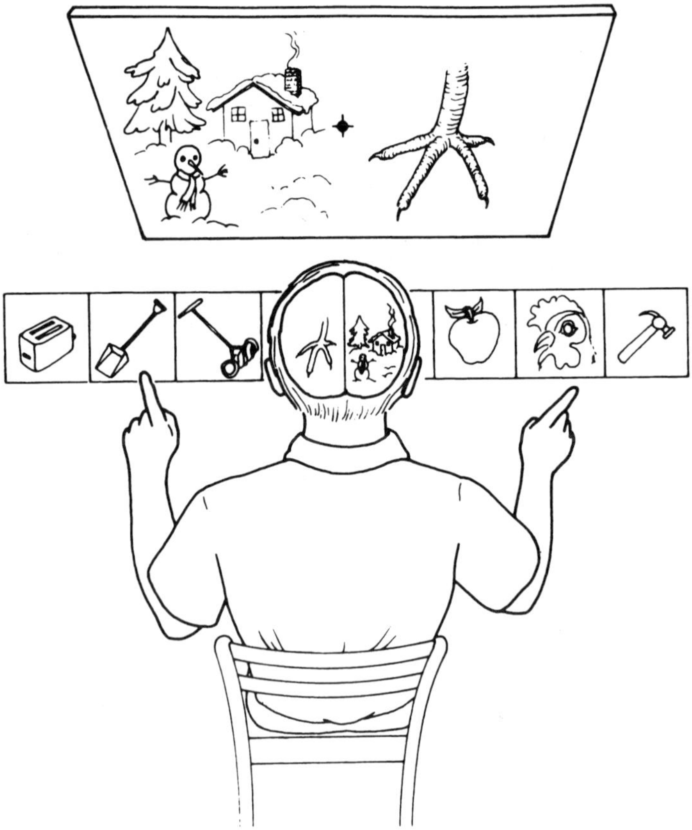

FIGURE 19. Simultaneous presentation of two different tasks, one to each hemisphere, to split-brain patient. From *The Integrated Mind*, by M. S. Gazzaniga & J. E. Le Doux, 1978, New York: Plenum. Reprinted by permission.

nation without any knowledge of what the right hemisphere has experienced. In the norm, the left hemisphere analyzes the content of the right hemisphere's experience and reconstructs it according to its own cognitive mechanisms, that is, interprets the content according to its mode of information processing. Thus, our knowledge of the right hemisphere's experience generally comes to us through the left hemisphere's interpretation. When the left and right hemispheres are disconnected and the left does not have access to the right's experience, the left continues to interpret without the complete information that it would normally operate with (i.e., data from both hemispheres). The left in this experiment gives an answer that resembles a *rationalization*, a term used in psychiatry to explain the process of "offering rational explanations in an attempt to justify attitudes, beliefs, or behavior that may otherwise be unacceptable" (Kaplan, Sadock, & Grebg, 1994, p. 257).

Interesting data come from a series of experiments performed on patients with right hemispheric damage (Wapner, Hamby, & Gardner, 1981). Patients were told several stories, each emphasizing different elements, and then asked to retell the stories. Stories emphasizing emotion used words and phrases to convey emotional states of the characters, for example: "The little girl began crying; her heart pounded as she crept in." The patients characteristically did not recognize the emotion implied in the situation, making inferences about the emotions of the character which logically could have been involved, but which were not. In recalling the story of the little girl, a patient stated, "The little girl did not express any opinion or feelings except being excited. She did not wet her panties, she did not kiss anybody around, and she did not hug anybody" (Wapner, Hamby, & Gardner, 1981, p. 25). We can compare these experiments to those with the split-brain patient cited earlier, in which the left hemisphere interprets information without access to right hemisphere cognitive mechanism. In this case, with a damaged right hemisphere, the left hemisphere does not have available the whole situation with emotion included, which is usually provided by the right cognitive mechanism.

In the norm, the right hemisphere's experience is "covered" by the left's interpretation; it is only in pathological conditions that exaggerated or distorted right hemisphere experience overcomes its counterpart's interpretation and comes to the surface. It is through this exposure of exaggerated right hemispheric experience that we may have a glimpse of its contribution in the norm.

5.2.2. Subjective Experience of Time and the Right Hemisphere

> Yet there is a big difference between the forward and backward directions of real time in ordinary life. Imagine a cup of water falling off a table and breaking into pieces on the floor. If you take a film of this, you can easily tell whether it is being run forward or backward. If you run it backward you will see the pieces suddenly gather themselves together off the floor and jump back to form a whole cup on the table. You can tell that the film is being run backward because this kind of behavior is never observed in ordinary life. If it were, crockery manufacturers would go out of business. [Hawking, 1988, p. 144]

Literature regarding unilateral brain damage shows that disorders of time perception arise mostly when lesions are located in the right hemisphere. These are various disorders in the tempo of events, passage of time, and unusual experiences of time (Cutting, 1990; Dobrochotova & Bragina, 1977). They are disorders in the subjective experience, the "feeling" of time, whereas knowledge and concepts about time are intact. They might be characterized as disorders in perception of the "flow of time," with intact understanding of relative sequences. The latter are based on the left hemisphere distinguishing the signs of temporal relationships that are embodied in language; for example, temporal preposition (before, after), ordinal numbers (first, second). Returning to Bernstein's system, recall that in the horizontal dimension of brain differentiation, it is the anterior brain that is concerned with time/movement, whereas the posterior brain is connected with space/afferentation (see Table 1). In Bernstein's vertical dimension, each level represents a relatively autonomous functional system, and each level is characterized by its own time. However, in actual brain functioning there is an integration of levels in which the highest level is the leading one and the one that determines what the role of the lower background levels will be (Bernstein, 1947). We will later revisit Bernstein's system specifically regarding time at the different function levels, but here we will extend and expand upon his basic concepts in the area of time and space, including interhemispheric differences.

5.2.2.1. Hypothalamic-Midbrain Level. We will start with Bernstein's level A. Our revision of this level is summarized in Figure 20. According to Bernstein, time of functional level A is a simple rhythm that is a temporal pattern of this level's movements—it can be registered as a simple sinusoid during tonic activity of striated muscle (muscle tone). Recent studies suggest that time of level A may have another, subjective aspect: the sensation of time intervals. Examination of time perception in patients with Parkinson's disease provides interesting data in this vein. Parkinson's disease involves damage to the substantia nigra, which

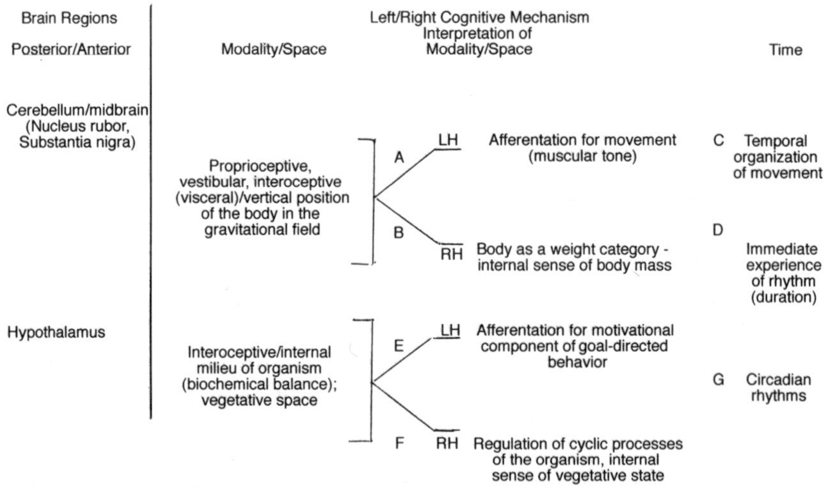

FIGURE 20. Hypothalamus-midbrain function level and right hemispheric time.

Bernstein considered as an effector center of level A, the affector center being the cerebellum, which receives proprioceptive and vestibular input from the trunk and head. Recent studies have shown that patients with Parkinson's disease could not accurately judge time intervals (interval clock) (Gibbon et al., 1997; Meck, 1996). When these patients took L-DOPA, a dopamine-stimulating drug, their interval clock ability was restored. In animal experiments, rats trained to recognize specific intervals of time could no longer do so following damage to the substantia nigra. Giving L-DOPA to the brain-damaged rats restored their ability to accurately judge time intervals. Researchers suggest that the substantia nigra acts as a metronome, sending a steady stream of dopamine pulses to the striatum. The frontal lobe appears to complete the interval clock's neural circuit (discussed in the following section). Functional MRI (fMRI) of college students while they were performing tasks to judge time intervals has shown that the same circuits measured in rats were selectively activated.

This recent work presents a fascinating complement to Bernstein's findings. Bernstein was concerned with the temporal organization of movement in the brain and concluded that time at level A is a simple rhythm reflecting the pattern of muscle tone. We infer from these studies that the ability to judge time intervals—a subjective sense of time duration—is connected to the same efferent part of the A level in Bernstein's terms. It may be that the immediate experience of time duration reflects the subjective sense of that simple rhythm that can be pictured by the sinusoid of muscle tone (Figure 20, D). Interestingly, it has been mentioned in the literature that "muscle sensation" possesses a sense of rhythm that far exceeds that of the visual and tactile senses, and is comparable to the auditory ability (Agadzanjan, 1967, cited by Dobrochotova & Bragina, 1977).

In our attempts to extend the concept of Bernstein's level A beyond motor function, we need also to consider "space." Bernstein defined space of level A as the vertical position of the body in the gravitational field, serving as afferentation for movement of this level (muscle tone) (Figure 20, A). One might, therefore, characterize the body at this level as a physical object, a mass in the gravitational field. Again, as with time of level A, we believe that there is a corresponding subjective side of space for level A, the experience or sense of body mass (Figure 20, B). Thus, as time of the A level is the temporal organization of movement as well as the subjective sense of these rhythms, so space of the A level includes afferentation for movement as well as the internal sense of weight and body mass. As noted, Bernstein defined the afferent center for level A as the phylogenetically ancient cerebellar structures (vermis), receiving input from proprioceptive and vestibular pathways from the trunk and head. This was information utilized in Bernstein's schema for maintenance of position in the gravitational field. Inherent in this same sensory information, we believe, is the subjective experience of these sensations, which is integrated as the experience of body mass. Although he did not include interoceptive information into afferentation for level A, Bernstein noted that a branch of the vagus nerve, conducting information from the inner organs, ends in the same cerebellar areas. This interoceptive input is easily incorporated into our model of subjective experience of level A, providing a sense of inner substance, of fullness, of vitality. One is not normally aware of experience of this level (which Bernstein called the level of "background of backgrounds") not only because of its depth but because it is also intimately incorporated into and used by higher levels.

Selective disorder of body mass is manifested as feelings of heaviness or increased density or the opposite, feelings of weightlessness or hollowness, of the whole body or of its parts. Disordered feelings of body mass are very common in depression. Lukianowicz (1967) gives examples from several depressed patients: "(My head feels) as if it was made of lead. It feels so heavy that the muscles of my neck are unable to sustain the weight"; "My legs became so heavy, as if somebody poured melted lead in them"; "My body feels like an empty box with another empty box on the top, instead of my head"; "I have a most terrible feeling of a large open cave, of a sheer *emptiness*, an excruciating feeling of the cold, dark, hopeless *nothingness* in my chest." (p. 39) The regularity with which these symptoms are seen in depression leads us to believe that disorder of level A is involved in depression.

Although the extent of lateralization at this level is unknown, we believe that there are differences in how the right and left cognitive mechanisms utilize sensory information from this level. It seems likely that higher levels, through reciprocal connections with level A, select and modulate the information contained within this level for their own purposes and according to their own information processing strategies, right and left. Afferentation for movement, Bernstein's space of the A level, is essentially a function of the left cognitive mechanism. The subjective sense of internal body mass is subserved by the right cognitive mechanism (Figure 20, B). We think that it is sense of weight or mass of the body that this level contributes to the next, B level I-space. We have already discussed the fact that body scheme (body image) is primarily connected with the right hemisphere. Therefore, we infer that it is the probably the right hemisphere cognitive mechanism which is responsible for incorporating the sense of body mass into the whole body scheme. Body scheme is composed of several different parameters—shape, size, mass, position in space—normally integrated into a whole coherent image, but any one of which may be selectively disordered in pathology, allowing us to know of their existence in the norm.

In our expanded concept of level A, and in particular, time of level A, we have incorporated the recently discovered rhythmic processes of the hypothalamus, which we feel belong to the fundamental functions subserved by this level. Bernstein included the hypothalamus as part of the brain substratum of level A, but, as his focus was movement, did not explore its role any further. Before we discuss more modern findings regarding hypothalamic function, let us present

Bernstein's hypothesis regarding the phylogenetic roots of voluntary movement, the connection between muscle tone of striated muscle (the background of voluntary movement) and the most ancient "movements," contraction of the smooth musculature of the internal organs.

Bernstein called level A the level of *paleokinetic regulation*, by which he meant the ancient motor system of the inner organs equipped with smooth muscles and enervated by the autonomic nervous system and hypothalamus. In contrast, the *neokinetic system* is the muscular skeletal system equipped with striated muscles and enervated by the central nervous system. According to Bernstein, tonic activity of striated muscle (as opposed to contractions for movement itself), represents an old (paleokinetic) function implemented by a new substratum, the brain centers of which include the relatively "new" nucleus rubor and substantia nigra. It has a regulatory role providing that flexibility and baseline readiness for action which the neokinetic process itself is lacking. Bernstein postulated that the physiological tone of striated muscles is equivalent to smooth muscle activity. As noted, Bernstein included as part of the brain substratum of level A the hypothalamus and the central part of the autonomic (vegetative) nervous system. Interestingly, damage to the hypothalamus not only results in dysfunction of systems enervated by the autonomic nervous system but also may cause disorders of muscle tone in somatic musculature (Markelof, 1948, cited by Dobrochotova & Bragina, 1977). Markelof concluded that muscle tone may represent a common link between motions and vegetative functions. Again, Bernstein emphasized that the paleokinetic process, born in phylogenesis together with its specific substratum (smooth muscle), also operates on the neokinetic substratum (striated muscle). Tonic activity of striated muscle and typical contractions of smooth muscles are depicted by a similar curve—a simple rhythm, which in turn is similar to the alpha rhythm of EEG, according to Bernstein.

Thus, if we consider this level in the brain function hierarchy in general and not just as applying to movement organization, its scope is much broader, encompassing the hypothalamic and autonomic nervous system's regulation of all the rhythmical periodic vegetative functions of the organism. We may call this level the level of the organism (vegetative self). Some authors called the elementary automatic regulation of the periodic functions of the organism implemented by the hypothalamus *vegetative time* (Shmaryan, 1949), the rhythms of one's own organism, its physiological processes. Damage to the hypothalamus leads to disorders of vegetative time, which are expressed in various dysfunctions: disorders of the rhythm of sleep/awakening, food intake, and various other metabolic and endocrine dysfunctions (for example, diabetic symptoms with polydipsia, polyuria, and change in glucose curve; disorder of thermal regulation; disorder of menstrual cycle; disorder of libido, and so on).

The ability of a living organism to generate rhythm depends on the so-called biological clock. Gastings (1961, cited by Dobrochotova & Bragina, 1977) indicates that living organisms possess a precise and fully autonomous biochemical oscillating system whose typical frequencies correspond to the natural geophysical frequencies. This thesis is important because it allows us to understand the essence of A level time, in which the organism's own apparatus maintains periodic processes in the organism in correspondence to oscillatory processes of the outside world. This beautifully parallels space of the A level, which is the maintenance of position and direction of the body in response to gravitational forces. There is as yet no division of *I-space* and *non-I-space*, in contrast to level B, for which space is strictly *I-space*, and level C, in which space is exclusively *non-I-space*.

The circadian timing system, recently examined in detail, encompasses regulation of the organism's cyclic processes, or vegetative time (Figure 20, G). "This system may be defined as a set of central neural structures, the primary function of which is circadian rhythm generation and regulation. Circadian rhythms have evolved as an adaptation to the solar circle of light and dark and have two principle characteristics—generation by endogenous pacemakers and entrainment by the light/dark circle. That is, the rhythms are generated in the absence of the light/dark circle, but the exact period of the rhythms and their phase relationships are set by environmental events. The pacemaker of the system is the suprachiasmatic nucleus … localized in the anterior hypothalamus" (Moore, 1991, p. 13). Anatomically, the suprachiasmatic nucleus belongs to the A level in Bernstein's system. Functionally, circadian rhythms, their nonindividually-specific but species-specific character (compare with the unique-for-one's own body propriomotor rhythm of the B level), their twofold nature—intrinsic generation and tuning on environmental events—come under the concept of A-level time.

Finally, it is likely that there are other nuclei within the hypothalamus, such as the mammillary bodies, which contribute to level A time. Some authors have proposed, for example, that the immediate experience of time duration is disturbed in Korsakoff's syndrome, which is manifested as a severe impairment of memory of recent events with relatively intact immediate recall and remote memory. These authors explain Korsakoff's amnesia as a consequence of temporal disturbances related to damage to the mammillary bodies (Dobrochotova & Bragina, 1977).

Curiously, some authors noticed that the disorders of vegetative time associated with hypothalamic dysfunction are found frequently in patients with right hemispheric lesions. Dobrochotova and Bragina (1977) described a surprising constancy in the combination of symptoms found with right and left hemispheric lesions. In patients with right hemispheric damage, in conjunction with specific psychopathological symptoms, disorders of periodic physiological processes were regularly observed: disorders of sleep–wake cycles, food intake, cyclic hormone levels. In contrast, according to Dobrochotova and Bragina, the clinical picture with left hemisphere–damaged patients was usually free of rhythmic dysfunction; these patients did, however, frequently manifest some disorders of arousal, the so-called state of wakefulness, in addition to characteristic psychopathological symptoms. From these findings, Dobrochotova and Bragina concluded that the right hemisphere had closer connections with the hypothalamus, and the left with the reticular activating system. What these differences may be in strictly anatomical terms, we do not know; however, we believe that the sensory information from these regions is utilized differently by the right and left cognitive mechanisms. We know that the hypothalamus provides information about the internal milieu of the organism, the organism's needs, and in the left cognitive mechanisms this information is utilized as "afferentation" for the motivational component of goal-directed behavior (Figure 20, E). It looks as if this same information on the right may be used to provide an internal sense of the vegetative (biochemical) state, and the regulation of these bodily biochemical processes (circadian rhythms) is more a right hemispheric cognitive mechanism (Figure 20, F).

We mentioned earlier that disorders of sense of body mass are frequent in depression. Disturbances of vegetative functions (appetite, sleep, libido, hormone levels) are also generally found in patients with depression, providing further evidence that level A is involved in depression. These two groups of symptoms (sense of body mass and vegetative signs) are dissociable, indicating that there are two relatively independent subsystems within the A level.

5.2.2.2. Thalamic Level. Level B for Bernstein is the level of possession of one's own body. Exhaustive proprioceptive information from the body musculature provides the spatial coordinate system that, for the left cognitive mechanism, provides afferentation for movement. Bernstein emphasized that space for level B is the body spatial coordinate system, and uniquely for level B among all the levels, is purely one's own body space, without external influence. Movements of level B are completely introverted, depending solely on afferentation from the musculature, and reflect precise, complex coordination of the various muscle systems, a kind of huge synergistic chorus of simultaneous involvement of different muscle groups. These movements are by nature repetitive and rhythmic, representing changing body positions over time—multilinked orchestrations exactly and characteristically performed with a striking sameness of pattern. Bernstein called this *propriomotor rhythm*, which in fact reflects time of this level and which is a complicated, individually specific pattern. For example, in walking, the highest necessary level is level C, but the rhythmical act of stepping as such is implemented by the B level's synergetic orchestra. The temporal organization of the stepping movements is quite characteristic and essentially unchanging for a given individual, as are the unique rhythmic patterns that are expressed in the movements of dancing, playing an instrument, sports, and so on.

Bernstein's construct of level B, in our view, is primarily left hemispheric cognitive function. We believe that there are other aspects to the propriomotor rhythms, reflective of right hemispheric cognitive processing of this information. For the left cognitive mechanism, space of the B level is the spatial coordinate system or afferentation for movement; for the right, it is I-space—one's own body space that is "filled" and its borders defined by proprioceptive sensation. This is the physical self, for which time is the unique-for-one's-own body pattern of coordinated propriomotor rhythms. One's individual rhythmic patterns of this level form the background upon which higher levels build their time, both for the right and left cognitive mechanisms. As the coordination of movement over time of level B forms the background of temporal organization of the more complex movements of the higher levels on the left, so the unique rhythms of the physical self of this level are incorporated into time of the higher levels of the self on the right.

5.2.2.3. Cortical, Sensory-Motor Level. In contrast to the B level, whose movements are defined by and directed at the spatial coordinate system of one's own body, Bernstein's level C is fully extroverted and its movements projected onto external space. The character of space of the B level, namely, the cyclical reiteration and alternation of elements in the coordinate system of one's own body (the multilinked pendulum of extremities) determines that movements of the B level are rhythmical and periodic. Space of the C level is aperiodic and homogeneous; movements of this level are also aperiodic, with onset and end, and goal-directed (i.e., have a point of destination). Again, in walking, for example, level C is the highest level necessary and determines the task: to travel in space to a definite destination point. The rhythmical act of stepping as such is implemented by the B level's synergetic orchestra. Level C adjusts the introverted B level's abstract walking to the external space, projecting the motor process onto the external spatial field with its forces and objects, and projecting the motion onto its destination (Bernstein, 1947). Temporal synthesis of the B level is "instilled" into the movement composition itself, embodying its rhythmical dynamics. At the C level, time becomes not rhythmical but linear, with one direction from past to present to future ("the arrow of time"). According to Bernstein, time of level C is temporal organization of movements in external space characterized by moment, speed, and duration; again, in our view, this is a process of the left hemispheric cognitive mechanism. The subjective sense or experience of the arrow of time seems to be a function of the right hemispheric cognitive mechanism. This is substantiated by the observations of many authors, who described numerous extraordinary and abnormal subjective experiences of environmental time in patients with right hemispheric dysfunction (Mullan & Penfield, 1959; Penfield & Perot, 1963; Dobrochotova & Bragina, 1977; Cutting, 1990).

We will give several illustrations from their observations. "For a moment I had a feeling that a woman I saw walking performed very slow movements, as though a film were running slowly, and I had a sense that time was moving down, or, more exactly, somehow the calculation of time was going backward." (Dobrochotova & Bragina, 1977, p. 89) "Suddenly it feels that everything is stopping, everything is motionless as in a photograph.... Time is stopping." (Dobrochotova & Bragina, 1977, p. 90) "Suddenly people are somewhat agitated, excited; they are fussing, moving abnormally quickly ... it happens just now but is seems that it was long ago, or it did not happen at all." (Dobrochotova & Bragina, 1977, p. 93) "Looking at the rug on the wall, the patient saw that violets and some other yellow flowers drawn on it became real and moved out of the rug as though they could be touched; they stirred, swayed, as from the wind. At this moment it was strangely as though time were stretching out" (Dobrochotova & Bragina, 1977, p. 92). Mullan and Penfield described a patient who experienced the nurses in the hospital as moving so fast he could hardly follow them with his eyes; he also reported feeling that time was stretching out endlessly (Mullan & Penfield, 1958). So we see various abnormal or unusual senses of time, distortions of the normally forward-directed arrow of time—time stops, time moves back, time stretches out, time is accelerating, and so on. Disorder of time is accompanied by distortion of the subjective sense of object movement, as these examples show. Distortion of the subjective sense of movement can range from the inability to detect movement or distinguish the fastness or slowness of motions, to movements seeming accelerated or slowed, to the inability to "catch" movement, progress in a general sense. The world can be deprived of its most essential feature—changeability or transformation (Dobrochotova & Bragina, 1977). Distortions of sense of movement parallel distortions of sense of time, accelerating or slowing conjointly.

The right hemispheric arrow of time is perceived through sense of movement of objects in external space. Time and movement are an intimate part of the visual scene-situation, which we have defined as the right hemispheric representation of action. Indeed, in patients with right hemispheric lesions, disorders of time occur as a rule in the context of a disturbed visual picture of the world. They are part of the derealization syndrome, where the surrounding world is perceived as unreal, unnatural, ghostly, unusual. The visual scene-situation represents that moment of real environmental time when it was perceived. The constant succession of moments of real time is a continuously changing impression of the outside world. We may see an example of exposed (unopposed by left hemisphere) right hemispheric constantly registered information from the outside world, or right hemispheric flow of consciousness, in an autistic individual's memory of his childhood: "Confusion and terror ... living in a frightening world presenting painful stimuli that could not be mastered.... Nothing seemed constant. Everything was unpredictable and strange. Animate beings were a particular problem. Dogs were eerie and terrifying. They were especially unpredictable. They could move quickly without provocation." (patient from Cutting, 1990, p. 384).

Pathology gives us an example of the splitting of time experience per se from its situational context, as in the schizophrenic patient described by Fisher who reported, "Life is now a running conveyor belt with nothing on it. It runs on but is still the same.... Outside everything carries on, leaves move, others go through the ward, but for me time does not pass.... When they run around in the garden and the leaves fly about in the wind I wish I could run, too, so that time might again be on the move, but then I stay stuck" (cited by Cutting, 1990, p. 268).

As we stated in the beginning of this chapter, the function levels in the right hemisphere are intimately interwoven, with lower levels' contributions incorporated into the higher levels' operations. Level A contributes the immediate experience of rhythm (duration), which is modified by the individually specific rhythmical pattern defined by one's body space (level B contribution). These two lower levels we suppose may constitute the so-called timing scale on which moments of real time are projected. Among disorders of time of the C level connected with right hemispheric dysfunction are also disorders of duration of events, such as past events' time getting "wrinkled," shortened, and the events themselves more densely located on the corresponding time's fragment (Dobrochotova & Bragina, 1977; Cutting, 1990). This is different from the level A time disorder of immediate sense of duration without any reference to the event (see previous in patient with Parkinson's disease).

5.2.3. *Visual-Situational Thought*

Based on studies of the psychopathological syndromes that follow right hemisphere damage, Dobrochotova and Bragina (1977) assumed that in the norm, right hemisphere images contain markers of that real space–time in which subject–perceiver—object–perceived contact occurred, when the object image is formed. Due to spatial/temporal marks, past experience is "written down" in successive order. In recollection, it is these marks that always relate the given images to the past.

As we assumed, in the right hemisphere mode of information processing images are perceived and stored as included within visual action situations (visual scenes), and this function is implemented by the posterior part (temporal-inferior and parietal-occipital cortex) of the corticocortical pathways. The visual scene is a "space" for the image; it is the equivalent of the term *spatial mark* used by Dobrochotova and Bragina. At the same time, the visual-action situation represents non-I-space, which, in the hierarchy of the function levels, would correspond to level C in Bernstein's terms. As we know, according to Bernstein, time of the C level is moment, speed, and duration.

Proceeding from this position, we will present the model of right hemispheric action at the stage of the right hemispheric cognitive mechanisms corresponding to functional level C. We then will support this hypothesis using examples from psychopathology that result from right hemispheric damage. The very moment that the image is being imprinted is the time unit of the C function level and is thus the image's temporal mark. The C level is completely extroverted and its subjective time reflects "real" environmental time, in contrast to subjective time of the introverted B level, which is the propriomotor rhythm of one's own body.

In the moment when the image is being imprinted, a single situation perceived as a separate space is marked. "Marking" the constantly changing impressions of the world establishes an order in the continuum of memory images. "Time points" are representative of real time, but on the other hand, these points of real time are internalized as an ordered succession of moments of experience in the external world. Thus, the single situation in the right hemisphere exists as a space closely united with the time in which this representation was formed. In other words, spatial and temporal synthesis are inseparable from each other as well as from situation-space's contents. The situation's content incorporates not only images and action but affect and motivation at the moment of its formation. The situations themselves, because they are momentary segments of real time, are as achronous and static as still shots in a continuously moving film strip. Each one of the situations, that is, actions of the real world, is included in the frame of its own space, and the number of spaces (non-I-spaces) in the right hemisphere corresponds to the number of the single situations. These separate spaces, gestalts of the right hemisphere, do not get intermixed or crossed, and the uniqueness of each situation-space is rendered by its temporal mark. In contrast, in the left hemispheric cognitive mechanism there is only one external spatial field, described by Bernstein in his definition of C level space, metric and geometric, which is the base for precise, goal directed movements. Owing to the temporal mark, situation-spaces are stored in their natural succession, from earlier to later layers of memory. They represent, in fact, one's past experience, or autobiography. Some experimental data support this hypothesis. Investigation of neuroanatomical correlates of remembering previously experienced events has shown that the recollection of past experience in young, healthy adults was accompanied by increased blood flow in the right dorsolateral prefrontal cortex (Brodman's areas 10, 46, and 9). Other principle regions of increased blood flow were situated around the left cingulate sulcus and bilaterally in the parietal lobes (areas 7 and 40). Major decreases in blood flow were situated bilaterally in the temporal lobes (areas 21, 22, 41, and 42) (Tulving et al., 1994). Neuropsychological experiments on autobiographical retrieval of incidental past events in patients with CT-assessed frontal lobe lesions has shown that poor autobiographical retrieval correlated significantly with disorder of executive performance (a test that is diagnostic for dorsolateral prefrontal syndrome). On the other hand, poor learning of new information did not correlate with disorder of autobiographical retrieval or executive performance (Della-Sala et al., 1993).

Memory for previously experienced events has been called episodic memory, in contrast to semantic memory (Tulving, 1983). Episodic memory is characterized as experiential; evoked events are infused with the strong feeling of intrinsic truth. The recollection of individual happenings and doings are accompanied by a "when and where" specification (Tulving, 1983; McKenna, Mortimer, & Hodges, 1994).

In contrast, semantic memory refers to recollection of information (facts, ideas, concepts), which does not have any personal "it happened to me" connotations and is devoid of any reference to the time and place at which it was acquired (Tulving, 1983; McKenna et al., 1994). "The organization of the store is conceptual rather than being ordered in time, when retrieved ... the knowledge does not evoke the feeling, and, in particular, the feeling that it must be true—rather, an intellectual judgment about its validity has to be made" (McKenna et al., 1994, p. 166). Tulving (1983) argued that episodic and semantic memory represent functionally distinct cognitive systems. Later, it was shown that they are subserved by different brain cortical systems.

As noted, episodic memory is impaired in patients with brain damage to the right prefrontal cortex. Selective disorders of semantic memory were reported in patients with posterior brain lesions; for example, in patients with herpes simplex encephalitis (damage to the temporal region). Episodic memory can be surprisingly well preserved in these patients (Hodges et al., 1992; DeRenzi, Liotti, & Michelli, 1987).

Two peculiar psychopathological syndromes, both found in patients with right frontal lobe impairment, give indirect evidence of the existence of a chronological autobiography, a sequential time line of lived experience. In the first syndrome, called *chronological regression* by Balonov et al. (1979), who described this peculiar distortion of consciousness, "evocation" of one's own time "pieces" was observed following selective administration of unilateral ECT to the right frontal lobe. During recovery of cerebral functions, the patient seems to return to a definite earlier fragment of his life, experiencing it as real and actual, with all its facts and events, with the knowledge, values, and judgments inherent in that earlier age period. A few minutes later, the patient might be in a later time period, with all experiences of that period of his life; three, four, and sometimes even more age periods may follow in succession. Every "return" is experienced as real and actual. Following full recovery of cerebral functions, the patient can retell the events of his life in succession, but denies having experienced certain episodes of his life as current. We conclude that in this syndrome the patient, due to some disturbance of the right frontal area, is returned to an earlier self in the autobiographical time line and is unable to dissociate from that self. His return-to-past-experience is perceived as current reality, because "revived" pictures of the past are inseparable from the time with which they are connected. The syndrome of chronological regression differs from so-called *flashes of past experience*, a condition described by Penfield and Roberts (1959) when the right temporal cortex was stimulated. Flashes of past experience are vivid, emotionally and sensory-colored recollections that coexist with the adequate orientation in current reality. In the state of chronological regression, revived past experience is devoid of emotion and imaginative saturation; it is indifferent registration of daily life events and facts pertaining to definite time periods. Thus, the clinical picture of the syndrome of chronological regression and its difference from the condition caused by right temporal cortex involvement support the hypothesis that the contribution of the right prefrontal cortex is separate moments of real time that, in their succession, make one's own biography.

The peculiar disorder *reduplicative paramnesia* illustrates impairments of spatial-temporal marks. The term was coined by Pick to describe a specific disturbance in memory characterized by subjective certainty that a familiar place or person had been duplicated (Pick, 1903; Alexander, Stuss, & Benson, 1979). It is usually observed as a tendency of a brain-injured patient to identify and name the hospital correctly but insist that it is located at a site connected with patient's past experience (often closer to patient's home) (Benson & Stuss, 1990). For example, a patient with brain injury described by Benson and Stuss consistently misplaced the hospital (which he correctly named) to a distant army base where he had been stationed a few years earlier. Paterson and Zangwill (1944) described two brain-injured soldiers who could tell the name of the hospital in which they were recuperating but insisted that it was located in their respective hometowns. Here we see that in fact the current situation receives a second "illegal" temporal mark from the past, which itself is still connected to its own place/situation from that earlier moment on the time line. Thus, the patient's current situation is connected with two temporal marks, each of which calls up its own moment of experience.

Temporal marks: Right frontal region	a	————	b
Situations: Right posterior regions	A		B

As we see in the norm in this illustration, each situation in the posterior brain is referred to the frontal lobe as a temporal mark, and we suggest that the mechanism of reduplicative paramnesia is a fusion of two or more temporal marks.

Neuroimaging studies of cases with reduplicative paramnesia have revealed evidence of either right frontal or bilateral frontal involvement in most cases, with some patients having right posterior lesions either alone or in combination with frontal damage. EEG and neurological exam data also indicated right hemispheric involvement (Malloy & Richardson, 1994). Kapur, Turner, and King (1988) reported one patient with demonstrable damage limited to the right frontal region, suggesting that frontal damage alone may be sufficient to produce a reduplicative paramnesia.

The *déjà vu* experience may be considered a transient condition of time duplication (additional temporal mark for one situation) and is usually observed when the right frontal-temporal region is involved. In *déjà vu*, every detail of present experience is perceived as being identical to an alleged experience in the past (Sno, 1994). Here we see an additional temporal mark for one situation with relatedness to the past. Sno also described a patient who had *déjà vu* who in addition claimed that he was living in a later year than the current one. Here we see an additional temporal mark related to the future.

Situation-action represents the first stage of right hemisphere cognition: *situational thought*. Although the singular situation represents the wholeness of object images within it, their spatial organization (non-I-space) and operations with them (action), we may attribute object images to the right posterior region, namely, the temporal-occipital area; spatial organization (non-I-space) to the right posterior region (parietal-occipital); and temporal organization (action) to the right anterior prefrontal region.

Each situation-action (non-I-space) has one temporal mark in the norm: it is a moment of real (environmental) time when the situation was perceived. These temporal marks form the sequence of the events of one's own life, or subject at that stage.

What is *subjective*? Non-I-space is a moment of environmental time that becomes internalized, forming a time line upon the rhythmic patterns of the lower function levels; in this way, "The moment is mine." The situation is saturated with affect; therefore, too, "The situation is mine." Increasing subjectivity will develop while I-space is interacting with non-I-space.

5.2.4. Symbolic Situational Thought

In this section we will discuss right hemispheric non-language symbolic thought. Traditional concepts about thought are concerned with the domain of the left hemispheric cognitive mechanism: categorical thinking and verbal-logical thinking. As far as we know, there is no description of the rules and regulations of nonlanguage symbolic thought, or visual-figurative thinking. Although neuropsychological research has elucidated information on right hemispheric specialization, this has been only at a perceptual level, regarding visual and visual spatial function, and not at the higher, symbolic level. Interestingly, symbolic systems, though in a completely different framework, are considered in the psychoanalytic literature (for example, Jungian archetypes).

Thus, the usual means of examination are of limited value in understanding right hemispheric symbolic thought; how, then, do we approach this area and attempt to build a coherent hypothesis? We have sought situations in which we believe the right hemispheric cognitive mechanism of the symbolic level is exposed in a more or less pure form, and have identified two such arenas. One is the study of specific cultures that, in our understanding, reflect and illustrate predominance of right hemispheric activity, and the second is specific forms of psychopathology, in which symbolic systems are exposed within characteristic phenomenology.

In the semiotic approach to the history of culture, there is an idea about the alternation of periods of relative dominance of left and right hemispheric consciousness in human cultures (Maslow, 1983). The reasons for these alternating periods of left and right prevalence lie in the limitations of each hemisphere's cognitive mechanism, and the essential incompatibility of their modes of information processing. Left or right consciousness might be captured in particular products of culture such as architectural style, for example, which may reflect left–right differences in information processing and, ultimately, brain organization. Thus, human construction of the world may be viewed as a projection of internal constructs of the brain.

Zeki and Lamb present an excellent example of the relationship between a product of culture and brain organization in their detailed structural analysis of kinetic art and the organization of the visual brain. They define kinetic art—exemplified by the work of Marcel Duchamp and others, beginning early in the 20th century—as art in which motion plays a dominant part or one in which the perception of motion is strongly induced by a static figure (Zeki & Lamb, 1994). Form and color are deemphasized and insignificant relative to motion. By analyzing what gives the viewer the perception of movement, even though it may not be there, Zeki and Lamb showed that the artist's technique is very specific for optimal and selective activation of small areas within the visual cortex specialized for visual motion. For example, lines oriented certain ways are exactly tailored for stimulation of orientation-specific cells and direction-specific cells located in this area. Zeki and Lamb concluded that in creating kinetic art, the artist "instinctively and physiologically" has unknowingly explored the organization of the visual brain. They state, "It is activity within specialized visual cerebral areas, not throughout the visual cortex,... which is particularly important in inducing the perception of motion in a stimulus in which there is no objective motion. It is as if, in studying the relationship between brain activity and the perception of at least some works of kinetic art, one should be enquiring not only into what the visual stimulus does to the cerebral cortex but also, and in particular, asking what the cerebral cortex does to the visual stimulus" (Zeki & Lamb, 1994, p. 623). Following this train of thought, how does the cerebral area specialized for visual motion interpret the world if isolated from other parts of the brain? It extracts motion per se, stripped of form and color (separate areas within the visual cortex are specialized for color and form, as mentioned in chapters 1 and 2). The artist creates a pure visual percept of motion that is another material version of the organization of the area specialized for visual motion.

Cultural evidence of brain organization of invaluable importance for us includes so-called collective representations of indigenous societies. The term was introduced by Levy-Bruhl, an early 20th-century French anthropologist (Levy-

Bruhl, 1930). It refers to a different attitude toward perceived objects in which objects themselves are not distinguishable from associated affects, motions, and actions. Levy-Bruhl emphasized that in collective representations, perceptions, feelings, objects, and subject are inseparably fused. We see here the wholeness of right hemisphere representation, where cognitive-perceptual and emotional components of consciousness content are inseparable. We think of collective representations as a unique natural model or cultural version of the right hemispheric cognitive mechanism that in our society is internalized and only able to express itself filtered through left hemispheric consciousness and interpreted by the left hemispheric cognitive mechanism. Here, right hemisphere symbolic systems are not only externalized and exposed but are embodied in societal rules, played out and reinforced in rituals and ceremonies. Levy-Bruhl indicated that another characteristic of collective representation is that things which do not have objective features in common are united in consciousness. For example, 19th-century observers of a Native American tribe recorded that tribe members claimed that, although they were human beings, they were at the same time red parrots. It is not that they believed that they would turn into red parrots after death; they believed that they were birds with red feathers in the present. It was not the name they gave to themselves and it was not that they were similar to red parrots: they *were* red parrots. Thus it is not analogy, not an association according to some features, it is *identification*. Levy-Bruhl gives another example from tribes of central Australia: they believed that each individual is himself or herself and at the same time is his or her reincarnated ancestor. Totem, the clan to which the individual belongs, is identified with this individual and with plants or animals whose name totem carries. Further, totem is identified with part of space whose borders are clearly cut and which is believed to be populated with spirits of totemic ancestors.

Thus, things which do not have common objective features, things which are different in their physical properties, have the same meaning (are identified) in these societies. Levy-Bruhl explained this identification by his *coparticipation law*, in which things are assigned some kind of mystical power in the consciousness of the aborigine, and it is this that things have in common despite differences in their physical features. We have attempted to understand this mystical power in light of our conception of the right hemispheric cognitive mechanism, and view it as the outside projection of one's own feelings and subjective sense, which may impart the same meaning to things having no common objective features. Let us outline our understanding of this process. The wholeness and indivisibility of the right hemisphere visual situation is characterized not only by the cohesion of spatial and temporal synthesis and situational content but also by the inseparability of emotional experience at the moment of the right hemispheric representation's formation. Association of the visual image with emotion is made possible by interconnections between the occipitotemporal (ventral visual) pathway with limbic structures (limbic emotion); in particular, with the amygdala, a small region in the temporal lobe identified as the source of the brain's emotional network (Harris, 1995). Thus, the final station of the ventral visual pathway—the orbitofrontal cortex—receives from the temporal region highly processed sensory information infused with emotional relevance. Medial frontal systems receive input from the hypothalamus, the latter providing information about the internal state of the organism. The extensive connections to and within the prefrontal cortex allow integration of information concerning the internal and external milieu.

Association of the visual object with emotion can be considered in a broader context, as a part of an emotional-motivational system that is connected with goal-directed behavior, the intentional selection of environmental stimuli based on the internal relevance that those stimuli have for the organism (Duffy & Campbell, 1994; Mega & Cummings, 1994). This goal-directed behavior is apparently in the domain of the left hemisphere cognitive mechanism. On the other hand, in the right hemisphere visual representations, objects are included in the situation and cannot be distinguished from it as separate entities. Similarly, in right hemisphere situational thought, emotion is projected upon the situation (non-I-space) as a whole and does not exist independently. In other words, in this deepest layer of the human psyche, affects are projected upon the outside world, giving meaning to objects. Remember that this layer also represents the sequence of situations (non-I-space) forming one's own life experience. Relating this to phylogenesis, we cite Kretschmer's statement about affectivity of aboriginal people: "Strong affects are not yet differentiated, they are diffuse and ... the affect is fully projected onto the external world. Affects are localized outside just as visual and auditory images are localized outside; indeed, affects are localized in theses images" (1927, p. 94). Outside projection of nondifferentiated affect onto the situations makes the different objects of the external world belonging to these situations equipollent. Through their affects, situations become associated and united into a symbolic system and in this way become imbued with meaning. Situations, through their unification, become multiple aspects of a continuous and indivisible whole, identical in their meaning yet at the same time unchanged from the instant they were formed. We see that the units for operations are the same here—the visual situation—yet at this step of right hemisphere cognition, they are reorganized into symbolic systems by affect.

In Levy-Bruhl's concept of coparticipation, he emphasized the continuity of things in collective representations united by mystical forces circulating in objects and creatures: "The thought of primitive man represents a continuum of magical forces, a continuity of vital principle," the aliveness of all inanimate objects. He stated that, "All animate and inanimate objects are imbued with continuous vitality resembling that vital power which they felt in themselves" (Levy-Bruhl, 1930). Indeed we suggest that the right hemisphere's senses of rhythm and the continuity of one's life (discussed earlier) are merged together and externalized much like affects in collective representations. In the preceding example of collective representation from central Australia, each individual may be somebody who lives now and at the same time is his reincarnated ancestor, and his own totem, and the animal whose name the totem carries, because they all are connected by the continuity of mystical power that circulates among them. In rituals such as animal dances of the totem (bison dance, snake dance, deer dance), the continuity–indivisibility of collective representation is reproduced in action, acted out; it is an intercourse in which the living individual, the ancestor transformed in him, and the animal that is the totem of the individual are merged together. Again, in collective representation individual–ancestor–totem is a single whole, indivisible and continuous (Levy-Bruhl, 1930).Thus, we view collective representations, as described by Levy-Bruhl, as externalized right hemispheric symbolic systems.

At the stage of situational thought, the right hemispheric unit, or continuous whole, is the singular visual scene-situation. Now the right hemispheric unit, or continuous whole, is the symbolic system. The symbolic system reflects the surrounding world to consciousness by simultaneous representation of singular situation-spaces. New simultaneous complexes, formed through uniting of singular situation-spaces by their affect, remain unchangeable. Because of their association through undifferentiated affect, situation spaces become identical, although the term identical should be understood here not as the sameness of two or more non-I-spaces, but as their indivisibility–continuity, in which one situation-space is simultaneously the second and the third one, and so on. Thus, situations are singular and momentary, whereas symbolic systems are simultaneous.

According to Bernstein's evolution of space and time, at each successive level there is reorganization from the physical, objective, metric, and geometric (remember Bernstein's space of the C level: the external spatial field, with main input from visual and auditory modalities) to semantic, meaningful, subjective space; in our interpretation, the abstract space of conceptual representations. All this refers to the left hemisphere cognitive mechanism. In the right hemisphere, we have reorganization from physical space to symbolic space.

The thinking described here represents the right hemisphere's reduction of a series of situations into compressed form, or symbolic situational thought (for details, see chapter 2). The classification of the external world characteristic of so-called primitive men described by Levy-Bruhl (1930) is an example of this thinking: "All things in nature—animals, plants, stars, parts of the world, colors, inanimate objects—in general are divided into the same classes as the members of the common group. If, for instance, the members of the given group are divided into a known number of totems, so the same division is maintained for trees, rivers and stars.... In their turn, parts of the world are tied with certain colors, winds, mythic animals; the latter are mythically tied with rivers, sacred woods, etc., to infinity."

At this stage, the individual worldview, the world perspective and intrinsic value system, is built upon symbolic systems. The organizing factor for one's self here is affect; situations are united by affect (compare with the temporal organization of situations at the previous stage). The same unit-situations are simultaneously organized at two levels working in parallel: in one, the units are organized by temporal marks; in the other, they are organized by affect. Structurally, this may reflect one group of neurons connected to two (or more) different networks. This view is consistent with current theory, which tends to assume that the same group of neurons code for many different concepts via different activation patterns (Hoffman & McGlashan, 1994).

The affect that unites situations into symbolic systems refers to two kinds of emotions: limbic (whose primary source is the amygdala [see Section 5.4]) and thalamic, which are integrated in the right frontal lobe. At this level, the self is liberated from the prison of time: situations located far from one another on the time line may be united into a symbolic system. However, the self is trapped by another prison, the prison of subjectivity that comes from affect. The tremendous input that the MD nucleus gives to the frontal lobe (Fuster, 1985) points to the significance of "thalamic emotion" in the formation of symbolic systems. Indeed, one of the key tasks of symbolic systems is to decode bodily feelings, to overcome one's body spatial borders.

In this vein, it is fascinating to speculate about brain mechanisms operative in psychodynamic psychotherapy. The importance of past experience in psychotherapy has empirically been understood; now we see more clearly the role of past experience, which is incorporated in brain networks as visual situations inseparable from emotion and time and united into symbolic systems at the symbolic level.

Psychodynamic psychotherapy primarily involves right hemispheric cognitive mechanisms of both the therapist and the patient. It "plays" with symbolic systems. In the patient's associations and narratives (stories, situations, past experience), the therapist "listens" for the underlying meaning from the patient's symbolic systems. The therapist's feedback and interpretations clarify the meaning/symbolic system to the patient. In the therapeutic triangle—current situation, past situation, transference—all three become incorporated into the symbolic systems, varying over time and united by the affect of the moment.

Regarding important aspects of the therapist–patient relationship, the terms *therapeutic* or *empathic resonance* are used to refer to a sense of being on the same wavelength, a sense of mutual transparency—of being fully heard by, and fully hearing, the other person (Orlinsky & Howard, 1986, cited by Rubin & Niemeier, 1992). Rubin and Niemeier note the similarity of empathic resonance with Winnicott's concept of the mother's preoccupation with her infant's feelings and needs. In a study of therapists, Larson (1987, cited by Rubin & Niemeier, 1992) determined that there were six specific aspects of resonance: intense concentration; therapist–client synchronization; therapist–client alignment; momentary merging of therapist–client selfhood boundaries; therapist's nonverbal understanding of the client; and specific sensation and feelings somatically and kinesthetically perceived by the therapist. These deeply felt, bodily, and psychologically perceived sensations characterizing resonance or empathy strike us as coming from the deep propriomotor rhythms of level B.

5.2.5. Symbolic Object Thought

As we discussed in chapter 2, the right hemispheric gestalthaft mode of processing associates objects based on resemblance of their holistic form. Examples were given of patients with left hemisphere damage who cluster objects according to their shape and appearance rather than their taxonomic category or function. We mentioned the work of Blonsky, in which free associative process in response to tactile-kinesthetic, visual and verbal stimuli in normal individuals of the so-called visual type (i.e., prone to the easy emergence of visual images) were examined. Blonsky emphasized several conditions necessary for the emergence of visual images: a rested and relaxed state and the absence of movements (indicating that movements and images are antagonistic). Examinees were instructed not to make any effort to call forth images but to close their eyes and wait until the images appeared. These conditions suggest the prevailing importance of the right hemisphere in the visual associations obtained in Blonsky's experiments. In one example, the subject reported these associations when a coin was placed in her hand (without any visual input): "Round cat's muzzle, its ears stick up; no, this is a muzzle of some other beast; two unpleasant eyes. It is as though there's a column from two ears, there are two triangles up to it, one above another. It all turned into a tree with a round crown; its roots can be seen as though the tree is dug out. This all is seen on a background of copper-brown color. It is getting lighter … it reddens. Some animal … round muzzle, huge ears" (Blonsky, 1935, p. 53). Here we may see transformation of "roundness" (cat's muzzle, beast's muzzle, eyes, crown of the tree, roots of the tree, muzzle) and "triangleness" (cat's ears, column from ears, triangles, huge ears). The subject explained later that, touching the coin, she had an impression not only of a round object but also something of triangular form on the surface of the coin. Thus, the image "flows," constantly changing (in content meaning) and at the same time remaining itself (preservation of general form, perseveration of the primary image: cat's muzzle—some other beast's muzzle—crown of the tree—roots of the tree).

In the second example the stimulus was verbal: "little stick." The response was: "Conductor's baton … familiar singing teacher … composer … composer Glinka, his portrait in a small cap … Roman … looks like Nero. Roman palace, Roman in white dress is walking. Garden, a lot of roses, a large quantity of warriors. A huge tree, there is a design from wooden sticks (Christmas tree sticks) on it. White birds fly out of there. Shooting … bullets. I see bullets are flying. I see how bullets, or more exactly their traces, white, are shining. They turn into a beast's paw with white claws. They are crawling ... growing while running. This is a road. The road turns into a waterfall in the Caucasus" (Blonsky, 1935, p. 55). In this example, little stick is transformed into a conductor's baton, a lane of trees, a tree with design from sticks, white tracks of bullets, a paw with claws, a road, a waterfall.

In the third example, the stimulus was the word "scissors." The subject responded: "Shining snow, winter, people are going, covering faces, a boy is rolling on some long gray sticks. A lot of bluish-silverish luster which forms a fountain, stripes of light ... Muslin curtain ... light of the sun penetrates through it. Stripes of light are spreading from the sun. It grows dark. The sun's rays slowly turn into long, loose, flowing hair and a woman's head. Ophelia over a stream" (Blonksy, 1935, p. 57). Here, the luster of the scissors gives the following image of long gray sticks, silvery luster spray of water, stripes of luster, sun's rays, long hair. In the second and third examples we see not only perseveration of the primary image but also its multiplication: little stick—design from wooden sticks—white tracks of bullets—white claws crawling—waterfall (example 2); scissors—stripes of light—sun's rays—long, loose, flowing hair—stream (example 3). There is a situation behind the image, and in these examples we can see fragments of these situations, not just the image. Indeed, the subjects reported that they saw not objects but pictures. One described his experience: "The transforming picture becomes salient, its one point, nearest to me, is defined, giving one distinct figure, other parts becoming less clear" (Blonsky, 1935, p. 59). Compare this experience with the response to a verbal stimulus given by the patient mentioned earlier who had undergone partial posterior commissurotomy: "It's like things are moving around constantly, and I'm trying to narrow it down to something that will just stop. I'm seeing a whole general picture but one thing is almost right in the middle" (Sidtis et al., 1981). Indeed, the image is flowing through situations. Blonsky indicates that the process of image transformation has an obvious perseverative character: the image is "flowing," partially changed and simultaneously staying itself, multiplying. Blonsky concluded that what is seen as image transformation is not two different images connected by similarity but two different moments, two conditions of the same image. We think that the flowing of the image represents recognition of the image in these situations based on resemblance of holistic forms. One can see this in indigenous languages, in which a spoken word represents many different things of similar holistic form belonging to one symbolic system; in Aranta, one word means knee, bend in the river, earthworm (see Chapter 2).

Thus, the object image is recognized in simultaneous symbolic systems according to right hemisphere associations—resemblance of holistic forms. Through this recognition process, there is continuous image modification: the image flows, multiplies, reproduces itself, which is nothing but right hemispheric action at this stage of right hemispheric cognitive mechanism. The "flowing" of the image reflects continuous interchange with like images belonging to symbolic complexes of different content. It is in this very process of continuous modification that the image acquires multiple meanings and becomes a symbol unique to the individual. We have designated this mechanism of symbol formation itself, this operation with images, as *symbolic object thought*.

Let us delineate the rules or parameters of symbolic object thought. The process of right hemispheric associations, or right hemisphere thought, is neither analysis nor synthesis of object images based on their logical connections, their common features and signs. This is the left hemispheric mode of operation. Right hemisphere thought is not based on abstraction from images, analogy among objects according to their common features—again a left hemisphere process. Right hemisphere associations are actions in which the process of operating with images is not separable from the images themselves. Relations between images are repeatedly thought as images themselves; connections of representations are representations themselves (Levy-Bruhl, 1930).

Right hemispheric associations are characterized by:

1. Resemblance of holistic forms (modification of the same image)
2. Multiplication of holistic forms (multiplying of the same image)
3. Contrast of holistic forms (duplicity of the same image)

Duplicity of the same image requires explanation. Situations and objects within them are inseparable from affects. Affects themselves are undifferentiated; they include opposite emotional tones that are indivisible—the root of ambivalence. We refer again to Kretschmer's notion that in primitive societies "affects are localized outside, just as visual and auditory images are localized outside. Indeed, affects are located in these images" (Kretschmer, 1927, p. 94). This undifferentiated affect, according to Kretschmer, holds "The whole chord of feelings: sacred, worshiped, admired, exalted, unpleasant, dangerous, frightening-dreadful, hateful, impure" (Kretschmer, 1927, p. 93). The emotional complex holds a tension of opposite affects, and the image itself is imparted opposite feelings, bad and good. Flowing through the situation, the image will be split into bad and good through a series of identifications. Cave paintings, burials, and other rituals from the early stages of homo sapiens were distinguished by a surprising consistency of themes and symbols, which were grouped around several binary oppositions, usually related to one of two emotionally colored poles (Ivanov, 1978). For example, left (sinister) symbolized bad; right symbolized good. In primitive religions, twins were both sacred and feared. Complicated symbolic systems were formed from these oppositions. For example, a sign of the left hand in cave paintings symbolized the feminine and the right hand symbolized masculine. Oppositions in paleolithic art were connected with difference in colors: a group of animals on the left was painted red and on the right, black. Feminine signs were done in red and masculine in black, and red and black were also connected with paintings of horse and bison, respectively, which themselves were probably symbols of feminine and masculine (Ivanov, 1978).

The individual symbol, the interiorization of right hemispheric action, is the highest stage of right hemisphere cognition. Due to the continuity of the situation-spaces that form symbolic systems in the right hemisphere cognitive mechanism, the symbol is inseparable from its component images, all bearing different content, all being equipollent parts of an indivisible whole. For example, in an indigenous culture with a predominant right hemisphere, in the individual symbol different objects from situations belonging to one symbolic system might be identified and the symbol will represent the continuous whole. Here is an example given by Levy-Bruhl: the feather "is" the cloud, the cloud "is" cotton, cotton "is" the deer's white tail, the deer's white tail "is" the deer, the deer "is" wheat, and so on (Levy-Bruhl, 1930).

However, for analyzing and separating out the symbols' components, and for knowing that an object may symbolize another object but is not that object, interaction with the left hemisphere is necessary. As an example of the interaction between the right and left hemispheres at the symbolic level, let us examine a line from the poem "Sail" by Russian poet Juri Lermontov: "A lonely sail gleams white in the blue mist of the ocean."

We understand that the image *sail* symbolizes a man through the metaphor *lonely sail*, yet we also understand that the sail is not man. For right hemispheric thought sail and man are interchangeable, identical in meaning. Left hemispheric analysis of the words *sail* and *man* reveals that the categorical meaning of these words have only one categorical sign in common: *objectness*. Thus, these two perspectives are not confused and we comprehend both the literal (i.e., the image as such retains its meaning) and figurative meanings. So the symbol may be considered as a complex, shaped through participation of both hemispheres. Despite the importance of the interaction of right and left hemisphere's for understanding symbols, it is the contribution of the right hemisphere that confers its content: the symbol is unique to the individual, its "volume," its depth and breadth, its quantitative and qualitative aspects, depend on the number and scope of associated situations and varies tremendously according to individual differences in the right hemisphere. These individual differences in the right hemisphere depend on a number of factors including various cortical abilities to modify the image (the more symbolic systems the image enters the more volume the symbol being formed will accumulate) and cortical-subcortical abilities to unite visual situation-spaces into symbolic systems by affect. In the norm, one can observe marked individual variations in relative predominance and ability in the different aspects of right hemispheric thought: situational thought, symbolic-situational thought, and symbolic-object thought.

Knowledge of right hemispheric cognition should be an integral part of the rehabilitation of patients with aphasia resulting from left hemispheric damage.

Although it is known that right hemisphere abilities should be used, we emphasize that it is the rules of right hemisphere cognitive mechanism, not just nonverbal stimuli, which will be most productive in this work. Due to the great variability in type of prevailing right hemispheric thought, the initial step in working with these patients should be assessment of these premorbid abilities. In fact, individually prevailing right hemisphere thought as a rule plays a role in spontaneous compensation, defining the individual specificity of aphasic syndrome. In each case, methods of rehabilitation should be developed that take into consideration utilization of the individually prevailing form of right hemispheric thought.

5.2.6. Right Hemispheric Thought and Its Relation to Delusional Misidentification Syndromes

The defining feature of the delusional misidentification syndromes (DMS) is intact recognition and faulty identification, with subsequent belief in doubles and replacements (Weinstein, 1994). Several types of DMS are distinguished:

1. Capgras syndrome is a delusional negation of a familiar (often close) person. The patient does not identify the person, though recognizing his or her appearance and behavior; the patient believes that the person is replaced by a double. For example, in the case described by Enoch (1963), when a male patient was visited by his wife in the hospital, he claimed that the visitor was not his wife, believing her to be a double. He further believed that this double was an impostor who had assumed the appearance of his wife in order to deceive him.
2. Fregoli syndrome is the delusional belief that the same person (usually a persecutor) is simultaneously identified in several persons. The patient believes that the persecutor takes the appearance of other people, "to change faces, as the famous European actor Fregoli used to on the stage" (Enoch, 1963, p. 438). For example, Courbon and Fail (1927) described a patient who became convinced that her persecutors were Fregoli, who had incarnated himself within the bodies of the people around her to thus torture her. There was no disorder of perception, as the patient found no similarity between the various transformations of Fregoli.
3. The syndrome of intermetamorphosis is the delusional belief in the transformations between familiar and unfamiliar people. Courbon and Tusques (1932) described the patient who believed that the persons in his environment changed with one another: A becomes B, B becomes C, C becomes A, and so on.
4. The syndrome of animate or inanimate doubles is the delusional belief that a personal possession or a domestic pet has been replaced by an

identical double (Weinstein, 1994). For example, Anderson (1988) described a patient who believed that over 300 objects had been removed from his home by his persecutor and a proportion replaced with identical doubles.

5. The syndrome of subjective doubles is the delusional belief that one has another self, or that the original self has been replaced (Weinstein, 1994). For example, a patient described by Capgras and Raboul-Lachaux (1923) not only believed that her family had been replaced by identical doubles but also claimed that there were doubles of herself in the community.

Since Capgras reported DMS in 1923, many cases of Capgras syndrome and other variants of DMS were described in the literature, mostly as scientific curiosities as interest in DMS was based largely on its fascinating and dramatic clinical manifestations. In the last decade, there has been a renewed clinical interest in DMS, which in many ways have become a convergence point concentrating critical issues of modern psychiatry, presenting relatively simple models to explore various hypotheses regarding cerebral pathogenetic mechanisms of psychiatric disorders.

DMS occur in patients with typical functional psychotic illness and in patients with definite organic brain disease. Initially, Capgras syndrome was described in patients with paranoid schizophrenia (Capgras & Raboul-Lachaux, 1923; Davidson, 1941); later it was found in patients with manic-depressive psychosis (Stern & MacNaughton, 1945; Todd, 1957) and was considered a functional disturbance associated with paranoid psychotic states. Various psychodynamic explanations were propounded, and until recently, these formulations dominated theoretical approaches to the phenomenon (De Pauw, 1994). Among leading formulations was an ambivalence theory. According to this theory, the patient is strongly ambivalent toward the object that became duplicated. The duplicate, then, appears to be a psychological mechanism of denial and projection; hate is projected onto the duplicate "impostor" so that the patient may continue to love the object (usually a person close to him) (Enoch, 1963). Another formulation emphasized the pathogenetic role of depersonalization with subsequent utilization of the defense mechanism of projection. Stern and MacNaughton (1945) argued that the patient confronted with the altered self-perception (depersonalization) projected the changes onto others in the environment, resulting in the delusion of doubles (cited by De Pauw, 1994; O'Reilly & Malhorta, 1987): Still another hypothesis regarding the psychopathology of Capgras syndrome is regression to the archaic mode of thought, marked by loss of identity and fusion of self-representation and object representation. Stern and MacNaughton (1945) supported this explanation based on the observation of Levy-Bruhl that in the conception of the primitive mentality, objects, persons, phenomena can be themselves and at the same time something else.

As more and more cases of Capgras syndrome and other DMS variants in patients with obvious structural (organic) brain diseases (dementia, interictal psychosis in epilepsy, brain tumor, intracerebral hematoma, posttraumatic encephalopathy, and so on) accumulated in the literature, the pendulum swung toward organic explanations and formulations. Neurologists claimed that Capgras syndrome indeed has a neurologic basis in many or even most cases. Alexander et al. (1979) stated: "Capgras syndrome should not be relegated to the 'unusual psychotic disorders and atypical psychoses' of psychiatry textbooks." (p. 337) Among the neurologically based hypotheses offered were a loss-of-familiarity feeling (Mandler, 1980) and an inability to discern individuality (Cutting, 1990). Other studies have suggested that there is a connection between Capgras syndrome and prosopagnosia (disorder of familiar face recognition)—a typical right hemispheric syndrome. Later it was proposed that there are two information processing routes in facial recognition, which both reside within the right hemisphere. The ventral route, responsible for conscious recognition, runs from occipital visual areas to temporal and orbital regions. The dorsal route runs from the visual cortex to the dorsolateral frontal cortex via the parietal cortex and may be responsible for the emotional significance of the perceived face (Bauer, 1986). Ellis and Young (1990) hypothesized that this dorsal route was damaged in the Capgras syndrome. Others (Luante & Bidault, 1994) postulated that Capgras syndrome is closer to other types of agnosia typical for right hemispheric dysfunction: visuospatial agnosia, neglect, and somatoagnosia. It might represent a variant of visual spatial agnosia characterized by a lack of corporal and/or egocentric space awareness.

Psychodynamic formulations have been harshly criticized as generally post hoc and teleological in nature, and as proposing motives and defense mechanisms that cannot be observed, measured or refuted (De Pauw, 1994). There are also particular arguments regarding the inner logic itself of psychodynamic explanations. For example, Anderson and Williams (1994) believe that the presence of the delusion of inanimate doubles within the spectrum of DMS makes it difficult to accept love–hate ambivalence theory, because it is not easy to explain how ambivalent feelings of great power can exist in respect to mere inanimate objects. Anderson and Williams declare: "The acceptance that disorder of inanimate doubles is a variant of the classical DMS ... puts another nail in the coffin of psychodynamic explanations of delusional misidentification" (p. 224).

In light of what we now know—that the primary psychiatric disorders are rooted in brain disorders—such an attempt to draw a line between neurologic and psychiatric syndromes becomes scholastic rhetoric. The type of damage to the brain in schizophrenia and other primary psychiatric disorders may be different qualitatively from what is encountered in more conventional neurologic disease such as tumors, surgical tissue removal, dead tissue due to stroke, cellular degeneration, and many other frank pathologies (Heinrichs, 1993). Yet, involvement of the same brain area in schizophrenia and organic brain disorder may result in similar clinical syndromes, whose formation and specificity are determined by the functional properties of the affected area.

Brain imaging studies shed some light on localization of DMS. It appears that disordered neural mechanisms of DMS involve widely spaced networks and interaction of the hemispheres (Weinstein, 1994). Most reports of structural brain damage in DMS have been of patients with closed head injuries and dementia, conditions in which there is extensive damage of the frontal, temporal, and parietal lobes. Case reports with brain imaging of epileptic patients with DMS give analogous results: involvement of the frontal, temporal, parietal regions (bilateral or unilateral right sided) (Kim, 1991; Lewis, 1987b). A case study of patients with paranoid schizophrenia similarly showed more frontal and temporal lobe atrophy in those with Capgras syndrome (Joseph, O'Leary, & Wheeler, 1990).

The role of the right hemisphere as a necessary component of the formation of DMS is emphasized in the literature (Ellis, 1994). According to Joseph and O'Leary (1987), the majority of DMS is associated with right frontal, temporal, and parietal lesions. Alexander et al. (1979) thought that bifrontal and right temporal damage is necessary for reduplication of the person, although they could not explain the particular form of the phenomenon.

Despite these findings, many authors admit that the identification of brain dysfunction does not, in itself, explain Capgras syndrome, and that an appropriate cognitive theory of the mechanisms underlying the phenomenon is needed (De Pauw, 1994). The noted neurological models (organic formulation versus psychodynamic formulation) employing various types of right hemispheric agnostic impairments also are not satisfactory: they fail to explain the very nature of DMS. DMS are not perception disorders even though some patients with DMS may have facial agnosia from involvement of bordering brain areas. In order to build a neuropsychological model of DMS, one must be able to explain, in terms of brain function, the essential features of DMS:

1. Intact recognition but disordered identification
2. Reduplication
3. Replacement, the sense that an alternate to the original object exists and that the original has in some way disappeared or vanished (Signer, 1994)
4. DMS as an entity (DMS variants can coexist and interchange in one patient [Christodsoulou, 1991]
5. Multiplicity of doubles (multiple doubles are a common finding in DMS)

How do we understand and approach these phenomena? The key is that DMS are disorders of the higher symbolic level. Of all theories proposed as explanations for DMS, only the psychodynamic have approached DMS as a symbolic disorder, as a disorder of the self rather than of perception. In his interpretation of Capgras syndrome, Enoch indicates: "The 'illusion' is not produced at the site of the object ... it is a lesion of the 'affect' or feeling of the patient" (Enoch, 1963, p. 455). For example, when a patient with Capgras syndrome was asked whether he saw any specific differences between his real wife and the "double," he denied that there were any outward physical differences that he could describe, but added that "there is some subtle difference." On further questioning he gave the very revealing answer: "I *feel* that they are different" (Enoch, 1963, p. 455).

We suggest that the symbolic nature of DMS, disorder of identification with intact recognition, can be explained in terms of specific brain functions. We see a correspondence of this disorder with the operational rules of right hemisphere cognitive mechanisms described earlier in this chapter.

If there is a breakdown in right hemisphere symbolic systems, the object that previously belonged to one symbolic system may not belong to that symbolic system anymore. As a consequence of this breakdown of symbolic systems, the object will be recognized but will no longer be identified: it will lose its meaning. As in collective representations, if an object is drawn not on the ritual place, the aborigine will know what it is, will be able to describe it, will know how to use it, will name it; but if you ask him what this object means, he will tell you it means nothing. That same object drawn on the ritual place will have particular meaning that is dependent upon the ritual place. Thus, it is disorder of the symbolic system which may be applied to Capgras, Fregoli, the syndrome of subjective doubles, the syndrome of animate and inanimate doubles, bearing in mind that the symbolic right hemispheric self is not divided from the outside world, is not divided from other people in the group. Both the self and inanimate objects may well lose their meaning.

We spoke of misidentification; now we will attempt to explain the mechanism of duplication and replacement. In the right hemisphere, the symbol and visual image through which it is expressed form a single, integrated representation. The symbol is multifaceted: at any moment, you may see only one facet, through which you may see all the others. The image is physical, it preserves its own value, its physical features, its form; it is the messenger through which the content of the symbol is seen. Form is physical, content is symbolic because it represents multiple images, images flowing through situations, constantly changing and staying the same. It is the continuity of images that will give symbolic, meaningful content. On the other hand, in the continuum of images, the images are not separate.

In DMS, the original has disappeared in some way, yet everything is known about it. The impostor becomes important but anonymous. The impostor may not be one person but several or a group; it is actually a collective idea, vague and symbolic in its nature, which may change as the image constantly flowing through situations. This is expressed in examples from the literature of patients with Capgras. As Weinstein wrote, "Although a patient misidentifies a person there is often implicit knowledge of the other's identity. Even though he may state in interviews that a nurse is an old high school classmate, he relates to her strictly in her professional role. While the Capgras patient misidentifies her spouse, she recognizes him, does not address him by another name, and does not mistake anyone else for him" (Weinstein, 1994, p. 132). When a patient described by Enoch was asked what could have happened to his wife, he replied, "She might have been replaced." When asked who would do that, the answer was "Some person of some sort, or a huge association.... An association of Jews would do it" (Enoch, 1963, p. 442). A patient of Silva, Leong, Weinstock, Sharama, and Kelin (1994) believed that his mother, father, brother, and sister had been replaced by robots that were physically identical to the original family members. Thus, we see an analogy in the relationship between the original and the impostor in Capgras and the relationship between form and content in the right hemisphere individual symbol.

Reversibility, perseveration, and multiplication of the image flowing through situations in the symbolic system may explain why there may not be one double but multiple doubles. It will also explain why the DMS variants may interchange within one patient. Indeed, because the images that constitute a symbol don't need to have physical features in common, being united by affect, the images may be self, other people, or inanimate objects. In this way, any separate delusional misidentification syndrome may be considered as a variant of the single entity—DMS are not really separate syndromes. In the instantaneous symbolic system, object A is B, B is C, C is D, and vice versa, which gives the basis for the syndrome of intermetamorphosis.

5.3. Mechanism of Left Hemispheric Thought

> It is obvious but not trivial to point out that to "think" is to order our ideas sequentially and toward a purpose. [Efron, 1963, p. 423]

5.3.1. Left Hemisphere and Temporal Order

> There is in space neither duration nor even succession in the sense in which consciousness understands those words: each of the states described as successive in the external world exists, and their multiplicity has reality, only for a consciousness capable of first conserving them and then juxtaposing them by exteriorizing them in relation with one another. [Bergson, 1912, cited by Brain, 1963, p. 392]

The right hemisphere reflects the real world; its cognitive mechanism is content-specific, with individual variations in right hemisphere cognition depending on individual variability in the quantity and quality of images and symbols and ability to operate with them. Left hemisphere cognition is the analysis of the world's events and phenomena as reflected and created by the right hemisphere. The cognitive mechanism of the left hemisphere operates according to a fixed program (Maslov, 1983), breaking down processed information with successive sifting of the variants. Individual variations in left hemisphere cognition depend on individual variability in relative development of the functional levels. Analysis in the left hemisphere consists of splitting of subject and object as well as the breakdown of the object itself. The right hemisphere's representation of "world movement" (movement of matter in time) is analyzed in the left frontal lobe by projection of the signs of the object (matter) onto the left hemisphere's own movement–time line, a vectorial unfolding of action (synthetic time, according to Bernstein). In the process of action unfolding, that time fragment being examined will be at the center of one's consciousness, the focus of active, voluntary attention. Like a point traveling along a line, the focus of active attention is moving along with the moving succession of representations (left hemispheric action). In other words, successive operating with units (the object's signs or features) in time represents left frontal function in general. Several studies have shown that temporal analysis of sequence, or *temporal order*, is a left hemispheric function (Carmon & Nachshon, 1971; Warren, Obusek, Farmer, & Warren, 1969; Efron, 1963). In Carmon and Nachshon's study, patients with unilateral lesions in the left or right hemisphere and controls free from neurological disorder were presented with sequential audiovisual stimuli. The ability of patients with left hemispheric lesions to recognize temporal order was severely impaired, whereas the performance of patients with right hemispheric lesions was similar to that of controls. Temporal ordering on the left does not depend on speech, whereas speech perception may depend on the sequential analysis of temporal order. In other words, the temporal ordering of events (such as are necessary for speech) is separable from the events themselves. This fits Bernstein's general concept about synthetic time as a framework of the anterior brain (the left frontal lobe, in our understanding), which is separable from the units with which it operates. Brain gives the example of the observation by Liberman, Delattre, and Cooper that the perception of a consonant depends in part upon the vowel that follows it, indicating that "the auditory stimuli representing the consonant and vowel are successive, and they are presented to consciousness as successive, but the quality of the earlier event, represented in consciousness by hearing the consonant, is influenced by the latter one, represented in consciousness by hearing the vowel. This would be explained if the neural state set up by the first stimulus had time to be modified by that set up by the second before it entered consciousness. Thus, in the mental present there is not only overlapping but mutual modification of representations of events, which in physical time are successive" (Brain, 1963, p. 394). Our understanding of the phrase "presented to consciousness as successive" is that this reflects the left cognitive mechanism's temporal ordering, which is an intimate part of left hemispheric consciousness, how we become aware of things. Brain went on to state that, "A consciousness which represents objects in time must itself be based upon time, and is unsuitable for dealing with reactions which must be immediate. We cannot afford time to think before adjusting our eye movements to vestibular information: hence such reactions are unconscious." (p. 394) We contend that "consciousness" here represents "awareness" and reflects left hemispheric ordering; yet in our view the right hemisphere also has a "consciousness," one that is immediate experience, a constant stream in which content and time are indistinguishable (flow of consciousness). This "immediate experience" of the right may in fact be very complex. Thus consciousness in our view has two aspects: immediate experience (right hemispheric) and awareness of experience (left hemispheric interpretation of the right hemispheric experience).

The left hemisphere's temporal ordering or "sequence analysis" reflects left frontal lobe operations with units presented by the left posterior brain. In this chapter, we will discuss these operations at two functional levels: gnostic praxic and symbolic.

5.3.2. Left Frontal Lobe and Inner Programming

5.3.2.1. Operating with Functional Signs. Paleoneurologic data derived from examination of brain macrostructure evolution in anthropogenesis indicate two new foci of intensive brain growth arising in the stage of synanthropus (Kochetkova, 1973). We speculate that the development of "left hemispheric action" in phylogenesis is connected with the development of these foci, the second and third foci of growth in human brain phylogenesis (the first focus, as mentioned in previous chapters, corresponds to cytoarchitecural fields 37, 39, 40). The second focus of intensive growth is situated along the lateral margin of the frontal lobe and corresponds to field 45 in modern man. The third focus is located in the premotor region and corresponds to field 8, which, in modern man, implements shifting from one motor act to another, "unfolding" movements into a serially organized row. Anthropologist V. Kochetkova (1973) indicates that development of field 8 might correspond, at that stage of anthropogenesis, to the progress in tool manufacturing which required several sequential operations. Field 8 (the so-called frontal eye movement field) allows redirection of attention, enabling concentration upon that part of an instrument that is being worked upon at a given fragment of time (stage of successive action). What we consider here is, in fact, in Bernstein's terms, temporal synthesis of the D level, which is concerned with object action, and here specifically applied to multilinked serial movements. As complex afferentation of these object actions includes spatial relations among the topological schemes of objects taking part in the action (mostly field 39 of the left hemisphere), the role of the frontal lobe involves the unfolding of the spatial scheme of action in time of the separate stages of object action (kinetic program according to Luria). As we discussed, afferentation for object action (praxis) is provided by the left posterior brain and it includes those properties of the object which determine its *usefulness* in certain situations (we defined these properties as functional signs). Combination of these functional features forms the topological scheme of the object, which implies all potential uses of an object. For example, in the case of a *cup*, topological scheme includes a handle (to hold the cup), solid bottom and walls (to contain liquid), open upper part (to drink from or pour out and empty of liquid). This will be a topological scheme of a cup in the visual modality. Correspondingly, in the kinesthetic modality, there will be positions of the hand that complement the object's functional properties or topological scheme: hand position to hold the cup, to pour out, and so on.

Afferentation (posterior brain) gives a *static*, spatial scheme of praxis (topological scheme of object and corresponding combination of hand positions). The program of action, or representation of object action, in the left frontal lobe is a *dynamic* scheme of positions of the object in the process of action with it along with the complementary positions of the manipulating hand. The dynamic scheme of positions of object and hand gives specific temporal order; it is the inner program of the action, left hemispheric action at the gnostic-praxic level. Figure 21 shows stages in the interiorization of left hemispheric action.

Functional signs of the object projected on the left hemispheric time line become signs of action—direction of object movement, speed of object move-

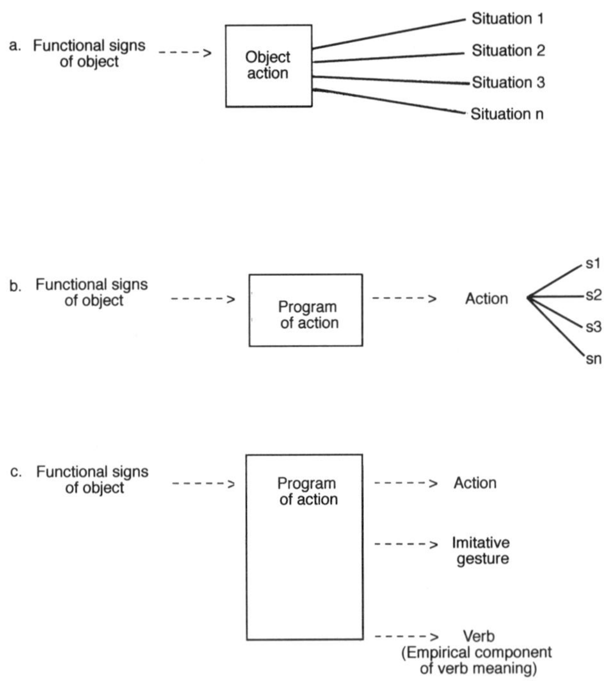

FIGURE 21. Development of inner programming of action.

ment, degree of approximation, approaching–departing. Coming from functional signs of the object, signs of action at this level remain connected to the object and are thus still indirectly connected with the situation. This internal program of object action allows the next step in left hemispheric cognition necessary for communication in the process of object action—gesture language (Figure 21, *c*), and in particular, imitative gesture, which is the symbolic performance of object action (hand movements without the object). Imitative gesture will be replaced by the sound word with further development; however, the word will not be a *noun* (as in replacement of indicative gesture, see chapter 2) but a *verb*. In other words, just as nomination of the object (noun) replaces indicative gesture in the phylogenesis of language development, so nomination of the action (verb) replaces imitative gesture (see Figure 21, *c*). Studies of patients with aphasia and presumed left frontal damage (Micleli, Silveri, Villa, & Caramazza, 1984; McCarthy & Warrington, 1985; Zingeser & Berndt, 1990; Caramazza & Hills, 1991) have shown greater deficits in verb retrieval than noun retrieval. This observation has been confirmed by recent findings of Damasio and Tranel (1993) using neuroimaging techniques. They found a selective deficit in noun retrieval in two patients with a lesion in the left temporal lobes, in contrast to another patient with a left premotor lesion who had a selective deficit in verb retrieval. The authors hypothesized that the systems essential for verb retrieval were in the left frontal cortices (Damasio & Tranel, 1993).

In the historical stages of phylogenetic language development, the verb which replaced the imitative gesture was not a verb in our contemporary meaning, in which the word is an abstract concept of action. In the earlier stages of the verb, the action is tied specifically to the object, and indirectly, to the situation. As an illustration, we turn to the history of language. Levy-Bruhl (1930) gives examples of nonwritten language, in which different words designate the same action performed with or by different objects. For example, in one language there are more than 30 words that correspond to the verb "to go," but there is no concept of walking as such. Each word corresponds to some special type of walking depending on the moving subject: The step of a leggy person who throws his legs forward; the walk of a portly person who steps heavily; the fast and hurried step of a small person; the running about of small animals such as mice or rats, and so on. After the detailed analysis of several nonwritten languages, Levy-Bruhl concluded that the extreme specialization of verbs is a natural consequence of the role that the hand motions have played in intellectual activity. So, at the described stage, the representation of action was connected with posture, position, the direction of motion of the object in the process of action, and so on. The combination of these signs represents nothing more than the *empirical component of verb meaning*. Interestingly, the empirical component of verb meaning (action with object) was intuitively employed in investigations using the Computerized Visual Communication System (C-VIC) (Weinrich, Steele, Kleczewska, Carlson, Baker, & Werte,

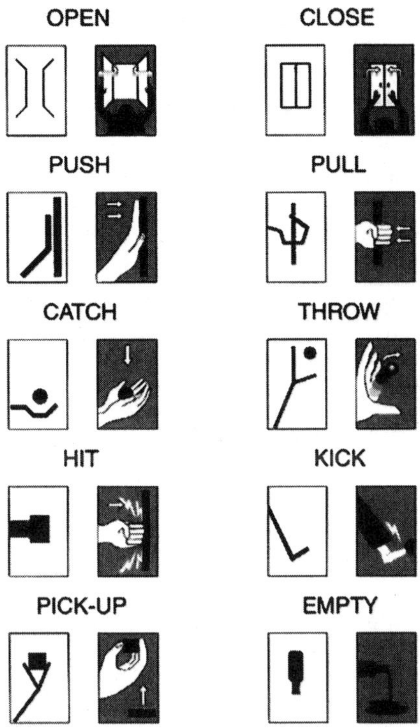

FIGURE 22. Visual symbols for verbs. From "Representations of 'Verbs' in Computerized Visual Communication System," by M. Weinrich et al., 1989, *Aphasiology 3*, pp. 501–512.

1989) to rehabilitate patients with severe anterior aphasia. In Figure 22, we see visual representations of verbs constructed in this study: two visual symbols for each verb, one more realistic and the other more schematic. Indeed, both verb symbols are visual representations containing spatial interrelations of objects in an action, the directions of their motions relative to each other. This is not the abstract idea of action as such; instead, this is a very compressed, symbolic representation of object action, indirectly connected with the situation and very similar to what we call the empirical component of word meaning.

In the C-VIC system, a small box was used as a demonstration *object* for all verbs. The trainer demonstrated the action—*open*, *close*, *push*, *pull*, and so on—with the small box and pointed to the appropriate symbol. Next the trainer selected the symbol and the aphasic patients were asked to demonstrate the action; the trainer then performed the action and the patients were asked to select the appropriate symbol (Weinrich et al., 1989). The aphasic patient would learn the connection between the action and the symbolic representation of the action, which is the empirical component of verb meaning. From our perspective, this actually reconstructs a step from language history: object action as a precursor of the empirical component of verb meaning. After several training sessions, patients were tested: in the standard context using the familiar small box; in the standard context using an unfamiliar object; in a novel context with different objects in a videotape demonstration, for example "to pick up" was illustrated by an actress picking up a chair and involved very different motions than picking up a small object. The authors indicated that patients experienced most difficulty in generalization when the "tool" of object action was changed; for example, in a symbol for "to cut" in which a *knife* is used compared to situations describing the action using *scissors*. This is understandable from the point of view of our concept—empirical verb meaning which is (literally) shaped by the functional signs of the object, its toolness.

In the schematic symbol, we see the transition from the more realistic visual picture to a topological scheme. So, this symbol involves more situations and it is less dependent on metric and weight signs of objects. It explains the results: that patients generalized better with more schematic verb symbols.

In this study, patients with anterior aphasia learned and generalized best with the verb "empty," which was represented by position of object, but they had difficulties learning verbs whose visual representations included mostly dynamic components, such as direction and magnitude of force. Patients made errors most frequently with direction, the sign needed to distinguish "pull" and "push," "open" and "close," "catch" and "throw," and with force, used to distinguish between "push" and "hit."

The emergence and further development of serial actions leads to a transition from immediate action in the situation to the action mediated by the inner program (interiorization of action). As noted, we believe that the structural basis for action interiorization in phylogenesis had been foci of intensive growth in the left frontal areas that correspond to cytoarchitectural fields 8 and 45. Field 45 holds a special interest for us in regard to language. This posterior-frontal (in Luria's terms) field is located between premotor and typically prefrontal areas. Damage to this field results, according to Luria (1966/1980), in a deficit of the inner programming of a sentence. We may speculate that subsequent specialization of field 45 in anthropogenesis for speech programming was determined by the fact that interiorization of action allowed double exteriorization—namely, of both the action itself and communication of information about it (see Figure 21, *c*).

5.3.2.2. Operating with Categorical Signs. In modern man, the highest forms of inner programming subserving thought are connected with the prefrontal area. It is in this direction—the further development and increasing complexity of structural organization of the prefrontal cortex—that human brain phylogenesis progresses. According to Kochetkova, paleoneurologic data indicate that the location of the most turbulent endocrane mold reorganization in Neanderthal man lay in the area of what would be, in modern man, fields 44, 45, 10, and 47. Neanderthal man showed differentiation of operations in the manufacture of tools, which, Kochetkova believes, reflected the ability of this human predecessor to establish mental connections between objects and phenomena.

Formation of the left hemisphere cognitive mechanism is intimately connected with two other dimensions of cortical differentiation: vertical and horizontal. It is expressed by distinguishing of the functional signs of objects implemented by the left tertiary cortical fields of the posterior brain. With development of the left tertiary cortical fields of the anterior brain, *operating* with the signs becomes possible. Operating with the signs involves their organization into hierarchical, continuous series or sequences. In the process of operating (interiorized action), the comparing of different series of object signs occurs and common features of groups of objects are recognized, allowing the formation of categories. Relations of objects according to their signs are established. Sequential sifting of the variants allows the ordering and grouping of the signs in a definite direction: from the general to the particular or vice versa. Thus, a new type of classification of the world emerges—categorical classification, which both characterizes objects as such and defines their interrelations based on common categorical features (inner connections among objects). We believe there must have been a close connection and interdependence between categorical thought (tertiary prefrontal fields) and categorical signs/categorical classification (tertiary cortical fields of the posterior brain, mostly field 37) during their phylogenetic formation. Paleoneurologic data are consistent with this assumption: two main foci of intense growth, in the temporal-parietal region and in the frontal region, coincide in time with the formation of the superior longitudinal fascicle, which connects these brain areas in modern man (Kochetkova, 1973). Emergence of categorical classification signifies distinguishing of the category objectness and therefore is closely related to the formation of the phonological shape of the word and categorical component of word meaning (see chapters 2 and 3). This, in turn, allows development of concepts, in which the categorical component and phonological shape of the word form the word as a concept characteristic of the given language. Operating with concepts constitutes conceptual, verbal-logical thinking (thinking with words). Both types of left hemispheric interiorized action—categorical thought and conceptual thought—are necessary for the possibility of sentence formation.

We presume that these processes describe the basic directions in human brain functional evolution. Brain development in the Neanderthal man had undergone fundamental changes that laid the groundwork for the basic, nuclear components of thought and language at their initial stages. Further evolution, accompanied by the continuing differentiation of the prefrontal fields—which Neanderthal man already had—allowed for the development of thought and language of modern man. Thus, as evolution of the modality-specific cortex of the posterior brain resulted in distinguishing the signs of objects, evolution of the cortex of the anterior brain initially connected with movements resulted in inner programming (Table 5).

Word and sentence are connected in their phylogenetic formation and development just as articulate speech is connected with categorical signs and categorical thinking (see chapter 3). In the linguistic literature, the connection between logico-grammatical categories and deep syntax is discussed in detail by Katznelson (1972, 1986). We will use Katznelson's concepts, which are largely consistent with our understanding of left hemispheric cognitive mechanism, in our interpretation of the cerebral organization of the sentence.

As noted, we have supposed that interiorization of action (inner programming) was connected in evolution with the focus of intensive growth in the cortical region corresponding in modern man to field 45. It is probable that in the further differentiation of the prefrontal region, the higher forms of interiorized action (subserved by field 45) "moved" to the anterior with the formation of tertiary prefrontal fields 9, 10, and 47. Categorical thought became the function of these fields, while inner programming of the sentence stayed connected with field 45. An additional sign indicating the emergence of articulate speech is the formation—together with fields 45, 47, 9, and 10—of field 44, which, in modern man subserves programming of articulation. Figure 23 illustrates different types of left hemispheric programming.

In language, operating with categorical signs is expressed in the unfolding of the sentence, which includes specific linear connections of categorical signs. The tertiary fields of the posterior brain of the left hemisphere distinguish categorical signs common for the given object group by analyzing whole images of the surrounding world represented in the corresponding parts of the right hemisphere. The left prefrontal region analyzes its subject matter: movement. Analyzing the right frontal region—"world movement": flowing images and symbols—the left frontal region distinguishes the main sign: the sign of change (action, state,

TABLE 5. Posterior-Anterior Brain System of Left Hemispheric Cognitive Mechanism and Its Relation to Language

I. Intrahemispheric differentiation	II. Interhemispheric differentiation	III. Language
Posterior brain, cognition of object world	Distinguishing of the signs of the object	Word
Anterior brain, evolution of movement	Operating with signs (inner programming of action)	Sentence

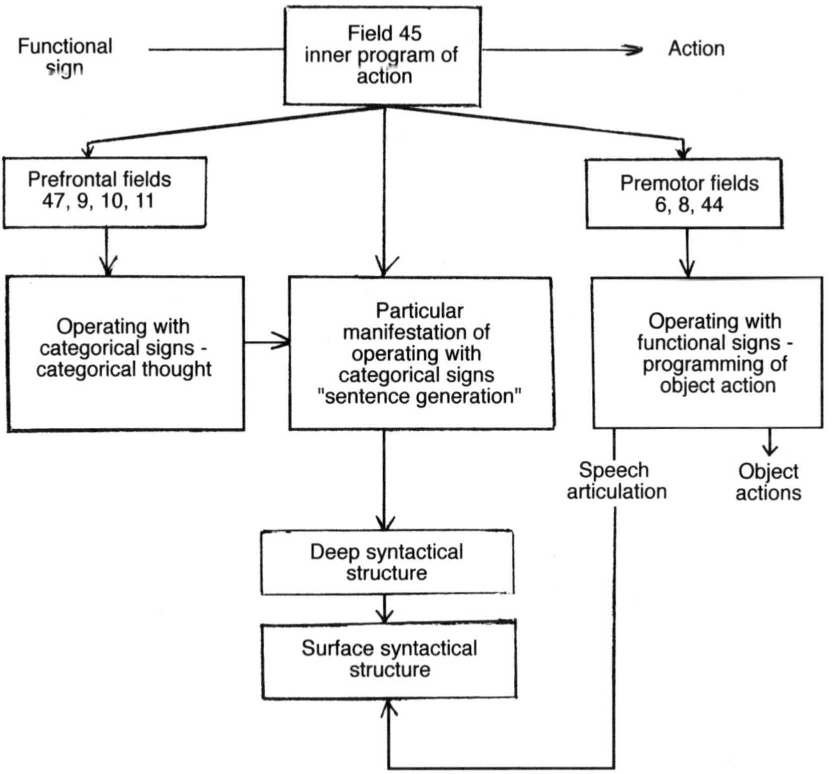

FIGURE 23. Left hemispheric programming.

process). This general categorical sign is superordinate to various more particular (specific) categorical signs. The hierarchical linear sequence of categorical signs of action forms the *categorical component of verb meaning*. For example, the categorical component of the verb "to dig," in addition to general categorical signs of movement, includes more specific categorical signs such as repetitiveness and external limit and goal (example from Katznelson, 1972). In parallel, the empirical component of this verb includes action connected with functional signs of the definite tool (object action) implemented in a few situations. Examples of more particular (specific) categorical signs of verbs follow: directional/nondirectional action; momentary/repetitive action; continuous/discontinuous action; voluntary/involuntary action; intentional/deliberate/forced action; sign of actions limits, goals, factuality, and so on. In categorical thought, these signs form a continuous series. In the categorical component of verb meaning there are a definite number of categorical signs covered by the phonological code of the given word and formed in the history of the given language. Verbs may have some categorical signs in common with one another; however, in each verb these common categorical signs are placed in different combinations of other categorical signs. For example, the categorical component of the verbs "to come" and "to leave" includes the following sequence: sign of movement → sign of active movement → sign of direction of movement. In contrast, the verbs "to wander" or "to roam" contain the categorical signs of movement and active movement but no direction of movement. The verbs "to go" and "to run" contain the same categorical signs of movement and active movement but different categorical signs of intensity. In the sentence "The soldiers retreated," the verb "to retreat" includes the following categorical signs: sign of movement → voluntary movement → direction of movement → forced movement. In the sentence, "He started in surprise," the verb "to start" includes the sign of not forced but involuntary action. The verb "to start" includes the sign of momentary action, which it has in common with, for example, the verb "to skip"; but the verb "to skip" includes categorical signs of repetitiveness.

In linguistics, the predicate plays a leading role in sentence formation, containing in itself the program of the future sentence. Katznelson (1972, 1986) indicates: "The meaning of the verb (predicate) is something more than just its lexical meaning. It is word meaning, but at the same time is a model of the future sentence" (1972, p. 88). The decisive role of the predicate is explained, according to Katznelson, by its selective ability to combine with words containing in themselves certain categorical signs (logico-grammatical categories). For example, predicates of active movements (to go, to wander, to crawl) may be combined only with a noun signifying animate objects (containing the categorical sign of animateness). Causative predicates may combine with nouns that include the categorical sign of agentivity ("Her father persuaded her not to go"). The categorical signs of the predicate, according to Katznelson, form vacant places or nests, valences which will be filled with corresponding categorical signs of other parts of the sentence. Hidden categories of the verb determine its valences, and hidden categories of the noun determine its ability to fill the vacant places attached to the verb. For example, the verb "to give" contains in itself categorical signs of possessiveness and factuality (to give means *to make* the object of possession pass from one person to another). Because of this, this verb opens three positions: one for subject (person), another for direct object (most frequently, the thing), and the other for indirect object (person)—"He gave the book to her." From a linguistic point of view, a sentence represents a combination of words bound by syntactic rules (Jakobson, 1970). Katznelson supposed that the building of a sentence starts with the formation of its deep, syntactic structure, which is the co-"articulation" or assembly of complementary categorical signs of words—the future parts of the sentence. The mechanism of this assembly—the model of explication—used often by linguists includes dynamic, progressive unfolding, the sequential connection of the elements in a definite direction, resulting in the linear structure of a sentence (Katznelson, 1972).

Some linguists have indicated that there is correlation between brain structures and linguistic processes (Jacobson, 1970; Caramazza & Berndt, 1978). Studies of sentence processing in aphasia showed a clear relation between the anterior portion of the dominant hemisphere and syntactic processes. In our model of the left hemispheric cognitive mechanism at the symbolic functional level, we see a strong correlation with Katznelson's model of rules governing sentence formation. We will reiterate Katznelson's model, interpreting from the standpoint of sentence generation as left prefrontal functioning.

The most general categorical sign of a verb is the sign of changeability, the basic *vector* independent from other more specific categorical signs. In other words, the categorical sign of changeability that is the result of the left hemisphere's analysis of right hemispheric world movement is an abstract representation of movement, or synthetic movement of the left hemispheric cognitive mechanism proper. As we have stated, in the categorical component of verb meaning as a hierarchical linear structure, there are general categorical signs and subsequent subordinate, more specific categorical signs (see previous examples). Regarding representation of movement in the left prefrontal lobe, we propose that on the basic vector there are marks or notches—more specific categorical signs upon which corresponding, complementary categorical signs of subject and object are projected (in this context, we use *subject* and *object* as grammatical terms). In the predicate, the basic vector corresponds to the formal, logical operations of the left hemisphere, whereas the categorical signs as such are semantic left hemispheric units (logico-grammatical categories). The predicate has a programming role, distinguishing within the categorical component of the subject's word meaning those complementary categorical signs that correspond to its own categorical signs (i.e., relevant to the given event) (Katznelson, 1972). Katznelson emphasized that the "vector" of the predicate, being directed at the subject, does not characterize the object any more; the logical ties between the categorical signs of the predicate and object are limited to the predicate permitting or not permitting the presence of the object in the sentence. Permission is due to the predicate's categorical signs, which open vacancies for the definite categorical signs of the object. For example, predicates of movement contain specific categorical signs of marked movement (to leave, to approach) which open a vacancy for an object that is a spatial point specifying the goal-directed character of the movement: "He left for New York; he came from New York."

The logic of the relationship between subject and object, according to Katznelson, is mediated through the connections of the categorical signs of each of them with the categorical signs of the predicate. However, there are also direct relations between subject and object that are realized at the surface syntactical level of the sentence. A sentence has two aspects: it relates the thought (event), and also reflects the situation at the moment of utterance and the emotional state of the speaker, his attitudes and values. Direct relations between subject and object are realized in the situation in which the event occurs (Katznelson, 1972).

5.3.3. Cerebral Organization of Sentence

We will build a model of cerebral organization of sentence unfolding (generation) based upon the linguistic concepts noted and the characteristics of left and right cognitive mechanisms and interhemispheric interactions (both complementary and reciprocal-alternative). Figure 24 shows our outline of the steps in sentence generation and the different cortical areas involved. A necessary precursor to sentence production is motivation, which originates in the right hemisphere and is intimately incorporated into the right hemispheric idea, underlying the future sentence. The right hemispheric idea is a condensed symbolic representation which includes the anticipated, motivating end result of the event. The idea will have various images unique to the individual, dependent on the vast array of individual symbols: thus, although sentences produced by different people may be identical, the underlying right hemispheric ideas, or symbolic representations, will vary.

The frontal area of each hemisphere can operate only with the "material" that is in stock at the posterior section of its hemisphere, and within the limits of its own cognitive mechanism. It should be emphasized that it is not that the right hemispheric idea is realized in the sentence by the left hemisphere; the sentence results from the explication or unfolding of those left hemispheric categories which correspond to the right hemispheric idea. The content of the right hemispheric cognitive mechanism is simultaneously represented as singular situation-spaces, as situation-spaces united into symbolic systems by affect, and as images constantly "flowing" through situation-spaces of the symbolic system, which, in the process of "flowing," acquire multiple meanings and become individual symbols.

At each stage of its activity regarding sentence generation, the left hemispheric cognitive mechanism analyses different aspects of right hemispheric content. In the individual symbol, there is an inseparable union of subjectively felt I-space and non-I-space represented by the constantly flowing image. The left hemispheric cognitive mechanism, distinguishing the most relevant event at the given moment, "fixes" the moment in the flowing image—a moment of world movement—and then "unfolds" it into a linear hierarchical structure of categorical signs of action in its own time (temporal order)—step 1 in our model (Figure 24). The fixation of the "flowing image" in itself signifies separation of object and subject, division of I-space and non-I-space. It is the realization of the separation of object and subject distinguished by the left hemispheric cognitive mechanism that

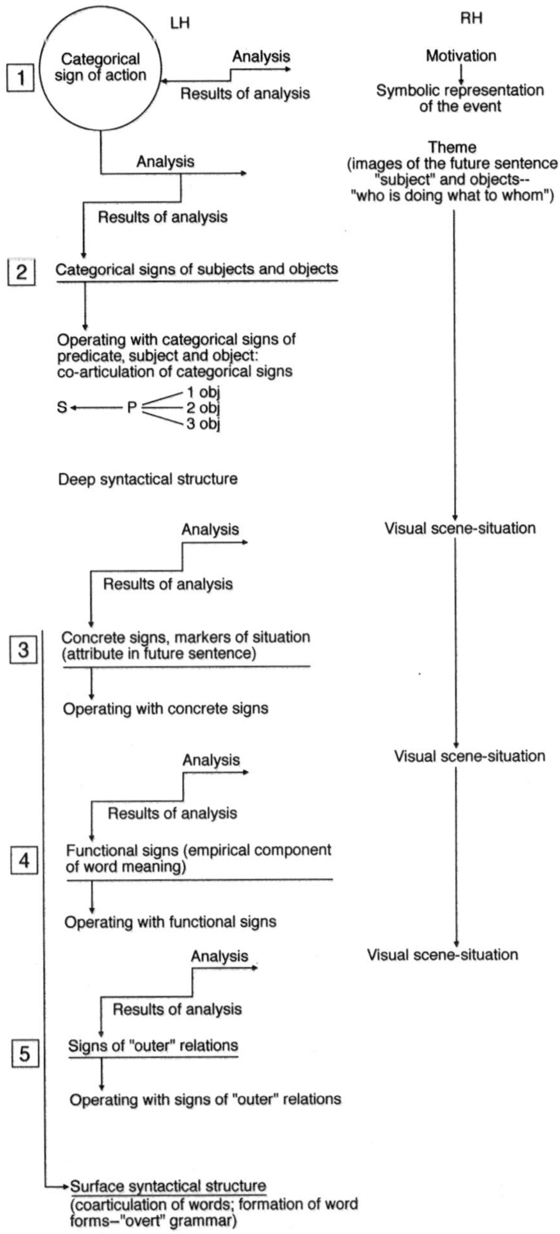

FIGURE 24. Cerebral organization of sentence formation.

results in realization of the existence of the separate, individual self—the awareness of "I," the awareness of "I" as an agent, awareness of the right hemispheric motivation to speak, "I want to say something." This represents the left hemisphere's contribution to self. Distinguishing the separate "I" means that in an individual symbol, images of objects that correspond to the subject and object of the future sentence get exposed (subject and object here being grammatical terms). The "supplier" of images is the temporal-occipital region of the right hemisphere. In turn, analysis of these images by the left hemispheric cognitive mechanism results in categorical and empirical components of lexical meaning of subject and "permitted" objects represented in the left hemispheric temporal-occipital region (field 37).

In the left frontal lobe, there is an operating with categorical signs in the direction in which linear connections between categorical signs of predicate and complementary categorical signs of subject of the sentence "delivered" from the temporal-occipital area of the left hemisphere are established (Figure 24, *2*). Thus, only those signs which are actual in a given event are used, not the whole categorical component of lexical meaning. For example, in the sentence "The grass grows" the categorical sign of the predicate, which may be defined as a sign of the growth process, makes actual the categorical sign of potential for growth in the meaning of the word "grass."

The sentence object is not defined rigidly by the predicate (Katznelson, 1972, 1986), and images of the sentence object in the right temporal-occipital area are represented by several variants. We speculate that there are common (for several objects) categorical signs, distinguished from continuous rows of discrete signs, which are complementary to the categorical sign of the predicate.

We have considered the stages in sentence unfolding from the position of its cerebral-cortical organization and the joint activity of the two hemispheres. From a linguistic point of view, articulation of complementary categorical signs of predicate-subject and the opening of positions for objects by the categorical signs of the predicate correspond to deep syntactic structure (Katznelson, 1972). The following stages in sentence unfolding correspond to transformation from deep syntactic structure to surface syntactic structure in linguistic terms.

Object images in the right hemisphere correspond to the topological schemes of objects in the left—the empirical component of word meaning. The empirical component of word meaning actualizes the phonological code of a word (see cerebral organization of word meaning, chapter 2). Because the categorical component of lexical meaning is not used fully while deep syntactic structure is unfolded, this incomplete categorical component cannot actualize the phonological code of the word. In other words, deep structure is that base upon which the empirical component of word meaning, together with word sound, is built.

Topological scheme, or the empirical component of word meaning, is fully used in sentence formation, because this scheme is, above all, a visual image, whole although synthesized from separate signs by the left hemispheric cognitive mechanism. Thus, this step in sentence unfolding from the point of view of left hemispheric action represents operating with functional signs (Figure 24, *4*). The left hemispheric cognitive mechanism analyzes singular situation-space, through which images of subject and object are flowing (Figure 24, *3*, *4*, *5*). Singular situation-space, achronous and unchangeable, is that steady form in which the image-symbol presents itself. Analysis of the situation by the left hemispheric cognitive mechanism leads to distinguishing of the concrete signs connected with this situation. In the sentence, these signs may be represented by attribute (*red* apple). Katznelson indicates that attribute is the marker of that momentarily steady situation in which the action occurs. Attribute distinguishes an object from the class of objects and by this specifies the visual scene-situation in which the event is happening (Figure 24, *3*).

The next step in sentence unfolding is marked by spatial analysis of the given whole visual scene-situation connected with the parietal occipital area of the left hemisphere. As a result, there are signs of outer relations, characterizing direct (i.e., not through the predicate) subject–object relationship in the situational context of the utterance.

Signs of spatial relations of subject and object are reflected in language by paradigmatic series in which nonroot morphemes are grouped (overt grammar). Operating with signs of spatial relations in the left frontal lobe is the next step in sentence unfolding (Figure 24, *5*). It results in the formation of the corresponding word forms. In this process, grammatical morphemes (nonroot morphemes and grammatical words) receive their sound expression and perform a double function: semantic, reflecting direct relations between subject and object (grammatic content); and formal-syntactic, serving "articulation" of words in linear sentence structure (grammatic form).

As we have already mentioned, operating with functional signs, concrete signs and signs of spatial relations results in formation of surface syntactic structure, in linguistic terms. Separate components of the sentence, explicated (unfolded) at different stages of its generation, appear in speech flow as a whole. In pathology, patients with left frontal lobe lesions may have selective disorders at the level of deep or surface syntactic structure unfolding. Both disorders of sentence generation should be defined as aphasia, and we refer to them as syntactical aphasia I and syntactical aphasia II. Two types of sentence generation disturbance, which correspond to our definition of disorders in operating with either categorical signs or functional, concrete and spatial signs, were described by Achutina (1975). Achutina called these disorders "disturbance of inner (semantic) programming of the sentence" and "disturbance in sentence grammatic structuring". We will here illustrate the different speech patterns of the syndromes of syntactical aphasia I and syntactical aphasia II with examples from patients' speech, and describe these syndromes in greater detail later.

In syntactical aphasia I, generation of a sentence is blocked at its initial and crucial stage of predicate vector unfolding (categorical signs of action). Speech aspontaneity will be observed in the absence of any sensory or motor speech difficulties. However, patients will be aspontaneous but not amotivational: they are able to give short answers to questions, usually in a so-called telegraphic style, in which they replace a sentence by a word. Patient K, who was observed by one of us (Glezerman, 1986) and who will be described in detail later, was a 27-year-old, Russian-speaking, right-handed male engineer with a diagnosis of dynamic aphasia (Luria's classification) as a result of a stroke in the distribution of the left middle cerebral artery. He responded as follows when asked to explain the meaning of the following words: *impede*—"the river and rapids, stony"; *spend*—"a desert and deficit of water." When asked to explain the expression "Much ado about nothing," he replied: "Trifle ... dust." We think that the production of words in this example is not a manifestation of the defect itself but a manifestation of spontaneous compensation: unable to produce a sentence, the patient used his intact ability for single word processing. The patient is not just saying a word: he cannot unfold his thought about some event into a sentence but he has a symbolic image of the event (right hemispheric idea) and *names* it. The mechanism of this compensation is illustrated in this schema:

	LH		RH
	Word meaning		
Phonological code ←	┌ Empirical component	⟵	Object image
			↕
	└ Categorical component	⟵	Symbol

As we see, the right hemispheric cognitive mechanism is exposed here (not opposed by the left hemispheric cognitive mechanism, which is impaired), initiating the word (see direction of arrows in this schema). Naming the event conveys the meaning in a condensed form. Telegraphic style is not just a phenomenon of pathology: poets use it quite often to strengthen emotional impact and make meaning more deep and polysemantic: "Night, street, street lamp, drug store, meaningless, and dim light around" (A. Beok in V. Smith, 1991, translation).

Telegraphic style in aphasia patients is a heterogeneous phenomenon: content words (mostly nouns) in initial forms may be names for events, *but* they also may describe a visual-action-situation. For example, to the question "What would you do if you were lost in the forest?" patient K replied: "Moss ... Northern side ... stars." It is the individual premorbid characteristics of the right hemisphere's abilities that determine what kind of compensation the patient will use. Patient K, as we see, used both types.

In syntactical aphasia II, generation of a sentence is blocked at the stages of surface syntactical structure formation. It will be expressed by so-called anterior agrammatism. Such patients will not be aspontaneous and will be able to build the basic structure S $\leftarrow$ P, but they will have difficulties in usage of grammatical morphemes (operating with signs of outer relations). Patient WH, described by Maher, Chatterjee, Gonzalez, Rothi, & Heilman (1995) and Chatterjee, Maher, Gonzalez, Rothi, and Heilman (1995), was a 66-year-old, left-handed college professor who had had a stroke in the distribution of the right middle cerebral artery involving frontal, parietal, and temporal regions. He presented with a selective syndrome of anterior type aphasia, disorder of syntactic processing with intact single word processing (both phonemic and semantic) and absence of sensory and motor speech defects. Following is an example of WH's narrative of the story of Cinderella:

> Once upon a time Cinderella, a beautiful young daughter and two ugly selfish of the sister and brother and, uh, she's .... Cinderella tends to, uh, shove aside and the preferred of parent, uh, of the sisters preferred, uh, but she shunted off and neglected, uh, Cinderella ... and she does she she she jewelry and finery and she dominate the Cinderella of the sister and she's ... uh ... a a great ball of the prince and royalty and she expected to be the, uh, older sister to to to the ball and to royalty look forward but Cinderella she said no you chose to the Cinderella I'll stay home and we'll all very enjoy and she's all alone and very unhappy. [Maher et al., 1995, p. 109]

Patient WH was not aspontaneous—actually, his rate of speech was normal and he could build a sentence and use main verbs, but his sentences were agrammatical. His speech production was characterized by unfinished sentences, omissions, and substitutions of grammatical morphemes and grammatical words (auxiliary verbs, prepositions).

Finally, within our model there are several steps within both deep and surface structure formation. Theoretically, we suggest the possibility of selective disorders within each of these syntactical processes. Many authors have found that the deficits manifested by agrammatic patients differ considerably, describing dissociations between the omission of grammatical morphemes and word order defects (Saffran, Schwartz, & Marin, 1980; Kolk & Van Grunsven, 1985; Berndt, 1987), and between production of bound versus unbound grammatical morphemes (Saffran, Berndt, & Schwartz, 1989, cited by Maher et al., 1995). Authors suggest that agrammatism probably represents a variety of disorders rather than one syndrome (Maher et al., 1995).

Patient WH was studied extensively by Chatterjee et al. (1995) and Maher et al. (1995). We will present a detailed account of his performance on a word order task adopted from the syntax comprehension task of Schwartz, Saffran, & Marini (1980), in which stick figures of a circle and a square represented various actions (Figures 25 and 26). Stick figures were used to eliminate animacy and semantic plausibility cues, shown in previous experiments to help the patient generate the correct sequence of words. All the target sentences were reversible because the circle figure and the square figure were thematically interchangeable, with the action occurring from left to right in half of the pictures and right to left in the other

FIGURE 25. Stick figures depicting actions from left to right (top) and from right to left (bottom). From "Agrammatic Sentence Production," by Maher et al., 1995, *Brain and Language*, *49*, pp. 105–124. Reprinted by permission.

half. The patient was asked to describe the picture. He performed well when the picture depicted the action going from left to right, but poorly when the action was drawn going from right to left, reversing the thematic roles. In general, in his sentence constructions WH uttered the grammatical subject first and matched it to the item on the left in the picture (in Figure 25, "The square is shooting the circle," "The square is kicking the circle"). WH was presented with the same action pictures (e.g., circle shooting square), in which the direction of action was left to right or right to left, and was asked to then choose which of two pictures depicted

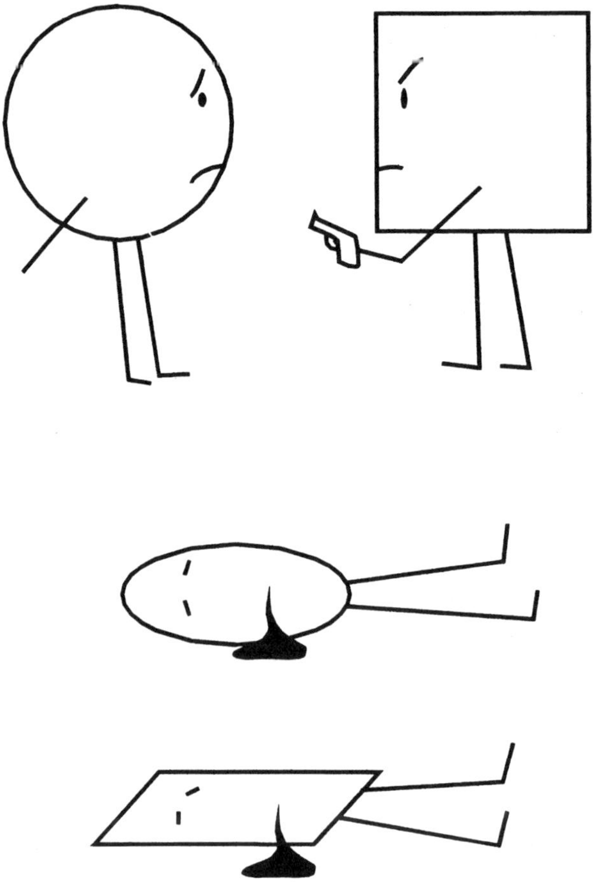

FIGURE 26. Examples of nonverbal stimuli used to test conceptual knowledge of thematic roles. From "Asyntactic Thematic Role Assignment," by A. Chatterjee et al., 1995, *Brain and Language*, *49*, pp. 125–139. Reprinted by permission.

the correct potential consequence of the action (with a circle or a square lying in a pool of blood—Figure 26). Here, with nonverbal stimuli and nonverbal response, the patient performed correctly, regardless of the direction of the depicted action. Under these conditions the direction of the depicted action did not influence his performance, and the authors concluded that WH was able to comprehend the thematic relationships.

In another experiment, WH was read a number of active and passive sentences and then asked to match each sentence to one of two pictures. In both pictures, the action was going in the same direction but one was the correct depiction of the read sentence and the other a drawing of the opposite thematic relationship. For active sentences, WH made the correct choice when the action was going from left to right; if the action proceeded from right to left, all his choices were incorrect. In the passive sentences, WH performed correctly when the action was depicted right to left, and incorrectly if drawn from left to right. Thus, WH would match the first noun that he heard to the drawing in which it was placed on the left, regardless of its grammatical role (subject or object) in the sentence. When WH was given the same task but read the sentences himself, he performed in the same manner, matching the noun that he read first (which was also on the left in the written sentence) with the picture with that item on the left, regardless of its grammatical role.

The authors conclude that the patient demonstrated a word-order problem suggesting an inability to map thematic relationships onto the appropriate grammatic categories (i.e., subject and object in the sentence). In our model of cerebral organization of successive sentence unfolding, the patient's deficits correspond to disorder of the direct relationship between grammatical subject and object at the level of surface syntactic structure. In our view of the process of sentence formation, distinguishing the categorical sign of action (Figure 24, *1*) is followed by left hemispheric analysis of right hemispheric general nonverbal representation of a theme (who is doing what to whom), distinguishing images of subject and object. The next step of left hemispheric analysis includes distinguishing of and operating with the categorical signs of the subject and categorical signs of possible (permissible) objects, resulting in construction of the deep syntactic structure of the sentence. There are two aspects of syntactic rules at the level of deep structure. The first is the connection of complementary categorical signs of the predicate, the subject and the possible object (the content semantic component). The second is the vector unfolding in time, the formal logical left hemispheric mechanism of explication, temporal ordering. Due to this mechanism, the linear structure of connection of categorical signs will be unfolded successively from left to right: subject → predicate → object (formal syntactic component). Both these aspects of deep structure seem to be intact in WH. As we mentioned above, the authors indicated that animacy and semantic plausibility cues from the figures had aided the patient in generating the correct word order. In our terms, these cues correspond to connections between categorical signs of predicate, subject and object in deep syntactic structure. For example, a causative predicate requires a subject that includes categorical signs of agentivity and animacy. Thus, the content semantic component (according to our model) appears to be intact in this patient. Similarly, the formal logical component, his ability to unfold a sentence in a linear direction in temporal order subject → predicate → object was also intact. In fact, he used this formal logical rule to unfold from left to right, designating the item on the left as the first, subject position of the sentence without regard to the actual grammatic subject and object, even though he did appear to understand the thematic relationship between the figures in the drawing (who is doing what to whom). According to our model, it is the direct relationship between grammatical subject and object at the level of the surface syntactic structure of the sentence that is impaired in this patient. The building of the surface structure starts with the left hemispheric analysis of the right hemispheric situation-space, that singular visual situation in which the event (action) is happening (situational context). The right hemisphere visual gestalt containing the subject–object relationship in action is analyzed, and signs of their outer spatial relations are distinguished. In language, these signs are expressed in overt grammar (see chapter 4): in languages using a declension system, by the declension paradigm and word order (e.g., Russian); in languages without declension system, by word order (e.g., English). For example, in the Russian declension system, the accusative indicates that action is directed at the object, and the marker of it is the grammatical, nonroot morpheme, the inflexional ending of the accusative. In English, the direction of action at the object is expressed by word order: subject → predicate → direct object. Although nonroot morphemes are "packed" in paradigms and stored in the left posterior cortex (parietal-occipital), operations with them take place in the left frontal cortex during the building of the surface structure. If we claim that WH's disorder was at the level of surface structure, he should also manifest difficulty with grammatical morphemes, inflectional endings, used in English (plural forms, the possessive form of the noun, the simple past tense of the verb, the third person singular present indicative of the verb, the comparative and superlative of the adjective, the progressive "ing" form of the verb). His spontaneous speech (Cinderella story) did show omissions and substitutions of grammatical morphemes: did this represent disorder of understanding and selection of grammatical morphemes (left parietal-occipital deficit) or disorder of operating with grammatical morphemes (left frontal deficit)? WH did well on a test in which he was asked to select the correct morphological form from among three alternatives that completed a sentence (e.g., "He detect/detected/detection."). However, he was unable *to use* grammatical morphemes *to produce* the sentence. For example, he performed poorly when, after listening to short sentences, he was asked to complete the last sentence with the correct word form ("The millionaire bought a new horse. He now has a stable full of ________.") (Maher et al., 1995). WH was successful in evaluating the grammaticality of a set of sentences, half of which were agrammatic, even correcting morpheme and word order errors. The authors concluded that WH produced grammatical morphemes more accurately when not required to provide the context in which they occur. In our terms, his disorder reflected difficulty in operating with grammatical morphemes rather than their understanding and selection.

Finally, we emphasize again that the patient's nonverbal gestalt of subject–object relationship was intact: he easily comprehended pictures of who is doing what to whom and accurately predicted the appropriate consequences of actions.

WH's performance was not random, but reflected a pattern. His strategy was rather peculiar. In picture description (sentence production), he always chose the item on the *left* side of the picture as a grammatical subject. When WH was asked to listen to sentences and match them with pictures, he would match the *first* noun he heard with the *left* item on the picture. When asked to read sentences and match them with pictures, WH would match the *first* noun that was also on the *left* with the item on the *left* in the picture. Finally, in the purely verbal task, when WH was asked to read sentences (e.g., the circle shoots the square) and to match them with the verbal description of possible consequences (e.g., square is dead or the circle is dead), he would assign the *first* noun that was also on the *left* as the agent (Maher et al., 1995).

Again, this case illustrates that unfolding of a sentence is not a direct reading by left hemisphere of right hemispheric content. WH understood who was doing what to whom in the picture (see Figure 26) but would say, when asked to describe the same picture, "The circle shoots the square." The picture gives a visual-*spatial* image from which relations between two objects are understood. In language, the message about relations between two objects is represented by the *temporal* linear structure of a sentence. Let us consider the linguist's view of how speech—which goes in one, *temporal* direction—reproduces numerous relations of multidimensional reality. Katznelson indicates that language should have some additional means to overcome one-dimensionality and linearity. "To create an illusion of three-dimensional space on the plane, geometry uses projection; painting uses perspective. These means of projection and perspective serve to add depth which the plane is lacking. The means to overcome one-dimensionality in speech are 'projective' forms of language, which serve as though additional, multidimensional space were built upon the speech line, broadening its dimensional potential.... This 'additional space' is language paradigmatic systems" (Katznelson, 1972, p. 186). As we discussed earlier and in detail in chapter 4, signs of relations are organized in paradigmatic series of semantic-syntactic functions and stored in the left parietal-occipital region. When the sentence is unfolded, appropriate signs of relations are chosen from the "repository of all possible constituents' parts" (Jakobson, 1970, p. 243), the corresponding paradigmatic series, and projected upon the vector of the already formed deep syntactic structure. Returning to the concrete example of the circle–square test, the picture shows the spatial relations of objects in the process of action. In language, the spatial relations of objects in the process of action will correspond to the sign of direction of action, which determines grammatical subject and grammatical object and which determines position of words in sentence.

Thus, we propose that what WH did was to leave out the linguistic steps that accurately translate the reality of the spatial relationships in the picture into relations of linguistic units. Instead, he behaved with visual images in the picture as though they were already linguistic units, using the left hemispheric mechanism of explication and unfolding them in temporal order from left to right, from first to next. Remember that this patient was left-handed, with anterior type aphasia after stroke that affected his right hemisphere, from which we may conclude that some of his linguistic abilities were connected with right hemisphere. But here we see that he used visual images as right hemispheric units with left hemispheric strategy, suggesting a premorbid peculiarity of cerebral organization of mental functions in this patient, in which his right hemisphere represented both language and visual perception, determining the type of strategy that he used.

5.3.4. Categorical Thought, Conceptual Thought, and the Sentence

In contrast to sentence unfolding, a particular manifestation of categorical thought that operates with categorical signs in a linear direction, categorical thought in general is characterized by operating with categorical signs in multiple directions. The formation of the categorical classification of the outside world is connected with establishment of the hierarchy of general and specific categorical signs in the process of operating with them. Categorical thought presents a base not only for sentence generation but for conceptual thought as well. The units for categorical thought are categorical signs; units for conceptual thought are combinations of categorical signs that make up the categorical component of word meaning.

The categorical component of word meaning represents a hierarchical structure, from general to more specific categorical signs. An example of a general categorical sign is the categorical sign of *objectness*. The most specific categorical sign in the categorical component of word meaning is also a marker of the concept, because it determines the level of abstraction at which the given concept is "located." As the concept is being formed, the process of ever-more specific sign distinguishing (infinite in categorical thinking) becomes complete at the marker sign. Establishment of the specific marker sign in the categorical component of lexical meaning occurs in the given language in accordance with the formation of the phonological code of the word. Thus, in the concept, the hierarchical categorical sign configuration is locked up in the sound contour of the word.

Concepts themselves are related on the one hand to thinking proper, and on the other to language. In the latter case they are presented as logico-grammatical categories, whose categorical signs will form the deep syntactical structure of the sentence (the skeleton of the sentence). As we discussed, in operating with categorical signs connected with sentence generation, only event-related signs are used. In historical language development, activity of concepts as logico-grammatical categories, or their realization in action in the sentence, promoted the formation of the categorical component of lexical meaning (Katznelson, 1972). Regarding conceptual thought, concepts present as units (configurations of categorical signs fixed in the given language by the word's phonological form). Operating with concepts will represent verbal-logical thought (thinking with words), construction of deductions or inductions, judgments, verbal reasoning.

Categorical thought, conceptual thought, and generation of the sentence, all so closely connected in their origin and phylogenetic formation, have acquired a relative autonomy that might be related to functional diffentiation within the left prefrontal region. The relative autonomy of categorical thought and conceptual thought is supported by such facts as the universal character of categories and the idioethnicity of concepts. We have already illustrated that in different languages, concepts that correspond to similar objects and actions may have differences in the categorical component of word meaning (see chapter 2).

We suppose that these three mechanisms of mental activity—categorical thought (operations with categorical signs), conceptual thought (operations with words-concepts), and sentence generation (mechanism of syntactic explication) have different representation within the prefrontal area. Because development of separate cytoarchitectural areas may vary independently of one another (Blinkov & Glezer, 1964), it seems clear that, in the norm, there may be unevenness or dissociation in the level of development among these components of thought in one individual. This is more apparent in disturbances following focal brain damage, in which the involvement of separate areas within the prefrontal region may result in selective disorders of categorical thinking, or conceptual thinking or impairment of sentence generation. In clinical practice, extended foci of damage are generally encountered, usually including several areas within the prefrontal region, in which case all enumerated disorders will present, constituting a syndrome complex. Nevertheless, even in this situation of complex frontal syndrome, it is sometimes possible to distinguish a prevalent disorder.

Within the whole range of functions different in character and complexity performed by the left frontal area—from programming of movements to sentence generation to higher forms of thought—there is a common feature for all of them: left hemispheric action; that is, explication or unfolding in time line. Impairment of left hemispheric action is expressed in difficulties in shifting from one step of the unfolding process to the next one (inertness) and in sticking at one of the stages (perseveration). This basic deficit underlies disorders of movement, disorders of speech articulation, and disorders in sentence generation as well as disorders of categorical and conceptual thought, although each of these disturbances may arise selectively, related to selective damage of corresponding areas within the left frontal lobe.

Impairment of left hemispheric action in regard to conceptual thinking presents with disorder of operating with concepts (categorical signs of lexical meanings) as successive series. Failure to unfold these series results in a breakdown of hierarchical connections of categorical signs.

In previous chapters, we have discussed disorders of conceptual thought related to focal damage to the left posterior cortical regions. Now we can compare them with the peculiar deficit of conceptual thought that emerges when the left prefrontal region is damaged. In patients with left temporal-occipital lesions (cytoarchitectural field 37), narrowing of the categorical component of word meaning due to a deficit of categorical signs is observed, resulting in impairment of ability to abstract. As the phonological code of a word is disordered (damage to the left temporal region, field 21), there is diffusion of the categorical component of word meaning due to difficulty in distinguishing a specific (marker) categorical sign.

There is a reduction in ability to abstract seen in patients with left prefrontal lesions as well as in patients with left temporal-occipital lesions (field 37), yet the basic deficit underlying this disorder is completely different. In patients with left prefrontal lesions, in contrast to patients with left temporal-occipital lesions, the stock of categorical signs is preserved, but because of impairment in operating with them, it is the hierarchical structuring of categorical signs that is disordered, and clear differentiation between more general and more particular categorical signs is impossible. It leads to concretization of concept formation, inability to distinguish differences between supraordinate–subordinate categories: for example, patients cannot understand the different levels of abstraction included in the words "animal" and "domestic animal." General concepts are used, possible because the phonological code of these words is intact, but without their full abstract meaning any more. For example, the word "animal" is associated with the words "dog," "cat," "goat," but does not unite them into a common category. The words "animal," "dog," "cat," "goat" are located at one and the same level of abstraction now; the word "animal" is perceived as one concrete animal at a time and not as a designation of a class. Because of the disorder of categorical sign hierarchy within one concept and inert sticking in categorical sign variant sifting of different concepts, these patients have difficulties with analogy (similarity) tests, not infrequently indicating the difference between two objects based on their participation in different situations rather than choosing signs of similarity. Thus, patients use a lower functional level of the left hemispheric cognitive mechanism—concrete situational thought. There are similar conclusions in the literature regarding disorders of conceptual thought in patients with left prefrontal lesions. Lhermitte et al. (1971, cited by Caramazza & Berndt, 1978) reported findings on patients with anterior and posterior brain damage given tasks involving sorting cards with printed words: patients with anterior damage manifested mostly a disruption of the hierarchical relationship among words, whereas patients with posterior damage displayed narrowing or broadening of the semantic field of words, corresponding to our understanding of deficiency of more general or more specific categorical signs in patients with posterior brain damage.

In some cases, patients with left prefrontal lesions, being unable to implement mental operations of sequential scanning of categorical signs of different objects needed to perform the similarity test (extract common signs of a category to which objects belong) unite the objects (words) according to affective-situational context of right hemispheric origin. Aphasic patients with anterior and posterior damage as well as neurologically normal control patients were presented by Zurif et al. (1974, cited by Caramazza & Berndt, 1978) with three words at a time and asked to indicate which two were the most similar in meaning. The nonneurologically impaired patients combined the items in terms of shared species membership, discriminating among fish, reptiles, and mammals. The anterior aphasics, in contrast, generated two major clusters, one consisting of shark, crocodile, and tiger (all ferocious) and the other consisting of trout and turtle (harmless and edible). Caramazza and Berndt indicated that the semantic representation in anterior aphasia is more restricted in its range of conceptual integration, and in effect, verbal concepts in anterior aphasia appear to be more tightly tied to affective situational data.

The mechanism of sentence generation includes a formal component, the process of explication or linear unfolding. We have already stated that in the predicate, the mechanism of operating with signs (formal-logical operations of the left hemisphere cognitive mechanism) is closely interwoven with categorical signs as such, semantic units (logical-grammatical categories). Although we consider the formal component in sentence generation from the point of view of its cerebral organization, the left hemispheric cognitive mechanism rules of successive unfolding, these ideas do not contradict linguistic theories of sentence generation (Chomsky, 1957, and others). According to these theories, syntax is a part of language as a code system, or rules, structured along the lines of a generative-transformational system. Each individual possesses a finite and relatively small set of operations, acquired early in life, that make up the syntactic rules for the process of sentence generation. It is because of this primary system of rules that we can produce, and understand, an unlimited number of sentences. The speaker builds first the "core" sentence (deep syntactic structure) and then performs a number of transformations to develop an end product—surface structure. "In this system, no specific sentence structure need be represented in the human mind; instead, it is the syntactic rules that produce the sentence structure that are stored" (Caramazza & Berndt, 1978, p. 911). We also believe that there is a special "memory" for the formal component of sentence generation (explication). Yet there are many steps in the generation of the sentence, each, we believe, subserved by a separate cortical system specialized for a specific operation and its own memory of this information processing. The ability for explication varies among individuals, with selectively high as well as low levels of development. Disorder of this ability results in difficulties of spontaneous sentence generation in speech activity. In our model of sentence cerebral organization, the semantic component (categorical signs of the predicate) is included into syntactic processing (together with the formal component of explication) at the very core of sentence generation. This component depends on right hemispheric ability for non-language symbolic thought as well as left hemispheric ability to operate with semantic categories (categorical and conceptual thought). These abilities may vary in their development independently of each other, and therefore, may be at different levels in one individual.

Non-language symbolic thought, categorical thought, and conceptual thought participate *indirectly* in sentence formation. However, there is a semantic component which is *directly* involved in sentence generation: the interaction between complementary categorical signs of the predicate, subject and object in the sentence. Remember that in our formulation, only those categorical signs that are relevant to the given event are used, not the whole categorical component. Interestingly, linguists have arrived at a similar understanding, making a distinction between lexical and sentence meaning, which, according to Caramazza and Berndt "differ in the very nature of their semantic representation. Lexical meanings can be considered to have fixed representations, but sentence meanings are novel, complex representations constructed by combining the meanings of single lexical items ... The meaning of a lexical item undoubtedly takes different senses in different sentential contexts ... there is a combinatorical operation implicated in sentential semantics that is not present in any formulation of lexical semantics" (1978, pp. 910, 911). This combinatorical operation we think is essential for the semantic component in sentence generation and memory for deep sentence structure.

The next stages in sentence unfolding are characterized by operating with functional, concrete signs and signs of spatial relations, which results in the formation of surface syntactic structure. Although we distinguish two components at the level of deep sentence structure (formal-syntactic and semantic-syntactic), at the level of surface structure formation there is no division into formal and semantic components. At this level of sentence generation, grammatical categories are expressed through grammatical words, word forms and word order. Operating with functional signs (i.e., the empirical component of word meaning) directly leads to actualization of the phonological code of the word. During operations with signs of spatial relations, grammatical morphemes and grammatical words that link words in linear structure receive sound expression. We think that at the level of surface syntactic structure there is a memory for the most frequently used spoken sentences or phrases, which then lose their full meaning and become speech clichés or ready-made utterances, retaining their grammatic (syntactic) form.

As we have already discussed, the sentence may simultaneously include two contents: abstract-categorical (distinguishing and analyzing the event) and concrete (analysis of the situation in which the event occurs). What relative proportion these aspects have in the sentence is defined by the sentence task and also the relative development of abstract thinking, functional (empirical) thinking or concrete-situational thinking in the given individual (with resultant prevalent cognitive style). In general, different and independently varying abilities connected with sentence generation may be at different levels of development in each individual (intraindividual variability).

Sentences in speech communication do not always rely on categorical thinking; indeed, free conversation is characterized by both full syntactic processing (sentence generation) and retrieval from memory of ready-made utterances. The speech clichés or ready-made phrases are not generated per se; they are automatisms, yet they retain the features of left hemispheric representation (e.g., "How do you do?" or "Good job"). These fragments may be combined with other words or phrases, generated or ready-made. We believe that there is a separate ability for manipulating speech clichés. The speaker modifies, transforms, and combines the ready-made speech forms and produces habitual speech combinations. These sentence fragments, for example "The idiot!" or "Stuck in traffic" (also referred to as *ellipses*) seem to be overused by patients with anterior aphasia when they no longer construct complete sentences but rely on utterance types that require less capacity (Hofstede & Kolk, 1994). We believe that this reflects a spontaneous compensation for impaired higher-level sentence generation by lower-level ready-made fragments. In contrast to left hemispheric speech clichés, idiomatic expressions that we, and others, think are of right hemispheric origin always represent whole, nonseparable forms. These idiomatic expressions (for example, "Drag through the mud," "The straw that broke the camel's back"), generated initially by the rules (syntax) of the left hemispheric cognitive mechanism, secondarily become gestalts, whole right hemispheric representations (Ivanov, 1978). In these representations, speech form cannot be divided into component parts and it is also not separable from the visual situation (literal meaning) and symbolic (idiomatic) meaning.

As noted, clichés play a significant role in speech activity in the norm, with individual variability in the relative development (and use) of different abilities participating in sentence production. In addition to cases of individuals who easily use and combine ready-made speech forms but experience difficulty in creative sentence generation, there are examples of high ability to use syntactic rules, ability to explicate, coexisting with difficulties in the use of speech stereotypes, an impoverishment of speech clichés. Psychopathology gives us extreme examples of the dissociability of the different abilities needed for sentence production. One of these examples is the speech pattern that has been seen in some cases of so-called idiot savants, who are mentally retarded individuals but have one unusually high ability (e.g., music, calculation, speech). In cases of speech ability, speech is fluent, spontaneous, and grammatical but devoid of content.

5.4. Left Hemisphere Cognitive Mechanism and Limbic Emotion

In Bernstein's system, level C is concerned with movement in the external spatial field. These movements have a distinct onset and end and are goal-directed; as such, they represent the simplest model of behavior. Bernstein spoke only about the program of movement. Viewed in a broader context, goal-directed behavior must also include components of motivation and emotion. In order to understand the role of level C in this broader sense, we will extend and expand Bernstein's concept of level C, applying it not only to cerebral organization of movement but to cerebral organization of emotion.

In neurophysiology, the term *behavior* implies the complex of goal-directed biological reactions completed by a definite result (Valdman, Evartau, & Kozlovskaya, 1976). The components of behavior are subserved by different brain structures, vertically organized. Movement and goal-directed behavior are by and large the domain of the left hemisphere, and their separate, distinct units function according to the rules of the left cognitive mechanism. That these units are distinct and therefore provided by separate brain regions may be illustrated by the dissociation between the emotional-motivational and performance parts of goal-directed behavior, which we see in psychomotor seizures of temporal lobe epilepsy—automatisms ranging from purposeless simple movements to complex, though meaningless, behavioral patterns. According to Dobrochotova and Bragina (1977), psychomotor seizures occur in patients with left-sided temporal lobe epilepsy versus the purely psychosensory seizures in patients with right-sided focus. Flor-Henry (1969) also observed mostly psychomotor types of seizures in patients with left-sided temporal lobe epilepsy.

The triggering link in goal-directed behavior is information about the internal state of the organism, its vital needs and drives. This is conveyed by afferentation from the hypothalamus (level A). The next step involves assessment of the external spatial field and its objects that may satisfy internal needs. Afferentation of level C gives information about the objects occupying the external spatial field. This afferentation, subserved by posterior sensory cortical fields, primarily visual, in Bernstein's terms is connected with exact evaluation of the physical features of objects necessary for implementation of movements of this level. However, these posterior cortical fields are also players in the emotional-motivational functional system that determines the biological significance of objects as a source of satisfaction of vital biological needs. The key "effector" in this functional system is the amygdala. We spoke before about the ventral visual pathway (the "what" system), whose end point in the posterior brain is the inferior temporal cortical region. This region, which is concerned with object recognition (a higher visual function than just assessment of physical features), is directly connected anatomically with the amygdala. In this system, we think that the amygdala marks the junction of information from level A (internal milieu) and level C (external objects). At this junction, integration of information from levels A and C, reflecting evaluation of external objects from the point of view of internal need satisfaction, is manifested by emergence of subjective experience (emotion). Although there are many definitions of emotion in the literature, a common feature has been object-related subjective experience. For example, Valdman et al. (1976) defined emotion as "subjective experience which reflects one's attitude to the surrounding world and oneself"; similarly, Smirnov (1976) stated that emotion reflects the attitude of an individual (subject of needs) to valuable objects and events.

The amygdala is a complex of nuclei located in the depths of the anterior temporal lobe. It is part of the limbic system, a circuit of brain structures traditionally viewed as the morphological substrate of the emotional-motivational system. We will consider the role of the amygdala within our three-dimensional conceptual framework of brain structural-functional differentiation (vertical, anterior–posterior, left–right).

The amygdala complex is phylogenetically connected with the striatum and archicortex. This connection with the striatum makes us think that in the vertical functional hierarchy, the amygdala belongs to the C level. Considering Bernstein's anterior–posterior dimension, the amygdala is an anterior brain structure, and in this sense, emotional experience connected with the amygdala may be considered as an equivalent of movement.

The amygdala is also phylogenetically tied with the olfactory system, and part of the olfactory pathway terminates in amygdala nuclei. This leads us to speculate that emotional experience associated with the amygdala, which we call limbic emotion, is somehow connected with olfactory sensation, as thalamic emotion is connected with bodily sensations. The olfactory modality is the only sensory system that does not have its center in the posterior cortex. Of all the sensory modalities, olfactory sensation yields least to categorical classification and is most associated with the object and the emotional experience connected with the object. Olfactory memory is almost always emotional.

Clinical and experimental data over the last 40 years have substantiated the amygdala's key role in emotional experience.

Within the motivational-emotional functional system, the contributions from level A (the hypothalamus) and level C (amygdala) are in reciprocal relationship with each other. In numerous experiments, animals became hypersexual following bilateral excision of the amygdala: males attempt to copulate with males, with animals of different species, with females not in a state of heat, and even with nonanimate objects. The hypersexuality could be eliminated after destruction of the ventromedial hypothalamic nucleus. Destruction of hypothalamic nuclei in the presence of an intact amygdala led to hyperemotionality in animals. In this case, subsequent excision of the amygdala led to a significant decrease in emotionality. In the monkey, Kluver and Bucy (1937) described a syndrome of hypersexuality, strong oral activity, placidity, and loss of normal anger and fear after bilateral temporal lobe resection. The Kluver–Bucy syndrome has also been observed in humans with either bilateral or left hemispheric temporal damage. Lilly, Cummings, Benson, and Frankel (1983) reported on a 31-year-old female whose brain neuroimaging results following presumptive herpes simplex encephalitis showed left frontotemporal abnormality. She demonstrated markedly altered behavior that included inappropriate sexual actions, constant oral activity, and emotional placidity. In contrast, the interictal personality of patients with temporal lobe epilepsy may have features (emotional intensity and hyposexuality) that are the opposite of those observed in the Kluver–Bucy syndrome. This emotional intensity and hyposexuality has been attributed to hyperactivity of the amygdala (Bear, 1979). Sudden stereotypical feelings of overwhelming fear, embarrassment, anger, and depression without any connection to environmental stimuli may occur just prior to seizures in patients with epilepsy (emotional auras). The limbic system has been implicated as the location for seizures beginning with emotional symptoms (Gastaut & Broughton, 1972). In studies of patients with temporal lobe epilepsy, spontaneous seizures or direct stimulation of the brain indicated that the amygdala was responsible for the emotional component that gives experiential immediacy to perceptual and mnemonic phenomena originating in the temporal neocortex (Gloor, Oliver, Quesney, Andermann, & Horowitz, 1982).

Emotion is a component of goal-directed behavior and is distinguishable from the other links in this chain. Emotion can be considered as a behavior itself, serving a communicative, social role, enabling individuals to gauge one another's attitudes. Again, its social communicative role is derived from the object-related nature of limbic emotion. Limbic emotion, born in the depths of the brain, involves amygdala-frontal lobe pathways. There are distinctive qualitative domains of human emotional experience. For example, Izard (1991) delineated seven main groupings: joy, surprise, fear, sadness, anger, disgust/contempt, and interest.

Emotion includes both experience and expression. There are different neural regions responsible for emotional experience and emotional expression, but they are interconnected and work together in patterns of behavior fixed in evolution. There are two types of emotional expression: somatomotor, which is voluntary and includes facial expression and body language, and visceromotor, which includes the autonomic responses accompanying emotions (involuntary movements of level A: visceromotor and angiomotor and tone of somatic musculature). The pattern of somatomotor expression is stereotypic and universally recognized, and is necessary for communication with the object (level C). Ekman (1984) identified universal, specific human facial expressions for fear, surprise, anger, disgust, and happiness. The external manifestations of visceromotor responses are also universally recognizable (flushing, sweating).

Although we have been discussing limbic emotion, human subjective experience includes also so-called thalamic emotion. External indications of thalamic emotion, implemented by the half-voluntary movements of level B, are idiosyncratic, communicating indirectly about the state of the subject (Mona Lisa smile).

In the norm, emotional experience and emotional expression are subserved by different regions and realized in parallel based on phylogenetically fixed brain networks of goal-directed behavior. The links of goal-directed behavior are relatively independent from each other and can be elicited separately, which again demonstrates left hemispheric information processing. In animals, experimental activation or inhibition of either of these regions illustrates the dissociability of emotional experience and emotional expression (Valdman, Evartau, & Kozlovskaya, 1976). Clinically, dissociation of emotional expression and experience can be seen in patients with focal brain damage, who may exhibit pathological crying or laughing without feeling (Benson, 1994). In certain psychiatric disorders, dissociation of emotional experience and expression may also occur. For example, in schizoid personality disorder, intense subjective experience may not be accompanied by emotional expression. In contrast, in the histrionic personality, a disproportionate display of emotional expression relative to depth of genuine emotional experience is characteristic. The basic role of the visceromotor reaction that accompanies emotion is to maintain autonomic tone for the subsequent links of goal directed behavior. This complex of emotion as a subjective experience and vegetative reactions is provided by reciprocal connections between level A and level C, with the direction of the vector determined by the degree of activation of the corresponding structures. In intense, pathological emotion (e.g., depression), the accompanying visceromotor reaction is similarly increased. On the other hand, the so-called autonomic crisis state of extreme activation of the autonomic system may be accompanied by intense feelings of panic and fear.

Visceromotor reactions were selectively induced by electrical stimulation of subcortical effector centers in humans during stereotaxic brain surgery (Smirnov, 1976). Patients reported a series of motor, visceromotor, and angiomotor reactions that reflected quickly developed changes in the activity of internal organs. According to Smirnov, information about these visceromotor and angiomotor reactions goes to the sensory afferentation system, and as a result of this feedback, the patient perceives these reactions as sensations, labeled *secondary sensations* by Smirnov. Patients describe these sensations as quick contractions, twitches, vibration, squeezing, tension, stretching, feeling hot, feeling chilly, feeling a lump in the throat. In the corresponding body area, motor or angiomotor reactions could be observed (hyperemia, cyanosis, muscle tension, temperature change). Patients also describe sensations suggesting involvement of vestibular cerebellar pathways, such as nausea, heart sinking, fainting, bursting, feeling of falling down, feeling of heaviness. These secondary sensations are often accompanied by emotion, frequently negative. Patients are unaware of the motor origin of these sensations. In contrast to described phenomena, when Smirnov directly stimulated sensory nuclei of the thalamus, the *primary sensations* experienced by patients differed from secondary sensations in numerous ways. The primary sensations were modality-specific, localized in a particular body part and emotionally neutral.

In the psychiatric clinic, patients with depression may experience pathological sensations that are similar to secondary sensations elicited by stimulation of subcortical effector centers (motor phenomena of level A). These sensations should be differentiated from senesthopathy, which we have discussed in Chapter 4. Pathological sensation in depression and senesthopathy are related to two different levels, A and B. The first is a motor phenomenon (anterior brain), the second, sensory (posterior brain). In the subjective experience of senesthopathy, the sensational component is not divisible from the emotional component, whereas in depression, pathological sensations and emotions are separate phenomena. The pathological sensations in depression, although complicated and not simply modality-specific, still may be described in language (words) and are comparable to known somatic sensations. Emotion also can be described by depressed patients; for example, anxiety, sadness, apathy. In senesthopathy, neither the emotion nor the sensation is describable. Traumatic situations tend to increase the pathological sensations experienced by patients with depression (extroverted C level); in contrast, environmental events do not influence senesthopathy (introverted B level). Careful clinical analysis of the patient's descriptions of sensational and emotional experience will allow determination of functional level of the brain involved, which is central to our new way of understanding the brain basis of psychiatric disorders.

Although the amygdala may be the seat of basic, primary emotions, it is the amygdala–frontal lobe connection that is crucial for the social and cognitive aspects of specifically human emotion. This is inferred from the extensive literature on the loss of empathic, socially appropriate and civil behavior observed with damage to the orbitofrontal region (Duffy & Campbell, 1994; Mega & Cummings, 1994). The social aspect of human emotion derives from the object-related nature of limbic emotion, the evaluation of the object as the means of biological need satisfaction. Evaluation of objects is also the foundation of interpersonal relationships, the roots of what is judged "good" or "bad," the rules and beliefs of the group. This is a left hemispheric cognitive process in that emotion is related to the object that induced it but is separate from it, in contrast to right hemispheric subjective experience, in which the emotion is projected into the external object. The division of object and subject in the left hemispheric cognitive process allows realization of the self as a separate "one," an awareness of "I," an "I" that feels, an "I" that thinks, an "I" that acts (Descartes's "I think, therefore I am"). This is left hemispheric consciousness. At the same time as the left hemisphere is aware of "I," it can observe "I," evaluate "I" as an object—the representation of "I" as an object. In this connection, we cite data regarding the difference in body image representation in the left and right hemispheres. Nikolajenko and Deglin (1984) examined drawings of a man made by patients immediately following unilateral electroconvulsive therapy (ECT), when the stimulated hemisphere is transiently inhibited. They concluded that each hemisphere deals with a different body space: internal (subjective body space), right hemisphere; external (objective metrics of the body), left hemisphere. According to these authors, one's body is represented in the left hemisphere as an "object" existing in extrapersonal space analogous to other objects. Cutting (1990), after analyzing disorder of body scheme in patients with unilateral right and left hemisphere disorders, concluded that the left hemisphere is responsible for knowledge of the essential features of body parts rather than their spatial image. In the frame of the vertical organization of the left hemispheric cognitive mechanism, we see representation of one's body as an object in external space at the C-level; one's body as a functioning object (essential features of body parts) at the D-level; and at the symbolic level, the psychical equivalent of the body as a physical object will be left hemispheric "I" as an agent, left hemispheric observing self, left hemispheric referential self.

The object "I" is a member of the group with other individuals (left hemispheric classification). Group beliefs are also evaluation of the object, and acceptance of the group's beliefs by the individual reflects the left cognitive mechanism: reliance on rules and formal logic, being a representative of a group. The power of group beliefs was captured by Arthur Koestler, who stated, "The rule is that the man who goes to war abandons his territorial home and fights for imperatives which are not territorial but mostly ... abstract: the true religion, the righteous cause, the correct political system. Wars are fought for words in semantic space" (Koestler, 1969, p. 20).* One can see that left hemispheric limbic emotion evolves in the frontal lobes united with cognition, becoming beliefs: "The violence unleashed in war and persecution is also a secondary or vicarious type of aggression derived from identification with a group and its system of beliefs. It is a depersonalized, unselfish kind of savagery, generated by the group mind *which is largely indifferent or even opposed to the interest of the individuals who constitute the group*" (Koestler, 1969, p. 20, italics in original). The term *identification*, used by Koestler, is not one that we would use referring to this left hemispheric process of being a member of a group, a separate but related part. Identification, as we have used it, is a right hemispheric cognitive process, the collective representations discussed in the corresponding chapters. This process involves the subject's projection onto the group and all its members, animate and inanimate (totem), an indivisible, continuous whole imparted with meaning. The left hemispheric group implies evaluation of objects, comparing objects with each other.

Let us dwell more on Koestler's idea of secondary or vicarious emotion in humans. Koestler wrote: "You watch a well-acted film version of the Moor of Venice; you soon begin to identify with Othello or Desdemona or both; so, of course, you hate Iago and are quite prepared to strangle him with your bare hands. Your anger will produce all the physiological symptoms of a genuine emotion; yet the psychological mechanism which pumps adrenaline into your bloodstream is totally different from that which operates when you are facing a real opponent.... The adrenaline is not produced by any primary biological drive" (Koestler, 1969, p. 20).

*From an article adapted from a paper read at the 14th Nobel Symposium in Stockholm, "The Place of Value in a World of Facts." Nobel Foundation Copyright 1969. Copyright 1969 by The New York Times Company.

Secondary or vicarious emotion may be related to the left prefrontal cortex function of integration of emotion and concept. Modern neuroimaging data give very interesting confirmation of this. Drevets, Videen, et al. (1992b) measured regional blood flow in patients with familial pure depressive disease (FPDD) and found increases in the left prefrontal cortex and left amygdala. In psychiatrically well subjects asked to contemplate sad thoughts, an increase in blood flow was found in the left prefrontal cortex but not in the left amygdala (Drevets, Spitznagel, et al., 1992a; Pardo, Pardo, & Raichle, 1993). Change in blood flow in the left prefrontal region was also found in healthy individuals during performance of a verb generation task (Peterson et al., 1989). The part of the left prefrontal lobe with increased activity in patients with FPDD consisted of two areas: ventral (orbital) prefrontal and ventrolateral prefrontal. Both areas were activated in subjects during contemplation of sad thoughts; however, only one area, the ventral lateral (along with dorsolateral, which was not hyperactivated in depression and when contemplating sad thoughts) was activated in people during performance of a verbal generation task (Drevets & Raichle, 1995). Drevets and Raichle suggest that although the left ventral (orbital) prefrontal area is responsible for the bridging of emotions and concepts, the left ventral lateral is more connected with making associations as such, and may be responsible for ruminative negative thoughts in depression.

5.5. Syntactical Aphasia I and Syntactical Aphasia II

The basic deficit in syntactical aphasia I is disorder of sentence generation at the stage of operations with the categorical signs of lexical meanings of the predicate, the subject, and the objects. In patients with this disorder, motivation, general intention, and the idea of the utterance are intact, but the presence of aspontaneity may create an impression of impaired motivation. Left hemispheric analysis of the general right hemispheric "idea" is impaired: linear unfolding of the categorical signs of the action gets stuck at one of its steps (Figure 24, steps 1, 2). The categorical signs of the grammatical subject and object cannot be projected upon the complementary categorical signs of the predicate, because the predicate vector (the general categorical signs of action) is absent. Disorder in distinguishing of the categorical signs of action (the predicate; step 1) creates a block impeding sentence formation even if the subsequent steps (Figurc 24, steps 2 to 5) are potentially intact. This disorder of sentence formation at the stage of initial unfolding is a disorder of deep syntactic structure, in linguistic terms. The complementary hemisphere interaction is not realized; in particular, in the right hemispheric symbol, the subject is not separated from the object, and images corresponding to the grammatical subject and object are not distinguished. Still, the right hemisphere idea—both the symbolic image of the event and the visual situation connected with it—are primarily intact. Single word processing, both word sound and word meaning, subserved by the left posterior brain regions, is also intact. As with other forms of aphasia, the clinical picture of syntactical aphasia I is a result of the interaction of specific deficits and intact abilities. The characteristic telegraphic style of the patient with syntactical aphasia I is an attempt to replace the sentence with the word, using intact right hemispheric symbol/situation and intact left hemispheric word sound and meaning. Unable to unfold a sentence, the patient will name the symbolic image of the event or name the object from the situation in which the event is happening. In the first strategy, right hemispheric symbolic associations may push out the word that designates the event (war, fire, wedding). In patients with syntactical aphasia I, simple ready-made sentences may be found, due to relatively intact memory of surface structure; these ready-made sentences are stuck and become perseverations. Also, phraseological unities connected with the right hemisphere may flow up, such as "golden fall," "real man," "rotten apple," "the last straw," "pins and needles." Disorder of operations with concepts as logical grammatical categories is combined with disorder of operations with concepts as units of thought, which results in concretization of words-concepts. On neuropsychological examination, speech in patients with syntactical aphasia I is characterized by intact comprehension of grammatically simple conversation, with difficulties with more grammatically complex structures. Recognition of oppositional phonemes and word comprehension are unimpaired, although errors secondary to perseverations may be found. Naming and repetition are primarily intact, although again, perseverations may interfere with performance. Expressive speech is impaired by lack of speech spontaneity, telegraphic style, speech clichés, and perseverations.

In syntactical aphasia II, there is a selective disorder of those steps in sentence unfolding which are connected with transformation of deep syntactical structure into surface structure. Formation of deep structure is intact, which means there should be no difficulty in explication of categories reflecting the event. There is impairment in operating with signs of the event's visual-situational context (concrete, functional, and signs of spatial relations) (see Figure 24, steps 3, 4, 5). Disorder of operating with concrete and functional signs during sentence unfolding may lead to word-finding difficulties for sentence formation (lexical deficit), which in fact has been mentioned in the neurologic literature as a *pseudoamnestic word deficit* in patients with anterior aphasia. Why does a lexical deficit appear in a patient with a deficit of sentence formation, in whom word sound and word meaning are primarily intact? This may be understood as follows. In the norm, as we discussed, during the formation of deep structure (coarticulation of categorical signs of the predicate, the grammatical subject and objects; Figure 24, steps 1, 2) not all categorical signs of word meaning are used, only signs actual for the given event. To induce word sound, the categorical component of word meaning as a certain combination of categorical signs fixed and connected with phonological code in the history of the given language should be presented. It is operating with functional signs (empirical component of word meaning) during formation of surface structure that is important for the actualization of the word in its sound form in a sentence. The empirical component of word meaning represents a whole (topological scheme of the object) and is the stimulus for word sound. In the patient with syntactical aphasia II, formation of deep syntactical structure is intact but it will not help produce the words of the sentence because, at this stage, it is still a particular type of categorical thinking, not speech.

Disorder of operating with signs of spatial relations embodied in paradigmatic morphological language series results in a deficit of overt grammar, usage of nonroot morphemes. The deficit will be manifested mostly in expressive speech. Patients speak in simple sentences. Words will be frozen in their initial form but, if word forms are used, there will be no syntactical agreement. At the same time, patients will recognize oppositional phonemes, they will understand words, and they will be able to repeat words, although in all these tasks errors of perseverative origin will be observed. Detailed description of a patient with combined syntactical aphasia I and II (Glezerman, 1986) follows.

Patient K, a 27-year-old, Russian-speaking, right-handed engineer suffered a stroke in the distribution of the left middle cerebral artery. His full-scale IQ was 84 a few months following the stroke (lower than average), with verbal IQ of 76 and performance IQ of 98. The patient was able to understand everyday speech and simple sentences but had difficulties in comprehension of complex syntactical structures. K did not have difficulties in speech articulation. He could easily differentiate oppositional phonemes. Repetition and naming were impaired secondary to perseverations. His answers to questions were characterized by simple sentences, sentence clichés, perseverations, and echolalia.

K's attempt to build a sentence in response to a picture with simple actions showed deficiency in syntactic coarticulation between the words: words were frozen in their initial form. For example, he was shown a picture of a woman peeling a carrot. He responded "Zshenshina chistit morkovk*a*" instead of "Zshenshina chistit morkovk*u*." He replaced the correct inflectional ending *u* with *a*, substituting the inflectional ending of the accusative case, indicating the relation of the subject and object in action, with the inflectional ending of the nominative case, which is the initial form of the noun. He would often make another error when asked to describe the action in a picture: instead of naming the action, he would name the object.

The patient's speech in the task of retelling a story presented to him auditorily was agrammatic; nonroot morphemes were missing or incorrectly replaced and there were perseverations not only of separate words but also of phrases. In general, patient K could not produce a coherent story, yet his verbal response showed that he understood the main idea of the story and remembered the details well.

Patient K experienced the greatest difficulties when required to build a sentence without support from the visual situation (picture) or from an auditorily presented story. Asked to write a paragraph on the topic *north*, the patient could only produce "Les*u* sosni" (Forest, pine trees). The word *forest*, "les," was agrammatically frozen with the inflectional ending "u," which case in the Russian language requires both inflectional ending and prepositions. Indeed, it was perseveration from the previous task. Curiously, patient K arranged the two words vertically, one over the other, instead of from left to right in a horizontal direction.

In patient K's WAIS verbal scores, there was a marked discrepancy between his above-average score in the information subtest (memory of learned general knowledge) and the very low scores in the vocabulary and similarity subtests (measuring verbal-logical thinking). His performance on the vocabulary subtest fell in the range of moderate mental retardation (4 against the average 10). K's responses in word definition were characterized by perseverations, clichés, echolalia, and telegraphic style. As we discussed earlier, telegraphic style is a result of compensation, an attempt to replace the sentence (which he cannot generate) with words. Here are his responses when asked to give definitions of the following words: *winter*—"snow, blizzard"; *repair*—"apartment"; *detail*—"detail and machine tool; detail and pig iron"; *fortitude*—"detail or person"; *pity*—"man"; *disaster*—"volcano or ship"; *gather*—"detail"; *impede*—"the river and rapids, stony." We can see that he does not produce a sentence in his responses, nor does he give any verbs, which, as we know, constitute the axis of the sentence. In his attempts to replace a sentence with words, he uses nouns, which are actually names for images from the situation connected with the target word and also symbols connected with the event. His inability to use verbs and adjectives was also expressed by his transformation of the target words that were verbs or adjectives into nouns: *conceal*—"evasive person"; *domestic*—"hearth"; *brave*—"brave person"; *matchless*—"size."

On the similarities subtest, the patient is asked how two words are alike. Patient K's score on this subtest was 7 (10 is average), which corresponds to the range of mild mental retardation. Qualitative analysis of his responses showed discrepancies in his performance within this subtest. On some occasions he would give categorical answers. For example, *chair-table*—"furniture." In other cases, K would give a noun coming from an image associated with a visual situation. For example, *axe-saw*—"well, wooden ... log." Instead of defining the common features, K would at times describe each object separately according to its sensational, affective-situational context. For example, *orange–banana*—"orange smells and juicy, banana is soft."

K was given an object classification test, in which he was presented a set of picture cards and asked to unite similar objects into groups. He was able to unite objects according to categorical principle, uniting into one group a few objects belonging to a common category, but at the same time he was unable to unite all the objects belonging to the same category, making instead several small groups. For example, cat and dog were called animals and grouped together, as were elephant, fox, and goat, and so on. This inability to make a larger, inclusive group indicated that the abstract concept of the word *animal* did not exist any longer for him, although the word was simply associated with particular objects (dog, cat, horse, and so on). His performance demonstrated disturbance in the hierarchy of supraordinate and subordinate categories and difficulty in operating with categorical signs of a concept.

Patient K had no difficulties in nonverbal tests on visual object gnosis, visual spatial gnosis, and praxis. He was spontaneous and interested while performing tasks. His performance on the nonverbal WAIS subtest was in the average range. He stayed focused on the task and had good insight into his performance. He was able to shift from one task to another without perseveration. On tests designed to assess nonverbal logical thinking, block design and picture arrangement, K scored in the average and high average range, respectively.

In summary, patient K had a disorder of sentence formation, which corresponds in our terms to combined syntactical aphasia I and II. He was unable to generate his own new sentence when he was not given any clues (visual or auditory). When he attempted to build a sentence, he did so agrammatically, with disordered usage of grammatical morphemes, which he omitted or misused. There were perseverations of words and phrases (clichés). He had a disorder of conceptual thought related to his inability to operate with or shift between general and specific categorical signs, with loss of hierarchy of categorical signs. His general ability to categorize was relatively intact. He also scored in the average range on performance (nonverbal) subtests assessing nonverbal logical thinking such as block design (spatial analysis, constructional thinking) and picture arrangement (ability to organize situations, planning and logical sequential relationships). On these subtests, he demonstrated no difficulties shifting from one task to another and no perseverations. Thus, his frontal lobe deficits were very circumscribed and only included the part of the verbal sphere involved in programming of the sentence, with programming for speech articulation intact. Patient K's full-scale IQ was 84, which clearly seems to represent a decrease from his premorbid level. There was a marked discrepancy between his verbal IQ (76, borderline mental retardation) and nonverbal IQ (98). His specific neuropsychological deficits were revealed by qualitative analysis of his performance on verbal subtests, and his low scores on these tests were a function of these particular difficulties.

K's clinical presentation also showed that he used right hemispheric symbolic thought and situational thought as a means of spontaneous compensation, which was expressed in his telegraphic style of speech patterns. The patient's right hemispheric abilities should be assessed as one of the first stages in rehabilitation work, as they will form the most effective foundations for strategic treatment planning.

III

Conclusions, Reflections, Perspectives

6

Thought and Focal Brain Damage

6.1. Neurolinguistic Classification of Aphasia

In previous chapters, we gave the theoretical basis for distinguishing neurolinguistic forms of aphasia, with descriptions and clinical samples. In this section, we summarize the data and present our proposed neurolinguistic classification of aphasia (Table 6). The main postulates that underlie this classification are these:

1. Aphasia is a disorder of the symbolic function level subserving language.
2. Aphasia is a result of damage to tertiary cortical fields of the left hemisphere.
3. Different aphasia forms are connected with selective disorders of separate links within symbolic (language) level of the speech functional system.
4. Different aphasia forms are the result of damage to different tertiary cortical fields.

Each aphasia form has its counterpart disorder at the gnostic-praxic level that is due to a lesion of the secondary, modality specific cortical field over which the corresponding supramodal tertiary field was formed during phylogenesis. For phonological aphasia, the counterparts are acoustic speech agnosia and articulatory (kinesthetic) apraxia; for logico-grammatical aphasia, visual anomia; for syntactical aphasia, articulatory (kinetic) apraxia. The exception is morphological aphasia, a somewhat confusing syndrome that in our view is not strictly aphasia as we formulate it, but a disorder of cortical fields which are viewed as transitional between secondary and tertiary fields (fields 39 and 40). These fields, which mostly subserve nonverbal visual-spatial function, we believe have specialized areas in

TABLE 6. Scheme of Neurolinguistic Aphasia Classification

Form of aphasia	Disordered link in language level of speech functional system	Assumed localization
1. Lexical (phonological)	Phonological code	Left temporal region Field 21
2. Lexical (logico-grammatical)	Logico-grammatical code (covert grammar)	Left temporal-occipital region Field 37
3. Lexical (morphological)	Morphological code (overt grammar)	Left parietal-occipital Fields 39, 40
4. Syntactical I	Syntactical code (covert grammar)	Left dorso-lateral frontal Field 45
5. Syntactical II	Syntactical code (overt grammar)	Left dorso-lateral frontal Field 45

which signs of spatial relations are incorporated into the paradigmatic series of morphological language code. Thus, damage to these fields may result in symptoms of language code disorder (morphological aphasia) in pure form or accompanied by gnostic-praxic visual-spatial deficit.

Speech agnosias and apraxias and aphasias related to damage to a certain brain area (temporal-occipital, parietal-occipital, temporal, frontal) are most often encountered together, due to the proximity of the corresponding secondary and tertiary fields. Indeed, selective disorders of the symbolic or gnostic-praxic level are rather rare. Nevertheless, clinical descriptions of these disorders are quite well represented in the literature, although they are presented in various different conceptual frameworks. We emphasize that it is necessary to accurately classify speech disorders regarding the functional levels involved, as rehabilitation strategies will require specific techniques targeting the very different deficits.

6.2. Aphasia and Intellect

Our assumption that the roots of the language–thinking connection go back to the phylogenetic formation of tertiary cortical fields and their symbolic function, and our conception of aphasia as a disorder of the symbolic (language) level in the brain functional hierarchy automatically lead to the understanding that aphasia represents disturbance in both language and thinking. This is not a point of view commonly found in the literature; indeed, analyzing these data is difficult because, although early investigators did examine the question of aphasia and thinking, later work relied more on psychometric intelligence tests, which, in fact, measure a variety of abilities and a broader field than thinking per se.

Basic data regarding aphasia and intellect were obtained primarily using the Wechsler Adult Intelligence Scale (WAIS), which gives a resultant full scale intelligence quotient score (FS-IQ) derived from verbal IQ and performance (nonverbal) IQ subscores.

In general, the average FS-IQ in the aphasia population has been found to be lower than the FS-IQ in the general population; however, the range of individual scores is very broad, from higher than average to profoundly mentally retarded (Lebrum & Hoops, 1974). The most common trend observed was for lower verbal IQ than performance IQ (Kennedy & Wolf, 1936; Weisenberg, Roe, & McBride, 1936; Piercy, 1964; Hebb, 1942; Bauer & Becka, 1954; Heilbrun, 1959; Reitan, 1959; Anderson, 1951; Smith, 1965). However, the significance of these findings is not clear because studies of the general population have found that individuals in the low average FS-IQ range have, as a rule, a higher performance IQ than verbal IQ. It was shown that as FS-IQ decreases from 120 (and higher) to 75 (and lower), the proportion of individuals with performance IQ greater than verbal IQ increases from 21% to 74% (Smith, 1965). There are also contradictory data about the degree of impairment of performance IQ in aphasic individuals, with descriptions in the literature of patients with severe aphasia and normal or even high nonverbal IQ and other patients with impairment of nonverbal IQ (Lebrum & Hoops, 1974).

These investigations did not reveal a correlation between decrease in FS-IQ and type of aphasia. However, there were significant differences in IQ depending on the severity of the aphasic syndrome and the extensiveness of brain damage (Lebrum & Hoops, 1974).

Some authors deny any pathogenetic connection between aphasia and intellectual impairment. Zangwill (cited by Lebrum & Hoops, 1974) suggested that any impairment of intellectual function that may accompany aphasia is due to damage to neighboring areas of the left hemisphere. De Renzi and Spinnler (1966) and Piercy (1964) speculated that damage to speech zones may cause, in addition to aphasia, disorders of general intellect due to overlapping anatomical representation of different functions. De Renzi and Spinnler, in fact, suggested that there might be a pathogenetic connection between disorder of speech and disorder of intellect; in particular, abstract thinking. Using special techniques to evaluate categorical, abstract thinking (classification test) De Renzi and Spinnler examined four groups of patients: patients with damage to the left hemisphere with aphasia; patients with damage to left hemisphere without clinical manifestation of aphasia; patients with damage to the right hemisphere; and patients with bilateral brain damage. Patients with aphasia (group 1) had the highest prevalence of disorder of abstract thinking. This bears some similarity to Goldstein's concept that brain damage produces a disturbance of both speech and intellect, although in his view this was secondary to the loss of abstract, categorical attitude.

The cause of the confusion is not that intelligence tests are not adequate for understanding the relations between aphasia and intellect. The problem is the incorrect interpretations of test results. The meaning of IQ obtained in patients with focal brain damage—in aphasia patients, in particular—is different from the meaning of IQ in a person from the general population. Intelligence level in patients with focal brain damage changes relative to the patient's premorbid intelligence level.

IQ represents a numerical summary of a multitude of distinct abilities that constitute the so-called structure of intellectual functions. To understand the intellectual deficit in patients with aphasia, it is not correct to rely only on an IQ number, even with consideration of verbal and nonverbal IQ; it is necessary to take into consideration the structure of intellectual functions.

Table 7 shows the main specific factors (abilities) that underlie performance on WAIS subtests. Performance of different subtests relative to one another will give the structure of intellectual functions, an individual profile of cognitive functioning that may be graphically depicted. In our earlier work, we showed that local cortical dysfunction directly affects specific intellectual functions, which secondarily leads to a decrease in IQ (Glezerman, 1983).

A comparison of neuropsychological examinations indicating local cortical dysfunctions in children with learning disabilities with the WAIS individual profile of these children, and the findings of the unevenness of WAIS individual profiles of school-age children from normal populations, led to the conclusion that the uneven level of development of separate intellectual functions within one individual identified by the WAIS profile may indirectly reflect varying relative development of distinctly different cortical regions. With this in mind, the individual WAIS profile may be seen as a neuropsychological profile. Individual variability of cortical formations underlies numerous combinations of independently (from each other) varying components of neuropsychological structure that give a variety of individual neuropsychological profiles in the norm (Glezerman, 1983).

In patients with aphasia, the individual neuropsychological profile is a result of two factors: one is the premorbid individual neuropsychological profile, and the second is the effect of the particular type of aphasia on the profile. Within the types of aphasia syndromes, there will be differences in presentation based on the severity of the lesion and individual compensatory efforts, which indeed represent interaction between the deficit and the intact abilities of the premorbid profile. Another source of individual peculiarities of aphasic syndromes is the interindividual variability of cortical cytoarchitecture within the same region. Blinkov found a high degree of variability in the cortical cytoarchitectonics in the borders of regions, both in size of the specific field and in the extent and structure of the transitional region between specific fields (Blinkov, 1955; Blinkov & Glezer, 1964).

Thus, intellect, as measured by psychometric tests in patients with aphasia, represents the result of complex interrelations between deficit due to a particular focal brain lesion (aphasia form), individual peculiarities of the aphasic syndrome,

TABLE 7. Factors Involved in Performance of WAIS Subtest

Verbal subtests	Factors underlying performance
Information	Learned general knowledge Education, culture, environment Long-term memory
Similarities	Verbal-logical thinking (ability to abstract)
Vocabulary	Verbal-logical thinking (lexical, logical grammatical language code—word meaning) Lexical, phonological language code Nonverbal, symbolic thinking Learned general knowledge Education, culture, environment
Arithmetic	Verbal-logical thinking (discursive, sequential logical operations) Spatial ability (spatial analysis) Attention and concentration Short-term memory
Comprehension	Adequate emotional orientation in situations ("common sense") Mobilization of conventional concepts (test of social standards)
Digit span	Attention and concentration Short-term memory Spatial ability
Performance subtests	
Picture completion	Attention and concentration (organization and concentration of visual attention) Visual object gnosis
Picture arrangement	Nonverbal logical thinking—ability to organize situations, planning and logical sequential relationship Adequate emotional orientation in situations Visual "gestalt" thinking, simultaneous synthesis
Block design	Spatial ability (constructional thinking) Visual-motor coordination
Object assembly	Spatial synthesis at object level Visual motor coordination
Digit symbol	Attention and concentration Visual-motor coordination (motor speed)

and individual neuropsychological pattern that determined the structure of intellectual functions premorbidly. Because of this, broad statistical comparisons of intellectual function in the aphasic population are not only difficult to interpret but may even be misleading.

Our approach relies on distinguishing a specific neuropsychological syndrome in the individual patient and examining the influence of this syndrome on the WAIS profile. This was accomplished by qualitative analysis of individual subtest performance and scores on subtests that have common factors necessary for performance. We now present a detailed description of our analysis of the WAIS as a neuropsychological tool. Our assessment includes several components: the IQ (an interindividual parameter), intraindividual intersubtest analysis, and intraindividual intrasubtest analysis (qualitative analysis of performance within the subtest). Intersubtest analysis includes comparison of performance on the various subtests with regard to the specific abilities required for each. As can be seen in Table 7, there is an overlapping of abilities that underlie performance on subtests. Verbal-logical thinking, a left hemispheric ability, is a component common to the Similarities, Vocabulary, and Arithmetic subtests. In the Similarities subtest, verbal-logical thinking involves the ability to distinguish the common categorical signs of two concepts, the ability to abstract in a relatively pure form. In the Vocabulary subtest, which requires the subject to define a word, verbal-logical thinking is involved in concept formation, the ability to distinguish combinations of categorical signs, or categorical component of word meaning. The complex of factors utilized in performance on the Vocabulary subtest, in addition to verbal-logical thinking, includes another language ability—lexical phonological code. Nonverbal right hemispheric equivalents of word meaning are also involved: visual-figurative, symbolic thinking. It should be noted that the relative contribution of these abilities and their varying degrees of development differs widely among individuals, giving a specific flavor and characteristic pattern to performance. In addition, performance on the Vocabulary subtest is dependent on the individual's volume of verbal information, which reflects education, cultural environment, and learned general knowledge, factors that overlap with the information subtest.

In the Arithmetic subtest, verbal-logical thinking is involved in discursive, sequential operations; other factors underlying performance include spatial ability, attention and concentration, and short-term memory. Attention and concentration are also measured by the Digit Span subtest. In addition, the digit backward component of this subtest requires spatial analysis. Attention and concentration overlap with the subtest Picture Completion; however, this last subtest measures specifically visual attention. Performance of the subtest Comprehension requires adequate emotional orientation to the situation; this emotional orientation overlaps with the Picture Arrangement subtest, in which ability to organize situations is determined. Logical thinking makes this subtest similar to the verbal subtests. On the other hand, the Picture Arrangement subtest may be successfully performed using nonverbal, visual-figurative thinking, ability to "catch" the whole, the gestalt, sense of humor. The Comprehension subtest also includes in its more complicated task—proverb understanding—a component of right hemispheric, visual-symbolic thought. (There are data that understanding of metaphors and sense of humor are connected mainly with the right hemisphere [Gardner et al, 1975; Winner & Gardner, 1977; Wapner et al., 1981].) The visual image component is common to both the Picture Arrangement and Picture Completion subtests. However, in Picture Completion, gestalt organization will be at the level of visual object gnosis rather than visual symbolic thought. An important component of the Picture Completion subtest is attention, which it has in common with the Arithmetic and Digit Span subtests. In performance on the Block Design and Object Assembly subtests, the spatial factor is important. In Block Design, the schematic concept of space (constructional thinking) plays the leading role: it is necessary to parse mentally the presented model into its component parts corresponding to the units of construction (spatial analysis). In Object Assembly, direct visual-spatial perception plays the leading role: it is necessary to determine the whole and the interrelations of its parts.

Thus, performance on each subtest is determined by several factors. However, high achievement or impairment on performance may be secondary to prevalence or deficiency of one of the factors. Qualitative analysis of subtest performance will allow one to determine the contribution of a particular factor. Then, scores on subtests having common factors of performance are compared. For example, low (relative to other) subtests scores in Block Design, Arithmetic, and Digit Span (with intact digit forward but inability to repeat digit backward) may indicate deficiency in spatial ability connected with left parietal-occipital region (spatial analysis, constructional thinking, spatial relations underlying concept of number). A low score on the Block Design subtest with intact performance on object assembly may suggest left hemispheric disorder of spatial ability, although the opposite result, intact Block Design performance with disorder of object assembly, does not necessarily mean disorder of right hemispheric spatial ability. Spatial disorder characteristic of the right hemisphere is more severe and will involve both subtests. In this case, qualitative analysis of Block Design performance is useful. There are specific patterns of performance on Block Design in patients with right versus left and parietal versus frontal dysfunctions (Luria, 1966/1980; Glezerman, 1983; Kaplan, 1990). From a neuropsychological point of view, low scores on Object Assembly may be due to visual-spatial perception disorder, but they may also be caused by a different disorder—fragmentariness of the visual object image as a result of right temporal-occipital dysfunction (Kock, 1967). Thus, low results on Object Assembly with intact results on Block Design may be interpreted as right hemispheric disorder of visual object perception rather than visual spatial perception. Comparison with results on Picture Completion and Picture Arrangement will further specify the level of dysfunction—object gnosis versus visual gestalt thinking.

In previous chapters, in our descriptions of patients with the corresponding types of aphasia, we have already applied this strategy. In most cases, the IQ in patients with aphasia is decreased from the premorbid level. Moreover, we concluded that this decrease is secondary to specific neuropsychological deficit (aphasia type), although in any patient with brain dysfunction there will be some decrease in IQ due to nonspecific factors such as psychomotor speed and coordination. When dissociations in profile could not be connected with the specific neuropsychological syndrome, we assumed that they belonged to the premorbid neuropsychological profile. We will briefly reiterate the examples of patients with each type of aphasia according to our classification.

Patient R with the syndrome of logico-grammatical aphasia (described in chapter 2), demonstrated specific deficits on WAIS subtests assessing verbal-logical thinking, contributing to a lowering of IQ from his premorbid abilities. His very high spatial abilities without a great deal of right hemispheric visual-situational or symbolic thought reflect his premorbid neuropsychological profile.

Patient S, with lexical, phonological aphasia (described in chapter 3), manifested a low FS-IQ and an even lower V-IQ, which were clearly a decrease from his premorbid level. Analysis of subtests indicated that this decrease was a result of his specific neuropsychological deficits. Right hemispheric visual-situational thought rather than symbolic thought prevailed in his premorbid neuropsychological profile.

Patient N, with lexical, morphological aphasia (described in chapter 4) showed no evidence of influence of his specific aphasic syndrome on IQ. His right hemispheric deficiency was not related to his specific aphasic deficits and was understood as a peculiarity of his premorbid neuropsychological profile.

Patient K, with syntactical aphasia I and II (described in chapter 5) had an IQ that clearly represented a decrease from his premorbid level. Qualitative analysis of performance on verbal subtests revealed his specific neuropsychological deficits. Right hemispheric thought, both situational and symbolic, was prominent in his premorbid neuropsychological profile.

6.3. Thought and Focal Brain Damage

The analysis of language and thought phylogenesis that we performed in the preceding chapters from the viewpoint of tertiary cortical field formation enabled us to propose a neurolinguistic classification of aphasia. We also presented a neuropsychological profile of patients with each form of aphasia within the frame of this classification. In this chapter, we will discuss the relationship between aphasia as a result of focal cortical damage and disorder of thought.

We considered thought as a process of operating with symbolic units: signs of the object in the left hemisphere; situation and object images in the right hemisphere. The units are connected with the function of the posterior brain cortical regions, while operating with them occurs in the anterior, frontal cortex. The level of thought development depends on both "richness" and variety of units, and the ability to operate with them. The history of tertiary cortical field formation and, correspondingly, formation of two cognitive mechanisms (left and right), predetermines the different relatively autonomous types of thought (see Table 8).

TABLE 8. Types of Thought Connected with the Left and Right Hemispheres

Left hemisphere		Right hemisphere	
Type of thought	Units of operation	Type of thought	Units of operation
Concrete-situational thought	Concrete signs of the object	Situational and symbolic-situational thought	Singular situation-space
Empirical thought	Functional signs of the object	Object-imaginative thought	Object image
Categorical thought	Categorical signs of the object	Symbolic-object thought	Individual symbol
Conceptual thought	Concepts (certain combination of catetgorical signs "contained" in sound form established in the given language		

We have already discussed the types of thought that are presented in Table 8 in the preceding chapters. It is important to emphasize that abstract (categorical and conceptual) thought and concrete thought have common origin connected with the phylogenetic formation of the left hemispheric cognitive mechanism; both rely on distinguishing the signs of the object and operating with them, which is in essence abstraction from the object itself. The difference between these two types of thought is in the *level* of abstraction, which, in turn, reflects the hierarchy of the functional levels provided by the cortical structures of different phylogenetic age.

For the types of thought connected with the right hemisphere, the terms *abstract* and *concrete* in their traditional psychological sense are not relevant. In the complicated development of symbolic systems, with the volumeness and polysemy of the individual symbol, there is no abstraction from the situation-object image itself; even the relations between images are thought of as images themselves. As we have mentioned previously, operation with right hemispheric units, or right hemispheric action, is inseparable from the units themselves. Because the right hemisphere image symbol is perpetually flowing, constantly changing, it never becomes concrete (definite, certain). Finally, each type of right hemispheric thought corresponds to the degree of integration of I-space and non-I-space, which is the cerebral basis of the self. Thus, each type of right hemispheric thought is inseparable from the stage in self-formation.

The types of left and right hemispheric thought, having developed at various stages of anthropogenesis, represent the different phylogenetic layers in the formation of higher cortical functions and coexist in modern man. Individual variability in the development of each type of thought in the norm might be connected with unevenness in the development of different cortical formations. Due to this intraindividual variability in cortical formation, individuals manifest prevalence of different types of thought and varying abilities even within one type of thought: for example, a richness in signs but difficulty operating with them, or, in contrast, good ability to operate with a limited number of signs.

The historical development of human language reflects all stages of phylogenetic thought development. The earlier stages of thought formation (concrete-situational, empirical, symbolic-situational) correspond to the *word-message*, in which the image of the whole visual-action situation and separate object image coexist. With the formation of categorical thought, the categorical component of word meaning emerges. In a sentence, historically different types of thought are represented at successive steps of sentence formation: unfolding of an event (operating with categorical signs), and unfolding of the situation (operating with functional, concrete signs and signs of spatial relations).

Goldstein (1948) postulated that patients with focal brain damage "dedifferentiate" from their premorbid abstract attitude to a concrete attitude. By attitude, Goldstein meant one's general view of or approach to the world, including one's behavior, thinking, and language. We understand the term *attitude* as a general personality trend and distinguish it from thought proper. There has been a connection in the literature between abstract attitude and an introverted personality trend (orientation toward the subjective, inner world), and concrete attitude and an extroverted personality trend (orientation toward the objective, outer world). Carl Jung (1971), one of the original proponents of the existence of these two key aspects of personality, felt that although each individual possesses the mechanisms that enable the development of both introversion and extroversion, the relative predominance of one of these innate capacities will determine the ultimate personality trend. It is of interest that these concepts of introversion and extroversion fit so beautifully the model of cerebral vertical organization, even though Jung himself did not speculate about the connection between the mind and the brain.

Considering our earlier discussion of self-formation by integration of I-space and non-I-space, we suggest that displacement in the direction of extroversion or introversion depends on the relative development of the thalamic functional level (Bernstein's level B) and the striatal-cortical level (Bernstein's level C). At the example of cerebral organization of movement, Bernstein demonstrated the fully introverted character of functional level B: movements of this level are based upon possession of the inner space. On the contrary, C-level movements are fully extroverted and are based upon possession of the outer space.

Jung determined the mechanism of extroversion as movement of interest toward the object, and the mechanism of introversion as movement of interest toward the subject. By the subjective factor, Jung understands "psychological action, which, combining with the influence of the object, results in a new mental fact." In other words, in the case of introversion, there is a steady tendency toward the further processing of the primary sensory material. From what is said, it is understandable that sometimes concepts of introversion and abstract attitude on the one hand, and extroversion and concrete attitude, on the other, are drawn together in the literature. Again, we distinguish between abstract and concrete attitude and realization of the general personality trend—thought proper. The latter depends on two cognitive mechanisms, left and right. Because the terms *abstract* and *concrete* do not fit right hemispheric thought, we feel it will be more justifiable to say that the introverted type is a premise for the prevailing usage (in mental activity) of the *symbolic* functional level. What Jung meant by the increasing role of the subjective factor, Bernstein actually postulated regarding cerebral organization of movement afferentation (which we think may be applied to the left hemispheric cognitive mechanism). It is more subjective reorganization of space at levels D and E, compared with the most objective C level. Regarding the right hemisphere, the increasing role of the subjective factor in the development of the right hemispheric cognitive mechanism is expressed by increasing degree of integration between I-space and non-I-space (penetration of subject into object). With the presence of an introverted trend, it is the relative development of the left or right cognitive mechanism that will determine the prevalence of the categorical thought or symbolic-object thought in the given individual. Analogously, with a prevailing extroverted trend, either concrete-situational/empirical or object-imaginative/symbolic-situational thought may prevail. Thus, besides personality attitude, independent variables that determine individual profile are the relative development of left or right type of cognition, and, more specifically, relative development of separate cortical formations (structures) whose function contributes to the different types of thought. If high level of one type of thought is present with the discordant personality attitude (for example, high ability for abstract thinking and extroverted attitude), one may suggest (at least theoretically) that personality attitude may be shifted toward the opposite direction. One may propose that external environmental (social) factors may also play a role in displacement of personality attitude. The prerequisite for such a shift is the presence of both tendencies (introversion–extroversion) in each individual, having its cerebral base in the existence of functional levels B (thalamic) and C (striatal-cortical). We think that attitude, dependent on the resultant personality trend closely connected with the emotional-motivational sphere, may serve as a trigger to stimulate the development of one or another type of thought. At the same time, the level of development of this prevailing type of thought will depend on other factors, namely, richness and number of structural units of thought and ability to operate with them.

Goldstein believed that brain damage results in replacement of categorical behavior by concrete behavior. He did not distinguish between the concepts behavioral attitude and thought proper, and explained the shift to the lower level by disorder of abstract attitude with intactness of the more primitive concrete attitude in patients with focal brain damage, and, in particular, with aphasia. However, as was shown in the present work, it is not general behavior attitude but the possibility of its realization—that is, thought proper—that is disordered in patients with left-sided focal cortical damage accompanied by aphasia.

Categorical and conceptual thought of the left hemispheric cognitive mechanism (see Table 8) are the types of thought that are primarily affected in aphasia. Moreover, for each type of aphasia, there are accompanying unique disorders of thinking that reflect disturbance of specific aspects of the predominant type of thought involved.

Focal damage to field 37 in the left temporal-occipital region (lexical, logico-grammatical aphasia) leads to depletion of categorical signs, which is manifested in both speech and thinking disorder. The latter reflects a partial disorder of categorical thought: impoverishment of structural units of thought while the ability to operate with them remains intact.

Focal damage to field 45 in the left prefrontal region (syntactical aphasia I) leads to disorder of operating with the categorical signs within the concept in the formation of the deep syntactical structure of a sentence, an aspect of conceptual thought. This, too, has ramifications for both speech and thinking. In the partial deficit of conceptual thought (verbal-logical thought) that results, concepts themselves are intact but the patient cannot operate with them, a disorder of reasoning. In a pure syndrome of syntactical aphasia I, there is a dissociation between intact categorical and disordered conceptual thought. However, such cases in clinical practice are relatively rare, because categorical and conceptual thought, as well as sentence formation, are subserved by proximate areas of the left prefrontal region. In extensive damage to the left prefrontal region, a combined syndrome is observed that includes all the preceding components. In these cases, a global disorder of thinking results, in which the entire hierarchy of general and specific categorical signs of objects is impaired. This is comparable to Goldstein's notion of dedifferentiation to a lower level of functioning.

Focal damage to fields 39 and 40 in the left parietal-occipital region (lexical, morphological aphasia) leads to depletion of signs of spatial, outer relations. In speech, this will be manifested by disorder in comprehension of nonroot, grammatical morphemes. There is no significant thinking disorder in this form of aphasia, except for secondary diffuseness of the categorical signs in lexical meanings due to disorder of their outer markers.

Focal damage to the left frontal region, possibly posterior parts of field 45 (syntactical aphasia II), is connected with disorder of operating with signs of spatial, outer relations and operating with functional, concrete signs in the formation of surface syntactical structure of a sentence, aspects of concrete-situational and empirical thought. Here we have a speech disorder accompanied by disorder of concrete-situational and empirical thought, in contrast to the preceding, which involved categorical and conceptual thinking. Although we have asserted that aphasia involves disorder of the symbolic function level, in this case we have impairment in concrete-situational and empirical thought, primarily the gnostic-praxic level. As already mentioned, Bernstein indicated that there are possibly several transitional levels (sublevels) between levels D and E, a view supported by cytoarchitectural data. We consider lexical morphological aphasia and syntactical aphasia II results of disorder of levels transitional between gnostic-praxic and symbolic.

Focal damage to field 21 in the left temporal region (lexical phonological aphasia) leads to depletion of specific categorical signs within the concept, a result of the breakdown of language sound code. This is manifested in speech in disorder of word comprehension and in a specific and partial disorder of conceptual thought: a depletion of the structural units for thinking (concepts) with an intact ability to operate with them. A concept is a thought unit that includes in itself formal (sound) and semantic (categorical) sides. Depletion of concepts in phonological aphasia is due to selective disorder of their formal side. Therefore, disorder of conceptual thought in phonological aphasia may be considered as secondary to disorder of language.

7

Perspectives for Psychiatry

In the preceding parts of this book, we outlined our theory of brain organization and its utility as a three-dimensional framework for approaching disorders of brain function. In this section, we will focus on the application of this conceptual framework to understanding brain mechanisms in psychiatric disorders, and the particular usefulness of this approach for consolidating what is already known and for formulating directions for further investigation. We will examine in greatest detail schizophrenia, a mental disorder that has continued to elude understanding despite great interest and efforts over the last century.

Today, it would be difficult to deny that schizophrenia is a brain disorder. Two decades of brain imaging data have opened a new era of neuropsychiatry and schizophrenia research in particular. Several areas of the brain have been found to be abnormal in schizophrenia. The most consistent and frequently replicated morphological finding on imaging is enlargement of the ventricles, the fluid-filled cavities within the brain. This enlargement is considered to be secondary to brain atrophy or some other processes that results in reduced generalized or focal tissue density. Indeed, periventricular structures such as basal ganglia, hippocampus, medial temporal lobe, and medial thalamus have been found to be abnormal in patients with schizophrenia. The dorsolateral prefrontal cortex has been found to be abnormal, and involvement of the posterior cortex (superior temporal and inferior parietal regions) has also been found. A good deal is known about the function of the areas that seem to be involved in schizophrenia, yet damage to these areas resulting in clear symptom complexes in other verifiable neurological disorders does not produce a clinical picture consistent with schizophrenia (Heinrichs, 1993). In this regard, we quote Heinrichs: "If we try to construct a behavioral profile by linking damage to the frontal lobes, hippocampus, and basal ganglia the following emerges: a mentally rigid individual with a movement disorder and impaired memory. This is schizophrenia?" (Heinrichs, 1993, p. 228). It might be expected that if these areas are involved in schizophrenia, the clinical picture would reflect what is known about the function of these regions. As this is not the case, the meaning of these morphological findings for the pathogenesis of schizophrenia remains to be determined.

We should recall how this knowledge of brain region function was obtained: the main strategy has been lesion analysis, which identifies "brain regions involved in specific cognitive processes by seeing what is lost when damage occurs" (Andreasen, 1996, p. 698). It appears as if lesion analysis does not work in the case of schizophrenia.

Traditional neuropsychological tests were used in attempts to understand brain mechanisms of schizophrenia. Many cognitive deficits were found. Again, the problem of how to interpret these data in regard to schizophrenia remains unresolved. Neuropsychological tests are designed to assess cognitive dysfunctions in regard to their localization in the brain. For example, the popular clock test is designed to assess visual spatial deficit; two syndromes, right parietal and left parietal, are known. Right parietal syndrome is due to fragmentariness of visual-spatial perception and left sided neglect, while in the left parietal syndrome, linear directions in space are disordered. These specific disorders will be manifested in patients' drawings. Figure 27 shows a clock drawing by a schizophrenic patient. We see the severe distortions in the drawing; however, they do not appear to relate to either right or left parietal syndromes. Again, we recall that neuropsychological tests were developed based on then-current knowledge regarding localization of cortical functions. As we said earlier, this knowledge was obtained on the basis of lesion analysis, and lesion analysis does not work in schizophrenia.

Why is it that our current knowledge about localization in the brain does not work in understanding schizophrenia? There may be two reasons. The first is that our knowledge about localization is incomplete. The second is that brain organization in patients with schizophrenia is different.

The new era in cytoarchitectural studies is characterized by great development of myeloarchitectonics. Scientists are increasingly aware that not only a brain region itself but its connectivity pattern is crucial for understanding brain functions. The same region may be involved in several networks and appear with its different facets in different networks.

A growing body of data indicates that schizophrenia is a neurodevelopmental disorder (Selemon, Rajkowska, & Goldman-Rakic, 1995). The absence of gliosis indicates that schizophrenia is not a neurodegenerative disorder (Andreasen, 1996). The neuropathology of schizophrenia shows a variety of subtle abnormalities in the cytoarchitecture of several brain regions. Indeed, authors think this may be due to disorder in programmed cell death, neuronal migration, neuronal differentiation, and neuronal connectivity during the neurodevelopmental process

FIGURE 27. Clock drawing by a schizophrenic patient. From "The Process Approach to Neuropsychological Assessment," by E. Kaplan, 1990, *Journal of Neuropsychiatry and Clinical Neurosciences, 2*. Reprinted by permission.

(Pakkenberg, 1990; Selemon et al., 1995). All current theorists emphasize the role of abnormal connectivity or disregulation of connections in the pathogenesis of schizophrenia. Further, a disturbance in the normal projections from the thalamus to cortex could result in a variety of cortical changes (Rakic, cited by Lyon, Barr, Cannon, Mednick, & Shore, 1989); overall, what could emerge is a new pathological organization in the brain.

Neuropsychological tests assess higher cortical functions, basically left hemispheric, such as disorders of memory, language, selective attention, and so on. Dysfunctions are observed and objectively measured. Even though many of these functions are disordered in schizophrenia, how do they relate to schizophrenia itself as a psychiatric disorder? Schizophrenia, like all psychiatric disorders, is a disorder of subjective experience; "[T]his world is unaccessible to direct observation and can only be grasped through the patient's report" (Bovet & Parnas, 1993, p. 181).

Schizophrenia is characterized by a multitude of symptoms (delusions, hallucinations, disorganized speech, bizarre behavior) and from the beginning, it has been known to be a heterogeneous disease, one "that, paradoxically resists subdivision" (Heinrichs, 1993, p. 221). Its various symptoms have been the basis for categorization of subtypes (e.g., paranoid, undifferentiated, catatonic), but relatives of schizophrenic probands of one type may have any type of schizophrenia. In addition, one individual may have more than one subtype over the course of a lifetime. All other attempts to divide schizophrenia into separate entities have ultimately failed.

Yet schizophrenia has been recognized for centuries and across cultures; it seems likely that there is some core entity which runs through all these variations. We believe the entity that is the core disturbance and is the source for all peripheral signs and symptoms is *disorder of the self.*

We will analyze subjective experience in patients with schizophrenia and other psychiatric disorders for correlation with brain function. Brain function will be considered in the basic framework of brain organization (hierarchical, intrahemispheric, left–right). Within this conceptual framework, a pattern of schizophrenic organization may be formulated.

7.1. Cerebral Organization of the Self and Schizophrenia

We have constructed a theoretical model of the cerebral organization of the self in the norm by integrating findings from several disorders that we view as selective disorders of the self, ranging from body schema disorders to depersonalization to delusional misidentification syndromes. These selective disorders of self were "mapped" within our three-dimensional conceptual framework of brain structure–function differentiation: hierarchical organization, intrahemispheric specialization, and interhemispheric specialization.

We will present this process, first reviewing briefly the three-dimensional framework of brain organization and then expanding on particular aspects relevant to the self. The dimensional concept of brain organization is summarized in Table 9, which represents our expansion of the work of Nicolai Bernstein on cerebral organization of movement (see chapter 1, and Table 1).

The hierarchical dimension is composed of five function levels of different phylogenetic age. The first level, which we have labeled hypothalamic-midbrain, and which roughly corresponds to Bernstein's level A, is concerned with the most basic regulation of rhythmic, homeostatic processes of the organism and is subserved by proprioceptive, vestibular, and visceral sensory modalities. The second level, the thalamic, corresponding to level B of Bernstein, provides definition of internal space derived from spatial coordinates of one's body and is subserved mostly by proprioceptive and tactile modalities. Levels 3, 4, and 5 are all cortical levels. Level 3, the striatocortical, corresponding to level C of Bernstein, is concerned with the external spatial field and subserved mostly by visual, auditory, and vestibular modalities. In the cortical hierarchy, level 3 is the lowest, the sensory-motor level. The next in the cortical hierarchy is level 4, the gnostic-praxic level, subserved by visual and auditory modalities and corresponding to Bern-

TABLE 9. The Three-Dimensional Model of the Brain

Dimension	Description	
Hierarchical	Five levels of different phylogenetic age:	
	Level	Major modality
	1. Hypothalamic-midbrain (Basic regulation of rhythmic homeostatic processes of the organism)	Proprioceptive/interoceptive Visceral
	2. Thalamic (Internal space)	Proprioceptive
	3. Striatocortical (External spatial field)	Visual (auditory, vestibular, proprioceptive, etc.)
	4. Cortical (Gnostic-praxic)	Visual, auditory
	5. Cortical (Symbolic)	Supramodal
Intrahemispheric	Posterior–anterior (see Tables 10 and 11)	
Interhemispheric	Right–left	

stein's level D. Level 5 is the highest cortical level, the symbolic, and operates with symbols, a supramodal function. Our expansion and extension of this level, which had been mentioned but not described in detail by Bernstein, is the subject of the present work.

The dimension of intrahemispheric specialization refers primarily to the general difference in the mode of operation between the anterior and posterior brain within each hierarchical level. Posterior areas are concerned with spatial organization of sensory information (simultaneous synthesis) and the corresponding anterior areas are concerned with temporal organization of the processed information (temporal synthesis).

The third dimension, interhemispheric, refers to the differences in mode of information processing between the hemispheres, the left in our conceptualization being analytic, and the right, holistic. In our present discussion of the self, we will focus primarily on the role of the right hemisphere.

In general, as we have said, the right hemisphere operates with discrete combinations of whole continuous images, whereas the left hemisphere functions with continual combinations of discrete signs. The whole continuous images, or units for operation, are provided by the posterior brain (spatial synthesis) whereas operations themselves take place in the anterior.

Figure 28 represents the regions, or nodes, and pathways involved in our model of the self. These known pathways may subserve many functions, but we have identified these regions and the interconnections between them as the core network (pattern) that constitutes the cerebral organization of the self. A detailed description of the contributions to the self of the involved regions and pathways follows.

The contribution of the right posterior brain to the self can be considered on several levels, presented in Table 10, which have identifiable roles yet combine intimately in forming the self. At the hypothalamic-midbrain level, there is no I-space or non-I-space. This level of "background of backgrounds" (Bernstein, 1947) contributes indirectly to the self, through forming and maintaining the subjective experience of body weight, sense of heaviness and fullness, and feeling of being alive, provided by propriovestibular and visceral modalities. Disorders of body mass in either direction (feeling of heaviness or of weightlessness or of hollowness of the whole body or of its parts), often described in the literature in patients with depression, we think may be attributed to disorder of hypothalamic-midbrain level. A patient with depression complained that her head felt "as if it were made of lead. It feels so heavy that the muscles of my neck are unable to sustain the weight of my head. Even if I rest my head on my hands, my head seems to slip through them, as they cannot take its leaden heaviness" (Lukianowicz, 1967, p. 39). Another depressed patient, quoted earlier, complained of the opposite: "I have a most terrible feeling of a large open cave, of a sheer *emptiness*, an excruciating feeling of the cold, dark, hopeless *nothingness* in my chest."

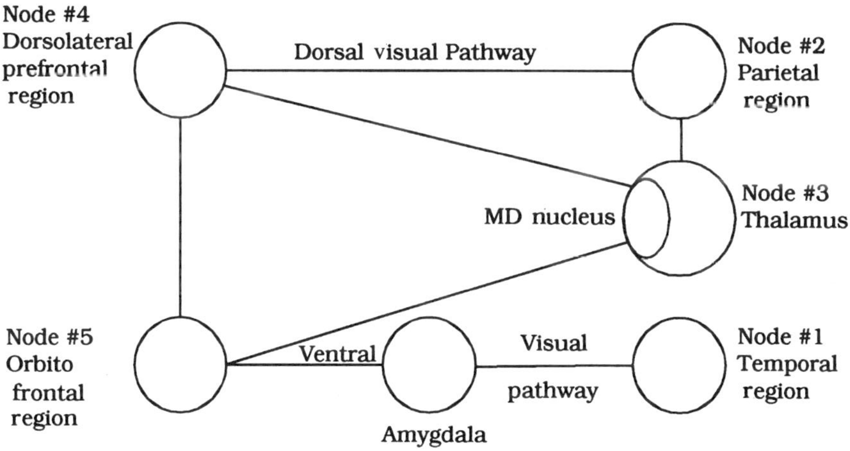

FIGURE 28. Model of cerebral organization of the self.

At the thalamic level (see Node 3, Figure 28), the constant inflow of proprioceptive sensation is integrated according to one's spatial coordinate system, forming the continuous image of one's own body space. This includes the subjective experience of one's boundaries, both defined and filled by somatic sensations, tactile and kinesthetic. This is I-space, the whole global entity of one's own body space as such (form) and its sensational content. We hypothesize that the pathological body sensations known to occur in schizophrenia may represent a disorder of this system. An illustration is presented by one of Angyal's patients describing his

TABLE 10. Contribution of the Right Posterior Brain

Level	Modality	Contribution to self
Hypothalamic-midbrain	Proprioceptive vestibular visceral	Body as a weight category in the gravitational field (experience of one's body heaviness and fullness)
Thalamic	Proprioceptive	I-space (subjective experience of one's body continuity and boundaries)
MD nucleus	Integrative	Subjective attitude to bodily feeling, "thalamic emotion"
Cortical		
Inferior-temporal and amygdala	Visual	Visual scene-situation (non-I-space) infused with limbic emotion
Inferior-parietal	Visual-spatial	Integration of non-I-space and I-space

experience: "My body feels too short for me; my whole body feels like a woolen suit which was left in the rain and became wet, then dried out and became short. When you move it stretches and binds" (Angyal, 1936, p. 1036). Here we see a disorder of this whole entity.

The thalamus is composed of many nuclei with different functions in the processing and integration of sensory information. The mediodorsal nucleus (MD) is the most intriguing to us: it evolved in conjunction with the frontal lobes and, indeed, the frontal lobes can be considered as a projection of the MD (Fuster, 1985). Although little is known with certainty about the role of the MD, data from multiple sources (electrophysiology, neuroanatomy, pathology) indicate that the MD receives from other nuclei multisensory information that has already been integrated—that is, is no longer separated by modality—and that results in a new whole somatic sense that is different from and more than the sum of its parts. Direct electrical stimulation of the MD may cause unusual and incomprehensible, virtually indescribable sensations accompanied by some emotional tone (Smirnov, 1976). Although we normally have no conscious awareness of this subjective somatic sense, we may find distortions of this bodily sensation in a pure form in otherwise healthy individuals, as well as included in the complex psychopathological pictures presented in patients with schizophrenia and psychotic depression. These experiences are again bizarre, almost impossible to articulate, and best captured only by metaphor: "A lot of snakes inside my back biting me all the time"; "I feel mentally sick; my body is filled with poison" (our observation). Here, I-space has been transformed into a new subjective sense of bodily feeling. The emotional tone, *thalamic emotion*, is perhaps understood as a subjective attitude to bodily feeling. This emotion is not separable from complex bodily sensations and thus is located in body parts. Although the otherwise healthy individuals and schizophrenics will both give these strange metaphors—which we feel reflect attempts to describe abnormal bodily sensations that there is no language for—the otherwise healthy individual, presumably with a more circumscribed deficit, understands that this is a feeling, though extremely strange and unspeakable. In the schizophrenic, this feeling will be incorporated into the whole psychopathological picture, emerging as a delusion, a false belief or explanation for the bizarre feeling.

At the cortical levels, the whole continuous images with which the right hemisphere operates are whole visual situations, singular unique "snapshots" of objects within a scene that are each separate and unchangeable. The separate visual situations are images of the external world, which we call non-I-space in contrast to the preceding I-space. Non-I-space is formed by the visual modality whereas I-space is formed by complex somatic sensations. Within the dimension of intrahemispheric specialization, there are two visual systems of the posterior cortex, both originating in the visual occipital cortex and then diverging, the ventral pathway to the inferior temporal areas and concerned with object recognition, and the dorsal pathway to inferior parietal areas and concerned with spatial relation of objects. The further fate of these pathways is determined by their connections, the ventral connected with the amygdala and then orbitofrontal cortex, and the dorsal with the dorsolateral frontal lobe (Mishkin et al., 1983). The parietal area also has interconnections with integrative nuclei of the thalamus, although not the MD, forming the thalamoparietal system (see Figure 28). The parietal cortex in general is concerned with spatial function, providing a supramodal framework for both I-space and non-I-space. Its emphasis in I-space is more purely on spatial coordinates (form split from sensational content): size, shape, position, spatial relationship of body parts—body schema and the integration of I-space with non-I-space. Damage to the right parietal area will give a spectrum of disorders that we view as a continuum, from partial body schema disorders to whole body schema disorders to depersonalization. Spatial disintegration at these different levels will be accompanied by the subjective experience of estrangement, ranging from detachment and estrangement of body parts, whose spatial image is distorted, to estrangement of one's body, to estrangement of one's self. For example, the spectrum of partial body schema disorders includes the following, taken from different patients with right parietal lesions observed by and cited in Cutting (1990): "My left arm is bigger" (a disturbance in size); "I feel as if my fingers are detached, decomposing, disintegrating" (detachment); "I felt as if instead of my left leg there was something that did not belong to me, a piece of meat, as if I had no leg" (estrangement) (pp. 192, 194). The closer we are to general body schema disorders, the more difficult it is to draw a boundary line between traditionally neurological (body schema disorders) and psychiatric (disorders in the sphere of one's self, depersonalization) symptoms. Examples of disturbances in whole body schema disorder are described by patients who feel as if their body is "dead" and doubt their own existence; others have felt themselves to be "a casing," "a cover," as if their "I" had been separated and located outside their body, which is depersonalization characterized as estrangement of one's own body space from one's self. There are clinical examples that seem to illustrate disintegration of I-space and non-I-space. Patients described by Dobrochotova and Bragina (1977) reported two real spaces having arisen, one in which the whole world remains and the other a space containing themselves. Patients use the word "space" trying to express the complex subjective experience of their existence relative to the surrounding world. This is another type of depersonalization that is a result of estrangement of one's own self from the surrounding world. This last example illustrates the overall contribution of the parietal region to the self: the integration of I-space and non-I-space.

The ventral inferior temporal pathway, concerned with object recognition, is connected with the amygdala, a central structure of the limbic system, the brain's emotional network. Emotion connected with the amygdala we refer to as *limbic emotion*, in contrast to *thalamic emotion*. This limbic emotion evaluates external objects or stimuli as potential sources for satisfying the organism's needs. Thus, the association of visual object with emotion might be considered in a broader context of an emotional-motivational system. In the left hemisphere, this system is linked with goal-directed behavior. In the right hemisphere, this system is concerned with the subjective experience of the object based on this hemisphere's holistic mode of information processing, in which the object, included in the visual situation, becomes permanently infused with the emotion experienced at the moment when the situation was perceived. This may be understood as projection of emotion upon the situation—emotion projected upon non-I-space—which we view as the contribution of the right inferior temporal region to the self. The subjective experience of a schizophrenic patient gives us an example of exposed projection of emotion into a situation: "When I am sitting next to someone in the ward I feel as if all my feelings seem to drain out of me into them" (example taken from Cutting, 1990, p. 281).

Table 11 presents the contributions of the right anterior brain to the self. We will focus here on the two cortical levels. The general function of the right prefrontal cortex is temporal synthesis, operating with visual situations or non I-spaces, units presented by the posterior brain. We will consider operations with these units at the two cortical levels, which we assume operate in parallel. At the first level, the situations receive a temporal mark, a fixed marker reflecting the moment of real time when the situation was imprinted. The situations represent momentary segments of real time, achronous and static as still shots in a continuously moving film strip. They are internalized as an ordered succession of moments of experience of the external world—temporal synthesis at this level. This, we think, is the contribution of the situational cortical level to the self. A disorder of temporal marks is illustrated by the syndrome of reduplicative paramnesia, in which a place or person is duplicated and linked to past experience (see chapter 5). Recent neuroimaging findings indicate that this syndrome is most commonly associated with damage to the right frontal lobe. Deja vu might be considered another time-duplicative experience, in which a present situation receives a transient, second temporal mark related to the past.

The internalization of moments representative of real time should be under-

TABLE 11. Contribution of the Right Anterior Brain

Level	Temporal synthesis	Contribution to self
Hypothalamic-midbrain	Simple rhythm of muscle tone; circadian rhythm	Immediate experience of rhythm and duration (interval clock)
Thalamic	Propriomotor rhythm; complicated pattern	Highly individually specific rhythmical pattern defined by singularity of one's body space
Cortical	Temporal sequence	Continuity of one's time line
Cortical symbolic	Semantic sequence	Individual symbol

stood as incorporating temporal synthesis of lower, subcortical levels of the anterior brain: rhythm of the midbrain-hypothalamus level, which gives immediate experience of duration, and complicated and highly individually specific rhythmical pattern of the thalamic level defined by the singularity of one's own body space. The result of this internalization gives continuity of one's own time line (Table 11). Disturbance of time experience in schizophrenia has been noted in both classical and recent literature; Cutting (1990) indicates that disordered temporal awareness was experienced by half of schizophrenics examined by him. Jaspers (1963) described discontinuity of one's own time line and discontinuity of past experiences in patients with schizophrenia. More recently, a disproportionately defective memory of temporal context, in which patients remembered that an event had occurred, but not when, has been reported (Rizzo, Danion, Linden, & Grangé, 1996). Another disorder of time in schizophrenia described in the literature we interpreted (see Chapter 5) as a splitting of time experience per se from its situational context: "Life is now a running conveyor belt with nothing on it" (example taken from Cutting, 1990, p. 268).

The internalized sequence of environmental events makes one's own past experience. Environmental events are non-I-spaces (visual situations) processed differently by the two visual pathways, dorsal (inferior parietal) and ventral (inferior temporal), because of their different functions and their different connections. The dorsal pathway (parietal-thalamic system) integrates non-I-space with I-space. The ventral pathway (temporal-amygdala system) infuses the visual situation with limbic emotion. Thus, non-I-spaces here are not simply non-I-spaces anymore, but subjective experiences, subjectively felt, the self at this level. Non-I-space from the posterior brain is presented to the frontal cortex as moments of environmental time, moments of one's past experience. I-space is presented to the frontal lobe not only indirectly, through the parietal cortex, but by an independent pathway through the MD nucleus of the thalamus, which is part of the thalamofrontal system, not the thalamoparietal system. We assume that it is thalamic emotion which is presented to the frontal lobe: the attitude, or emotional tone, to bodily sensations.

The experience of the patient with schizophrenia offers an eloquent illustration of disorder of this level of the self: "I am now living in eternity. Outside everything carries on, leaves move, others go through the ward, but for me time does not pass.... When they run around in the garden and the leaves fly about in the wind I wish I could run too, so that time might again be on the move, but then I stay stuck." We presented this quote when discussing level C time as an illustration of splitting of the visual situation from its temporal context; at this point, we emphasize that the visual situation and temporal context are also split from subjective sense.

The second cortical level concerned with the self is a symbolic system (see Table 11). Visual situations here are united into a symbolic system through their affect, becoming suffused with meaning and becoming multiple aspects of a continuous and indivisible whole—all identical in their meaning yet unchanged from the instant they were formed. Within the symbolic system object images are identified and associated according to resemblance of their holistic forms, irrespective of content, a characteristic of right hemispheric cognitive mechanisms. In this stream of images, the object image both remains the same and undergoes multiple transformations through constant interchange with similarly appearing images, giving multiple conditions, multiple moments of the same image. Thus, a polysemantic symbol is formed, with an image acquiring multiple meanings as it flows through situations, accumulating and condensing multiple contents. But the right hemisphere cognitive mechanism is limited: objects within the symbolic system having different content are equipollent facets of an indissolvable whole. It is only through interaction with the left hemisphere that we can "dissect" the symbol's component parts and know that an object may symbolize another, but is not that other. Disorders of the symbolic layer of the self are well illustrated in delusional misidentification syndromes, phenomena which include intact recognition and faulty identification. Regions and pathways that have been shown to be associated with these syndromes (right hemisphere temporal, parietal and frontal pathways) are also structures that we have included in our model of the self in the norm. None of the theories regarding pathogenesis of DMS, other than the psychodynamic, address their symbolic nature.

Tables 12, 13, and 14 show the correspondence between right hemispheric cognitive mechanisms at the symbolic function level and the clinical manifestations of the various DMS.

Let us review our construction of the self in the right hemisphere as we have described it. The self at the first level is the temporal sequence of units; the units are non-I-spaces infused with affect and integrated with I-space. At the second level, these same units are organized not by their temporal marks but by affect, resulting in a symbolic system of meaning, a semantic sequence of units, in contrast to a real time sequence. Both cortical layers integrate the subjective sense of bodily feeling from the thalamus: at the second layer this subjective sense is intimately incorporated with the symbolic system, imbuing one's worldview. We can see that as the self develops, it becomes determined less by external events and real time and is increasingly defined subjectively, with an inherently developing system of values and view of the world. An example of disturbed bodily sensations as a trigger for disorder of the symbolic system was given by a patient with schizophrenia: "I have a bug in my back biting me all the time and giving me pain and a lot of snakes inside biting me all the time and giving me pain. I'm eating human bodies, snakes, and bugs on trays alive and they're giving stiffness and pain." When asked to draw how he feels, the patient produced the following pictures and writing (Figure 29). It seems to us that bodily sensation is transformed into a visual metaphor of the snake/bug biting inside, and immediately incorporated into symbolic systems. We see here a flow of images identified and transformed, his feelings projected onto

TABLE 12. Clinical Manifestations of Delusional Misidentification Syndromes and Right Hemisphere Cognitive Mechanism

Delusional misidentification syndromes (DMS)	Operational rules of right hemisphere cognitive mechanism at the symbolic level
Capgras syndrome:	
Patient does not identify a familiar person but does recognize his/her appearance and behavior. Patient believes person has been replaced by a double, an imposter.	The object included into situations belonging to different symbolic systems will be perceived correctly regarding its physical features but not identified as the same (have a different meaning).
Fregoli syndome:	
The same person (usually a persecutor) is simultaneously identified in several persons.	Different objects which belong to non-I-spaces united into a symbolic system are recognized as physically different but will be identified with one another (allotted the same meaning).
Syndrome of intermetamorphosis:	
Transformation between familiar and unfamiliar people. Persons in the environment change with one another: A becomes B, B becomes C, C becomes A and so on.	Temporal synthesis at the symbolic level (operations with images are inseparable from images themselves). Image is "flowing" through situations, multiplying, constantly changing and remaining the same (different moments of same image).
Multiple doubles:	
Multiple doubles may present in any variant of DMS.	

the outside world: snakes–dogs–cats–human bodies, becoming the same as a pig; meat and blood in the market; eating a pig is identical to eating human bodies alive. Finally, his worldview: "On top of it, there (sic) eating them alive, and there (sic) up higher than pain. The world stinks with that."

To summarize this section, we conclude that there is no such thing as localization of the self in the brain, but there is a cerebral organization of the self, the interconnection of separate brain areas in a specific pattern of connectivity. We believe that schizophrenia is a consequence of a specific pattern of disruption within this system and its components.

7.2. Bernstein's Concept of Function Levels in the Brain and Psychopathology

7.2.1. Bernstein's Levels and Two Types of Emotions

The introverted B level (I-space) and extroverted C level (non-I-space) are relatively independent and work in parallel. They are both background levels that

TABLE 13. Delusional Misidentification Syndrome Variants and Right Hemisphere Self

Delusional misidentification syndromes (DMS)	Non-I-space in symbolic system	Contribution of right hemisphere to self (limitations of right hemisphere cognitive mechanism)
Syndrome of animate and inanimate doubles	Multiple non-I-space	Multiple "selves"; (no individual "I")
Syndrome of subjective doubles (duplication of self)	Integration with I-space	Self is not divisible from others (subject/object division)
Unusual variant of Capgras syndrome ("impostor city"):		
Patient states that there were eight impostor cities and he spent the last eight years wandering between them without finding the real one. Eight duplicates of his wife and children, each duplicate living in a separate duplicate city with a double of the patient (from Thompson, Silk, & Hover, 1980).	Space itself and visual-situational content are not separable.	Self is not divisible from outside world.

are integrated at the symbolic level. Within an individual, the relative contribution of each will determine personality trend: predominance of level B will result in a more introverted personality, whereas predominance of level C will lead to greater extroversion. In general, the relative development of levels B and C will give a continuum of personality types from theoretically pure introversion to pure extroversion.

TABLE 14. Capgras Syndrome and Right Hemisphere Cognitive Mechanism at the Symbolic Function Level

Capgras syndrome: Relationship between "original" and "impostor"	Right hemisphere individual symbol: Relationship between form and content
Although original in some way disappeared, everything is known about him. Impostor (replacement) is important symbolically but is anonymous. All actions are with replaced image as though it is not replaced.	A symbol and the visual image through which it is expressed form a single integrated representation. However, the visual image (form) preserves its own value. There are no monosemantic relations between form and content.

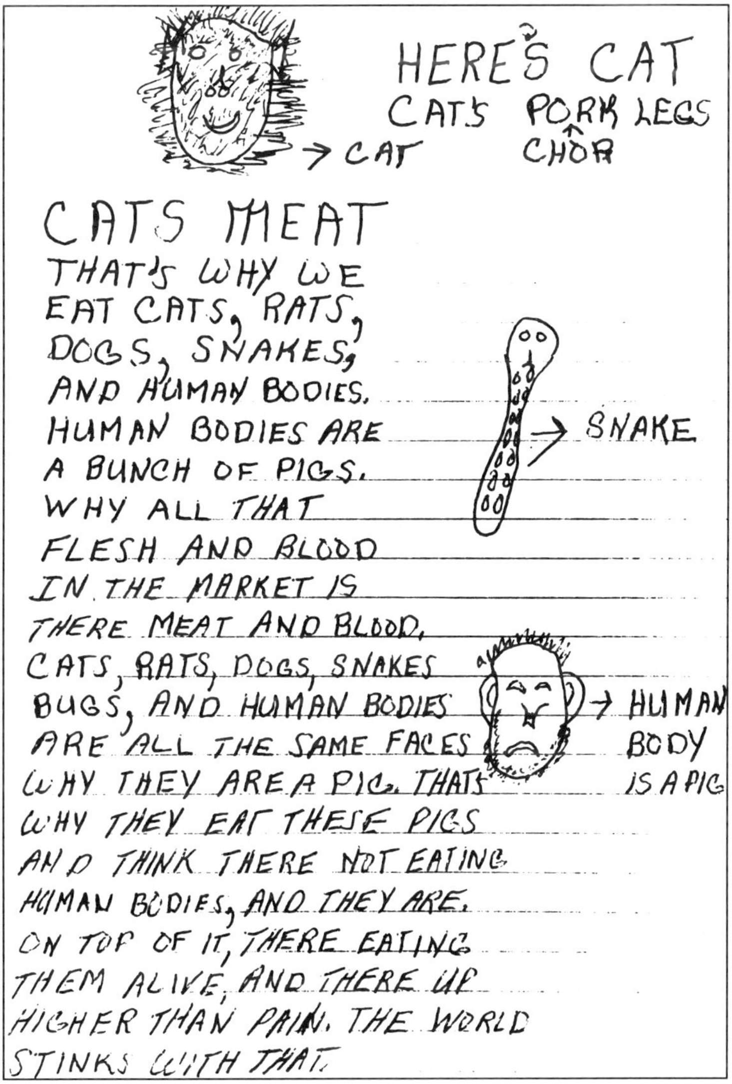

FIGURE 29. Drawing and writing by a schizophrenic patient.

One of the significant factors that determines personality type is the emotional sphere. In previous chapters, we introduced the idea of two types of emotions: thalamic, which we think originates in the MD nucleus of the thalamus, and limbic, originating in the amygdala. It is their integration in the frontal lobe that results in specifically human emotion. Thalamic emotion belongs to level B and limbic to level C. Table 15 summarizes the basic features of the two types of emotion.

Kretschmer (1936) proposed that human emotions are differentiated along two dimensions—mood and sensitivity (see Table 16). The mood dimension is represented on a continuum from joy to sadness, present in the individual in "diethetic proportion." The sensitivity dimension is represented in a continuum from sensitive to dull, in "psychesthetic proportion." These scales are relatively independent and function in parallel. Based on the relative contribution of one or another scale, Kretschmer distinguished two basic types of personality: cyclothymic and schizothymic. According to Kretschmer, extremes of either continuum represent psychopathology. Within the first dimension (mood scale), the continuum extends from cyclothymia (variant of norm) to cycloid personality disorder to manic-depression. Within the second dimension (sensitivity scale), the continuum includes schizothymia (variant of norm), schizoid personality disorder, and

TABLE 15. Function Levels in the Brain and Two Types of Emotions

Limbic emotion	Thalamic emotion
Extroverted	Introverted
Directed toward the object	Sensational feeling of one's own body
Belongs to function level C	Belongs to function level B
Not spatial	Spatial (localized in body parts)
Object-related	Not object-related
Motivated, conscious	Not conscious (becomes conscious only in pathology, senesthopathy)
Included into goal-directed behavior	
Plays the basic role in the formation of left hemispheric "I"	Plays the basic role in the formation of right hemispheric "I"
Directly connected with but distinguishable from psychomotor, goal-directed activity (behavior)	Embodied into symbolic thinking (in pathology, autistic thinking)
Dependent on external stimuli	Not dependent on external stimuli
	Dependent on phase of one's own B-level propriomotor rhythm
Attitude to the object of one's need satisfaction	Attitude to one's own bodily sensations
Expressive	Impressive
Brain substratum originated from the motor, effector formation (anterior brain)	Brain substratum is sensory formation (posterior brain)
Related to successive synthesis (time, movement)	Related to simultaneous synthesis (space)

TABLE 16. Mood and Sensitivity Scale according to Kretschmer

	Cyclothymia	Schizothymia
Psychesthesia and mood	Diethetic proportion: between elevated and depressed	Psychesthetic proportion: between hypersensitive and anesthetic (cold)
Psychic tempo	Wavy temperamental curve: between mobile and comfortable	Jerky temperamental curve: between unstable and tenacious, alternative mode of thought and feeling
Psychomotility	Adequate to stimulus, rounded, natural, smooth	Often inadequate to stimulus, restrained, inhibited, stiff

Note: From *Physique and Character* (p. 265) by E. Kretschmer, 1936, New York: Harcourt Brace. Adapted by permission.

schizophrenia. It is of interest that, in determining personality features, Kretschmer included not only emotional but also psychomotor patterns and psychic tempo. According to Kretschmer, the mood scale will fluctuate in full, regular, rounded waves between joyful and sad in accordance with the endogenous situation and the environment. The psychomotor pattern of cyclothymia varies from fast to slow. For example, when it is slow, it is a simple, even slowness of psychic tempo with a tendency toward depressed mood. The sensitivity scale (psychesthetic proportion) is characterized by abrupt, irregular variations, with all possible shading of psychic sensitivity alternating from mimosa-like timid fineness of feeling to a continual state of passionate excitation to psychic insensitivity, dullness, lack of spontaneity; in severest psychiatric cases it is "affective imbecility" (Kretschmer, 1936). Psychesthetic proportion, according to Kretschmer, fluctuates between excitability and dullness, oversensitivity and coldness at the same time, and that in quite different proportions; a continuous ladder from the mimosa-like extreme to the insensitive and cold extreme in the same person. The psychic tempo of the sensitivity dimension is characterized by abruptness. It will range between jerky and tenacious, with inflexibility broken by jerkiness, with a jagged curve, the extreme psychomotor pattern being catatonia.

Within the psychomotor pattern, Kretschmer included manner of relatedness to environmental stimuli. The cycloid personality, according to Kretschmer, is characterized by "giving up of himself to the external world, a capacity for living, feeling and suffering with his surroundings" (Kretschmer, 1936, p. 133). For the schizoid personality, it will be a lack of direct connection between the emotional stimulus and its motor reaction.

Clearly, we see an analogy between cycloid diethetic proportion of mood and its psychomotor pattern and limbic emotion, the rhythm of Bernstein's level A and level C movements directed at objects. On the other hand, schizothymic psychesthetic proportion is analogous to what we call thalamic emotion, and its psychomotor pattern is analogous to the propriomotor rhythm of Bernstein's B level, which is not influenced from outside, is individually specific and idiosyncratic, and at the same time stereotypical.

7.2.2. *Bernstein's Levels and Two Types of Psychopathology*

The two scales of emotion proposed by Kretschmer, mood and sensitivity, are consistent with our view of the dichotomy of brain organization of left hemispheric and right hemispheric self.

Here we will attempt to show that two major psychiatric disorders, schizophrenia and bipolar disorder (manic-depression)—which in Kretschmer's terms represent extremes in the sensitivity and mood scales—result from dysfunction of two distinct brain networks. Depression reflects disorder of the left hemispheric self and, connected with it, object-related limbic emotion. Schizophrenia reflects disorder of the right hemispheric self and, connected with it, subject-related thalamic emotion.

7.2.2.1. Depression. Recent neuroimaging studies have pointed to the involvement of the left prefrontal cortex and the limbic region (amygdala) in patients with manic-depression and unipolar depression (Drevets and colleagues, 1997; Mayberg and colleagues, 1995; Bench, Priston, Brown, Irackowiak, & Dolan, 1993; Dolan, Bench, Brown, Scott, Friston, & Irackowiak, 1992). The main feature of depression is a feeling of sadness, but there are a multitude of other characteristic feelings and symptoms, both physical and emotional, that occur with such regularity in depressed patients that they are also used as diagnostic criteria. Yet there is little understanding of, or attempt to integrate, the consistent cooccurrence of these additional signs and symptoms.

Common feelings described by patients with depression include emptiness, worthlessness, guilt, and badness. These are expressions regarding conception of left hemispheric self. Underlying and coexisting with these psychological feelings, one may see the physical roots that derive from the multileveled vertical hierarchy of the left hemispheric self. We spoke earlier of feelings associated with body mass and density, which in the norm impart sensations of well-being and vitality; patients with depression not infrequently experience disorder in the sensation of the density or the weight of the whole body or certain body parts (Lukianowicz, 1967). A continuum of gradual transitions may be traced from the physical feeling of weightlessness; to feelings of body parts' disappearance; to physical-psychical feelings of hollowness, emptiness, nothingness; to the experience of nonexistence (nihilistic delusion). Patients are often quite eloquent in their attempts to describe these feelings. For example, a depressed patient reported, "I have a most terrible feeling of a large open cave, of a sheer emptiness, an excruciating feeling of the cold, dark, hopeless nothingness in my chest" (from Lukianowicz, 1967, p. 39). This same patient also stated, "I often feel that it is not a head any more but an empty shell with nothing inside it and a lot of holes around ... my body feels like an empty box with another empty box on the top, instead of my head.... I am an empty and hollow nothing." Another depressed patient reported, "My spine doesn't seem to be there anymore. My heart is no longer. There is something beating in its place, but it is not my heart. My stomach has gone. I no longer feel any hunger" (Cutting, 1990, p. 351). In the nihilistic extreme, a patient reported believing that she was dead, and on one occasion described herself as "consisting of mere fresh air or atmosphere"; on another, she said that she was "just a voice and if that goes I won't be anything" (Young, Leafhead, & Szulecka, 1994, p. 228). We refer to this continuum as the *theme of disappearance*.

The second theme in depression that we identify is the *theme of the nonfunctioning object*. This theme also can be traced in the continuum of gradual transitions: from not-working internal organs and body parts to the not-functioning self. Examples of patients with depression reporting their experience follow. "All my internal organs are rotten and decayed.... I can feel my left rotting lung, but my right lung is completely dissolved and missing" (Lukianowicz, 1967, p. 39). Disorder of functioning self involves the concept of self as an agent, one which is not thinking, not feeling, not acting. Cutting (1990) observed that "most depressive patients complain of a change in at least one of the self's following activities—thinking, perceiving, feeling, acting, remembering, concentrating, or sleeping" (p. 355). However, he emphasized that objective psychological testing showed that the claimed dysfunctions are much milder than experienced. Disorder of the concept of self in patients with depression also involves self as a member of a group. Patients' estimation of themselves is devalued: "I have fallen below my ideals ... there is something wrong in my life which I must put right ... I have committed the unpardonable sin—the closed mind." "I am a washout ... I'm selfish.... I'm certain I'm disliked wherever I go" (Lewis, 1934, cited by Cutting, 1990, p. 352).

Cutting indicates that "ideas or delusions of self-reproach, worthlessness, and sinfulness can be considered as nihilistic ideas on the spiritual or moral plane, arising out of a disruption to the left hemisphere's representation of such abstract concepts. They are equivalent to the annihilation of the existence of body parts and outside-world objects on the objective plane" (Cutting, 1990, p. 352).

Interpreting this, we identify in depression disorders of functional levels participating in building of the left hemispheric concept of self. The physical feeling of body mass, as we showed earlier, is connected with level A. Afferentation of level A represents the body as a weight-object provided by the proprioceptive-vestibular modality. In neurological patients with verified left-sided lesions, anomalous bodily experiences have been reported that are similar to those we have described here in patients with depression. For example, patients reported, "I feel as if I no longer have any hands." "It's as if there was an emptiness vaguely on the right, that on that side everything is far away and empty." "Right arm keeps turning, as if it's breaking up into pieces. Hand feels all wet. I feel as if I'm in a hole ... as if I'm dead. I don't feel as if I'm living in the world" (patients from Hecaen and Ajuriaguerra, 1956, cited by Cutting, 1990). In these examples we see the contribution of the vestibular modality. Further evidence of level A's involvement in depression comes from the disorder of vegetative self that occurs in depression. Vegetative self disorders of rhythmical, cyclical processes of the organism (circadian rhythm) are well known in depression, and include disorder of the rhythm of sleep/awakening, rhythm of melatonin secretion, food intake patterns, disturbance in the normal pattern (rhythm) of cortisol secretion, dysregulation of the hypothalamic-pituitary-adrenal axis, and so on. Other supporting data that level A is involved in depression come from visceromotor and angiomotor reactions (movements of A level), which are the expressions of limbic emotion that are frequent accompaniments of depressed feelings.

Anomalous feelings of the body as a functioning object we attribute to disorder of levels C and D. Additional support for the extroverted level C's involvement, that we deal with disorder of body image as an object in the external world, comes from the fact that patients with depression who feel that their bodies are not working may feel that other people's bodies are not functioning either. As an example, one of our patients with psychotic depression believed that his grandson's internal organs were rotten. In the extreme, it will be what Kraepelin referred to as "ideas of annihilation." A patient reported, "The world has perished; there are no longer railways, towns, money, beds, doctors; the sea runs out. All human beings are dead, 'poisoned with antitoxic serum,' burned, dead of starvation.... No one eats or sleeps anymore" (Kraepelin, 1921, cited by Cutting, 1990, p. 348).

Finally, disorder of the functioning self (self as an agent and self as a member of the group) we understand as a disorder at the symbolic level. In the left hemispheric vertical hierarchy, one's own body is understood differently and acquires different meaning at each function level: level A determines one's body as a physical body in the gravitational field; level C determines one's body as a physical body, metric and geometric, in the external visual field (body image); level D, one's body as a functioning object; level E, concept of self (self as an agent, self as a member of a group). Thus, there are multiple meanings of one's body, but as in all vertical organizational systems that we have described, there is overall an integrated left hemispheric concept of self. This conscious concept of self at the symbolic level includes all lower background levels in an integrated fashion, giving a continuous line of the self. The pathological condition of depression illustrates disintegration of the integrated left hemispheric self, in which the patient may manifest disorder at any point along the vertical axis of left hemispheric self, from clearly physical proprioceptive vestibular and visual vestibular feelings of weightlessness, to highly symbolic feelings of disappearance and nonexistence.

We identify depression as a disordered interaction between levels A, C, D, and E, skipping level B. We don't know much about the primary site, or whether disorder at the symbolic level will impose dysfunction on the lower levels of this axis, but we are sure that depression is a heterogeneous condition regarding degree of relative involvement of regions of these different levels along the axis.

In fact, at least some data show that lower levels—in particular, the amygdala—may be of primary importance in the pathogenesis of depression. Drevets and Raichle used PET scans to measure regional brain blood flow, a reflection of functional activity in the brain, in patients with familial pure depressive disease (FPDD) (Drevets & Raichle, 1995). During the depressed state, blood flow was found to be increased in the left prefrontal cortex and the left amygdala. In contrast, blood flow was elevated in the left amygdala but not the left prefrontal cortex during remission of depression in these same patients. In additional studies of individuals without FPDD or other psychiatric illness, contemplation of sad thoughts resulted in an increase in blood flow in part of the left prefrontal cortex but not in the amygdala (Drevets, Spitznagel, et al., 1992a; Pardo, Pardo, & Raichle, 1993). In another type of depressive illness, bipolar depression, blood flow was found decreased in exactly the same area of the left prefrontal cortex where it was increased in FPDD (Drevets & Raichle, 1995). Drevets and Raichle emphasized that it is abnormal modulation within the anatomical system, rather than an increase or decrease of activity in any single structure, which may produce the major depressive syndrome. In general, these data confirm that brain mechanisms in the different types of depression may involve different patterns of pathological organization, with disordered interactions between relatively autonomous links along the vertical axis of left hemispheric self.

7.2.2.2. Schizophrenia. Examining body image disturbances in psychiatric patients, Lukianowicz (1967) came to the conclusion that they occur as a part of the basic psychiatric disorder, coincide with the onset of psychosis, and come to an end with the termination of the psychotic episode. Lukianowicz observed that in psychiatric patients, body image disturbance has its roots in "complex organic sensations, mainly of visceral, kinaesthetic, and labyrinthine origin, which become misinterpreted and secondarily elaborated in terms of a 'change' taking place in the *shape*, or the *size*, or the *mass* of the whole body, or its parts, or in its *position* in space" (Lukianowicz, 1967, p. 45). Lukianowicz emphasized that the role of abnormal bodily feelings in the formation of the clinical picture in the major psychiatric disorders is grossly underestimated: "In *psychiatric patients* their quite frequent complaints of various bizarre bodily sensations are usually ignored or brushed aside by their well-meaning physicians: in *schizophrenics* under the vague name of 'somatic delusions', in *depressed* patients as 'nihilistic delusions', in *neurotics* as 'hypochondriacal complaints' " (Lukianowicz, 1967, p. 42). In our experience, patients usually don't complain spontaneously about their abnormal bodily experience, which is an additional obstacle to seeing them as an integral part of the clinical picture.

According to Lukianowicz, schizophrenia is most often accompanied by bizarre disturbances in the *shape* of the body image, with further delusional elaboration. Lukianowicz gives several examples of patients with schizophrenia and complaints of disturbed body image (1967, pp. 33–34):

> "I began to have the *feeling* that my body shrank and changed into the body of a dog."
>
> "I ... *feel* that my hands and feet shrank, became covered with hair and turned into cat's paws."
>
> "I had crab's hands. I also had the *feeling* in my hands that they shrank and became hard like crab's claws."
>
> "One night I *felt* that I turned into the Evil One, with hooves like a horse, with a tail like a goat and with two small horns on my head. I can't see them in the mirror, but I *feel* them all the time."
>
> "For some time, I had a *feeling* that my breasts became large and full, and my buttocks became plump, like in a woman. I can see in the mirror that it is not so, but still I have this strange sensation in the corresponding parts of my body. I don't know how to describe it. It is a sort of pressure from the inside upon my skin, pushing it out, and stretching. The *feeling* you have when your body becomes swollen. It itches and stretches and pulls."

Our interpretation of these bizarre, complex bodily sensations in schizophrenic patients is that they represent a disorder of the sensational filling of I-space (senesthopathy). In other words, we see here distortions of the sensations that make up the physical self, disturbance of the introverted B level. However, as we stated in chapter 4, the B level has no direct access to language for expression and must rely on visual images. It is our opinion that, in schizophrenia, pathological bodily sensations are translated into visual metaphors and decoded at the symbolic level of the self (level E) through series of identifications (see description of patient with schizophrenia in Section 7.1 of this chapter).

We agree with Lukianowicz that disorder of body image in schizophrenia is secondary to distorted bodily feelings. It is our contention that these abnormal bodily feelings are thalamic in origin, and as such, represent a disturbance in the sensational filling (content) that makes the contours of the body parts (form) (see Table 4). This is different from the disorder of body shape seen in patients with cortical right parietal lesions, in which disturbance in spatial contours of body parts themselves (form) is observed (as we showed in Table 4, in the right parietal cortex, the coordinate system of one's own body [form] is split from its sensational filling [content]). In schizophrenia, sensorial feelings of one's own body, which are usually unconscious in the norm, come to the surface and become actualized. They are distorted, distressing, and exaggerated. In contrast, in patients with right parietal cortical lesions, sensations of body parts are less actualized. There are feelings of estrangement, of not belonging, of body parts whose shape is distorted, against a background of unconcern, anosagnosia, and neglect. In right parietal syndrome, there is a denial of belonging of body parts; they are "not mine." In schizophrenia, body parts are felt to become different, and the self becomes different. The subjective sense of feeling of one's own body has a distinct flavor of kinesthetic and tactile modalities (patients described feelings of pressure, stretching, pushing, shrinking). This distorted thalamic B-level emotion is decoded at the symbolic level of the self when it is presented to the right prefrontal region, through immediate involvement of symbolic systems. We see the same flow of identified and transformed images in Lukianowicz's patients with delusional metamorphoses as we described in our delusional schizophrenic patient. We conclude that the pathological organization in schizophrenia involves a peculiar pattern of direct B to E level connections, skipping level C. Anatomically, this may represent thalamic (MD nucleus) and right prefrontal cortex connections. Thus, the basic core disturbance in schizophrenia lies, in our opinion, in the thalamic-right prefrontal cortex system, not the thalamic-right parietal system.

In the literature, there are descriptions of patients with somatic delusions secondary to documented lesions of cerebrovascular origin. The localization of these lesions most often involves subcortical frontal white matter that may correspond to connections between prefrontal cortex and thalamus (MD nucleus), the connections that we believe are involved in schizophrenia. For example, a patient described by Flynn et al. (cited in Malloy & Richardson, 1994) with focal cerebrovascular disease involving right subcortical frontal, right splenial, and bilateral periventricular areas, complained of a crawling sensation on the left side of his head under the skin, which he believed was due to worms. It is interesting that in this case of organic delusions and in others, in contrast to schizophrenia, interpretation of abnormal bodily feelings stops at the level of the visual image without involving the symbolic system and a change in the self. The following is an example of schizophrenic symbolic reading of thalamic emotion at the onset of schizophrenia. A patient of ours with schizophrenia reported that his problems all started in his neck: "All the muscles on my back are pushed and pulled. Chemicals are shifted around. Something is pushing against my brain ... those senses. It was very strenuous." Asked to draw a picture of how he felt, the patient drew his spinal cord and brain, with small circles covering the spinal cord. He described his drawing: "I felt all these small things coming out of my spine into my brain, as though I tried to squeeze myself through a pipe, as though I was taken over, somebody controls me. Suddenly, I began to hear voices. All these small things went to my brain creating messages. I had a perfect experience, whether it is negative or positive." The delusion of being controlled by another and its opposite, the delusion of controlling others (omnipotence), so-called first-rank symptoms of schizophrenia according to Schneider, are specific for schizophrenia (Schneider, 1959). Both of these delusions have been recognized as resulting from the same root, from a disturbance in the border between self and nonself: "In both cases (delusion of omnipotence and delusion of control) the link between I and the environment seems qualitatively changed in a characteristic way. It is not the direction, that is, it is not the question of either I or the environment being experienced as the more powerful, nor is it the unilaterally concerned mineness of experiencing; it is rather the border between 'mine' and 'yours' (i.e., the outer world), and not the border only, but also the way in which it is overstepped, that is specific for schizophrenia" (Blankenburg, 1988, p. 187, cited by Bovet & Parnas, 1993). Bovet and Parnas refer to Bleuler's (1911) original identification of this overstepping of the me-not-me boundary, which he called *transitivism* (cited by Bovet & Parnas, 1993).

We think that this "overstepping" reflects an exposed abnormal symbolic system and externalized distorted symbolic self. Exaggerated and distorted bodily feeling, thalamic emotion, becomes conscious and is then translated into a symbolic system of meanings. In skipping level C, responsible for the continuity of one's time line and the organization of non-I-space in time, the schizophrenic self is missing that part of the self that has organized moments of experience of the external world (situational self). Without the internalized non-I-spaces, the self of I-space is exaggerated, exposed, and distorted; this self is projected onto the world, and the distinction between me and not-me is different than in the norm. This distorted symbolic self, split from the situational self (right hemispheric) and split from left hemispheric self, reveals the continuity of the symbolic system; at the same time, the individual's continuous time line is missing. Exposure of the continuity of the symbolic system is expressed in the reversibility of "I am controlled/I control," and consistent with the right hemispheric mechanism. Also consistent with right hemispheric cognitive mechanism is the fact that perception, feelings, and action are not divisible. Bovey and Parnas indicate that "in delusions of control and omnipotence, it is not the strength or weakness of the feeling of activity, nor the sense of power or impotence that shows specificity to schizophrenia. Rather, it is the 'immediacy' (*Unmittelbarkeit*) of the access by which the patient experiences his being and acting that qualifies such experiences as typical for schizophrenia. In the delusions of control and omnipotence, clinicians are struck by the fact that, for the schizophrenic patient, to be and to act fuse with each other" (Bovey & Parnas, 1993, pp. 590–591). We agree with Bovet and Parnas that delusions of omnipotence in patients with mania and schizophrenia are qualitatively different phenomena. In schizophrenia, delusion of omnipotence is a result of a qualitative disorder of self; in mania it is a consequence of an inflated self, a quantitative change in self. Just as in schizophrenia, delusions of being controlled and delusions of omnipotence are two sides of the same coin, so the delusion of omnipotence in mania is the opposite of the disappearing self in depression.

From what we have said, Bleuler's insight that autistic thinking is the core disturbance in schizophrenia can be more fully appreciated. Indeed, autistic thinking is, in terms of brain cognitive mechanisms, exposed right hemispheric symbolic systems split from right hemispheric situational thinking and left hemispheric thinking, again reflecting a distorted symbolic self. Bleuler's three other characteristic disturbances of schizophrenia (associations, affect, ambivalence) are all a function of autistic thinking. For example, a patient of ours with schizophrenia was asked by the examiner, who had a Russian accent, how he felt. He replied: "Yes, I have a cold. My aunt hypnotized me. She is half Russian. She had Russian eyes. I have visual paranoia. She looked at me and I was scared. I read a Russian book about Rasputin who hypnotized the Czarevitch." The patient was asked when he last saw his aunt; he stated that he had not seen her in many years. We see this as indicative of a gap in his time line and not linked with current reality, reflecting his autistic thinking. The patient's affect toward the examiner was included into the situation, and subsequent identifications were without any environmental connections. We see the flow of his right hemispheric associations: spoke with the examiner who had a Russian accent, felt uncomfortable, associated with an aunt who induced similar feelings, identifications with Rasputin and Czarevitch and examiner and himself.

To summarize, we say that what has happened in schizophrenia is by no means the same as what occurs in organic delusions but reflects a unique pattern of new organization that most likely involves abnormalities of both connections and brain regions. The complex pathological interrelationship between levels B (thalamic) and E (cortical) may have origins in early brain development, in disturbances in the processes of cell proliferation and migration. In a discussion of genetic and epigenetic regulation of cortical mapping, Rakic reported, "While cells migrate to the cortex, thalamocortical projections growing during the second trimester of gestation accumulate in the transient subplate zone, which is a 'waiting' compartment for the cortical afferents. These terminals, in turn, could determine the end points for the second wave of cellular migration, which produces corticocortical connections. Thus, changes in the input from the developing thalamus could alter the area of destination or reduce the thickness of superficial layers by preventing the second wave cells from reaching their proper position" (Rakic, 1989, p. 154). Other authors have recently expressed interest in the thalamus and thalamocortical connections in schizophrenia based on the extensive connections between these areas. Comparing the MRI of an average normal brain with an average schizophrenic brain, Andreasen et al. (1994) found that the thalamus was smaller in the schizophrenic brain. Similarly, in a study of PET and MRI of the thalamus in schizophrenics who had never been medicated, Buchsbaum et al. (1996) found a diminished metabolic rate in the right thalamus in schizophrenics, with a loss of the normal pattern of right greater than left asymmetry. Both authors view the thalamic abnormalities in schizophrenia as evidence of deficits in filtering sensory information, leading to an overloading or decreasing of information. This does not explain the clinical manifestations of schizophrenia. Our view of the role of the thalamus in schizophrenia more specifically correlates the core symptoms of schizophrenia and function of the thalamus, and derives from detailed examination of the importance of the thalamus to higher brain function and to the self in the norm.

7.3. Brain Mechanisms of Psychiatric Thought Disorder (Delusions)

Delusions are a cardinal feature of schizophrenia. A delusion is a disorder of thought, defined as a "false belief, based on incorrect inference about external reality, not consistent with patient's intelligence and cultural background, that cannot be corrected by reasoning" (Kaplan, Sadock, & Grebb, 1994, p. 305). Among the distinctive features of delusions identified by Jaspers (1963) are: (1) they are held with an extraordinary conviction, with an incomparable subjective certainty; (2) there is an imperviousness to other experiences and to compelling counterargument.

Delusions are symptoms. Although they are key features of schizophrenia, they are also present in other brain disorders with clear, localized pathology. Yet while one may use data about delusions from these localizable disorders to help decipher the complex puzzle of brain mechanisms in schizophrenia, this should be done with caution. First, although there are similarities in the delusions of schizophrenia and disorders with known localization, we propose that there are fundamental differences. The crucial difference is that delusions in schizophrenia represent an end point to a complex pathological process that involves disruption and dysfunction of the system of cerebral organization of the self. Delusions as a result of focal lesions are also an end point, but of a specific injury in an otherwise intact system of self. They may reflect in this case an attempt at compensation for focal damage, a process we have discussed frequently throughout this book, an interaction between intact and disordered links in brain functional systems. On the other hand, evidence of delusions in patients with focal brain damage can be useful in delineating those areas of the brain involved. It may indicate a piece of the pathogenetic mechanism in schizophrenia, a fragment of the shattered mirror that is the self, and a similar final common pathway with delusions resulting from focal brain damage. In the following sections, we will analyze the psychopathological phenomena of delusions with respect to specific brain areas (and hemispheres) and their interactions.

7.3.1. *Delusions and Left Hemispheric Self*

The definition of delusion in the beginning of this section is the standard textbook definition, which expresses the common notion that delusions belong to the class of ideas called *beliefs*. In light of what was discussed about the left hemispheric cognitive mechanism in this book, belief is in close connection with the *left hemispheric self*. Indeed, the belief system of an individual reflects a conscious concept of the self, the self as an agent and member of a group (see Section 5.4). In other words, it reflects the social self (social cohesion, shared values of the group). In our three-dimensional framework of the brain, left hemispheric self in a very distilled form represents integration of limbic emotion (levels C and D) and conceptual thought in the left prefrontal region (symbolic level E).

Belief is close to *judgment*, which also is not a purely intellectual category, as it includes not only units of information (left hemispheric units) and operations with them but also the evaluation of information in correspondence with one's needs (limbic emotion/left hemispheric thinking). Some authors (Fulford, 1991) prefer to define delusions as value judgements (rather than false beliefs) on a scale of good and bad. This again refers to left hemispheric self (see 5.4).

As with other left hemispheric phenomena, there is a continuum between normal beliefs (religious, political, and so on) and delusions. Jaspers (1963) distinguished among normal belief, overvalued idea, delusionlike idea, and primary delusion. Kurt Schneider spoke about delusional notions: "Notions such as those of religious or political eminence, or having special gifts, or of being persecuted or loved, which are not qualitatively specific, unique phenomena but found both in the norm and in pathology. The difference is in the degree of expression along a continuum from normal ordinary notions to overvalued ideas to paranoid psychosis" (Schneider, 1959, p. 107).

In terms of the left hemispheric cognitive mechanism, including the interaction between left and right hemisphere, we may ask the question: under which circumstances will false beliefs or incorrect interpretations of reality occur? As we discussed in section 5.2.1, it is the right hemisphere that gives an experiential sense of events. The left hemisphere analyzes the content of right hemisphere experience and reconstructs it according to its own cognitive mechanism, its own mode of information processing. Thus, the left hemisphere gives an interpretation of right hemispheric experience. It is this interpretation, not the primary experience, which is on the surface, the focus of our conscious awareness.

When the left hemisphere does not have access to the right hemisphere's experience, it continues to interpret, but without complete information, as in the classic split-brain experiments of Gazzaniga and LeDoux (see 5.2.1), in which the left hemisphere, surgically separated from the right, gave a logically plausible (though incorrect) explanation. As we mentioned, this left hemispheric explanation resembles a rationalization, one of the defense mechanisms in psychodynamic theory, which is defined as giving incorrect, justifying reasons for otherwise unacceptable feelings. Galin (1974) hypothesized that a temporary functional disconnection between hemispheres could explain the cerebral basis of defense mechanisms. Repression, another of the defense mechanisms, is characterized by unconscious forgetting of unacceptable ideas and feelings. Galin postulates that "in normal intact people mental events in the right hemisphere can become disconnected functionally from the left hemisphere (by inhibition of neuronal transmission across the corpus callosum and other cerebral commisures), and can continue a life of their own. This hypothesis suggests a neurophysiological mechanisms for at least some instances of repression ... [and] requires that parts of the transmission from one hemisphere to the other can be selectively and reversibly blocked" (Galin, 1974, p. 575). The defense mechanisms are part of the functions of the ego, in Freudian terms, which overall bears a strong resemblance phenomenologically to our concept of the left hemispheric self.

False belief occurs in circumstances in which left hemispheric interpretations are not consistent with reality. Several possibilities can result in this situation: (1) both hemispheres are intact but the left does not have access to right hemispheric primary experience; (2) the primary experience of the right hemisphere is distorted but the left hemisphere is intact; and (3) the left hemispheric cognitive mechanism itself is disordered. In regard to the situation involving disorder of the left hemispheric cognitive mechanism, it seems that left hemispheric thought, as we described it earlier as operations with symbolic units or signs, is basically intact in patients with delusions. Many authors have indicated that general intellectual abilities must be relatively intact in order to elaborate complex delusional beliefs (Moor & Tucker, 1979; Butler & Braff, 1991). Indeed, authors who described organic delusions noted that whether patients with organic brain disorder will or will not have delusions depends on the intactness of intellect (Cummings, 1985). Thus, if the ability to operate with categorical signs in pure form is not impaired, what is responsible for thought disorder (false beliefs)? As we postulated in the beginning of this section, belief/judgment is closely connected with left hemispheric self, and in this sense involves integration of limbic emotion and conceptual thought as such. If pure conceptual thinking is intact, thought disorder will be due to disorder of limbic emotion or disorder in the integration of limbic emotion and conceptual thought (left hemispheric self).

To conclude, we think that the brain mechanisms of the defense mechanisms, such as intellectualization and rationalization, and false beliefs are on same continuum of left hemispheric interpretations.

7.3.2. Delusions and Left Hemispheric Injury

Flor-Henry (1969) found that temporal lobe epilepsy with a focus in the left hemisphere is associated with schizophrenic-like psychosis; in particular, delusions of persecution. Flor-Henry indicated that both the psychomotor seizure and psychosis are directly correlated with left temporal involvement.

According to Flor-Henry, "The phenomenon of 'forced normalization' of the EEG during psychotic episodes ... is perhaps a reflection of the complex antithetical equilibrium which appears to underlie the organization of the temporal lobe: the normalization of the (scalp) EEG during a psychotic episode being the probable manifestation of increased disturbance in the depths of the temporal lobe. The presence and frequency of psychomotor seizures are inversely correlated with psychosis, suggesting that such seizures and psychosis are opposing manifestations of the same underlying disturbance of temporal lobe function" (Flor-Henry, 1969, p. 390).

Seizure phenomena in patients with left temporal lobe foci may be expressed by the following acute symptoms: anxiety, impending doom, apprehension accompanied by restlessness and psychomotor agitation. If hallucinations are present, they are simple, verbal auditory hallucinations such as calling names or voices (Dobrochotova & Bragina, 1977).

Chronic psychiatric disorder may accompany left posterior (temporal, parietal, occipital) brain lesions. Patients may present with tension, anxiety, worry, and constant vigilance and guardedness, carefully reading perceptions and feelings. These symptoms constitute a continuum ranging from watchfulness, to suspiciousness, to paranoid ideations, to paranoid delusions. Authors indicated that involvement of the temporal lobe of the left hemisphere predisposes to a paranoid-hallucinatory state, delusions of persecution, delusions of reference (Dobrochotova & Bragina, 1977; Cutting, 1990).

Patients with unilateral left hemispheric lesions often manifest a so-called catastrophic emotional reaction (acute apprehension, feeling of impending doom), whereas those with right hemispheric lesions generally present with the opposite, with apparent indifference (Goldstein, 1948; Terzian, 1964; Gainotti, 1969). It appears that the catastrophic reaction represents an irritation, or pathological activity of the left hemisphere, as opposed to a lack of functioning. Experimentally, when the left hemisphere has been briefly depressed by local anesthetic, a catastrophic reaction has often been observed not when the left hemisphere was depressed, but when emerging from this state. In this same situation, the right hemisphere gives a completely opposite reaction, of carelessness, euphoria, and complete lack of apprehension. Bruton's study (1988), cited by Cutting (1990), showed that five out of eight left-sided temporal lobe epileptics with psychosis showed marked improvement in their psychosis after a portion of the left temporal lobe was removed. This indicates that psychosis in left temporal lobe epilepsy is a consequence of pathological activation of the left hemisphere rather than absence of function. As we mentioned previously, some authors have speculated that the left hemisphere is more closely functionally connected than the right to those brain formations which regulate arousal and provide general functional tone for the cortex (reticular activating system) (see section 5.2.1). Left hemispheric damage leads to a disturbance in consciousness far more often than damage to the right hemisphere. Patients who received left-sided unilateral ECT lost consciousness earlier and recovered more slowly than patients who received right-sided ECT.

Irritation or pathological activation of the left hemisphere is manifested by increased arousal, which may be the physiological basis for the clinical symptom continuum ranging from increasing vigilance to paranoia. In an attempt to make sense of or justify the pathological anxiety or fear arising from damaged or disordered left posterior regions, the patient may utilize environmental cues for explanations. This may lead to misinterpretations and false beliefs resulting in delusions of persecution or reference, although by nature these delusions are generally fleeting and fragmentary because they serve the immediate purpose of containing the pathological anxiety. When chronic irritation of posterior regions results in fairly continuous anxiety and apprehension, delusions of persecution may be fixed; this has been described in patients with left hemispheric lesions. The interpreter of the pathological fear or anxiety is the left frontal region; the explanatory beliefs are a result of the interaction of the disordered (posterior) and intact (anterior) parts of the brain. This interaction between intact and disordered links in functional networks of the brain is a mechanism we have illustrated earlier in a less complex brain disorder, aphasia, in which the clinical picture is due to two components—the defect itself and an attempt at spontaneous compensation. We think that the mechanism of interaction between intact and disordered links is a universal rule in brain disorders that can be applied to psychiatric disorders.

7.3.3. Delusions and the Right Hemispheric Cognitive Mechanism

Kurt Schneider described delusions that he considered characteristic for schizophrenia, which he called *delusional perception*: "Delusional perception takes place when some abnormal significance, usually with self-reference, is attached to a genuine perception without any comprehensible rational or emotional justification" (Schneider, 1959, p. 104). To illustrate, we use Schneider's example of a schizophrenic patient who described the following strange experience with a dog: "A dog lay in wait for me as he sat on the steps of a Catholic convent. He got up on his hindlegs and looked at me seriously. He then saluted with his front paw as I approached him. Another man was a little way in front of me. I caught up to him hurriedly and asked if the dog had saluted him too. An astonished 'no' told me I had to deal with a revelation addressed to me" (Schneider, 1959, p. 105).

What Schneider described as delusional perception corresponds, we think, to exposed right hemispheric symbolic thought, in which the object perceived is immediately included into a symbolic system of meaning. In other words, it is an individual symbol, or the symbolic layer of the self, that is projected into the object, assigning it with meaning. Although Schneider used the term delusional perception, he really meant *thought*: "Perception itself is not altered but the meaning of it. Delusional perception belongs, therefore, not to disturbances of perception but to those of thought" (p. 104). It is remarkable, however, that in right hemispheric thought, operation with images (thought) is not divisible from the images themselves (perceptions). Further, Schneider's description of delusional perception corresponds to our understanding of projecting the symbolic self, not the situational self (situational layer of self). Schneider indicates that the meaning attached to delusional perception "almost always carries great import, is urgent and personal, a sign or message from another world. It is as if some 'loftier reality' spoke through the perception, as one of Zucker's patients described it" (Schneider, 1959, p. 104). Schneider emphasized the individual idiosyncratic character of delusional perception. According to Schneider, delusional perception does not derive from any particular emotional state; it cannot be explained by reaction to emotional stress or based on such emotions as anxiety, distrust, jealousy, and so on. As Schneider said, "When, for example, someone fearing arrest suspects that everyone who comes up the stairs is a police officer, the paranoid reaction is closely in line with the emotional background and is basically quite comprehensible. Here there is something quite different from the delusional perception of schizophrenia" (p. 106).

Again, Schneider's understanding of delusional perception corresponds to what we described as right hemispheric representation, which includes images, emotions, feelings, and actions within an inseparable whole.

Similar to Schneider's concept of delusional perception is Jaspers' *primary delusions*. Primary delusions are direct, immediate experience, in contrast to experiences which are mediated by thought and which "developed, evolved based on thinking and working through" (Jaspers, 1963, cited in Walker, 1991, p. 97). "The primary delusional experience is the direct, unmediated, intrusive knowledge of meaning ... not considered *interpretations* but meaning directly experienced"—the sense of presence (Jaspers, 1963; Walker, 1991, p. 98).

Jaspers's theory about delusions was derived from his examination of patients' inner subjective experience, a phenomenological approach that was without reference to brain function. It is a tribute to his great insight that his phenomenological analysis remains clinically applicable and is not only consistent with the framework of the right hemispheric cognitive mechanism but enhances our understanding of this mechanism in delusions.

The right hemisphere gives an experiential sense of events, and the left hemisphere interprets this by analyzing information from the right and rearranging it according to its own left hemispheric mode of information processing. Right hemispheric experience is covered by left hemispheric interpretation and is presented to consciousness through left hemispheric interpretations in the norm. As we assumed earlier, it is only in pathology that exaggerated or distorted right hemispheric experience breaks through or comes to the surface. May this be the case in delusions, with right hemispheric experience coming to the surface?

According to Jaspers, the experience of meaning is implicit in all perception, and it is the distortion of this implicit meaning which is the primary delusional experience (Walker, 1991). Jaspers meant that perception is intact but the meaning is distorted. In the light of right hemispheric cognitive mechanism we know that right hemispheric gestalt symbolic content is expressed through visual image and in right hemispheric thought there is no content (meaning) without form (visual image). As we have stated, in the right hemisphere, the symbol (meaning) and visual image (perception) form a single integrated representation. If a visual image is involved in different symbolic systems, the same visual image will have different meanings. For example, as we discussed in regard to Capgras syndrome, appearance (form, perception) is the same but identity (meaning) is different. It is change in the symbolic system of meaning or distortion of the symbolic system of meaning that will give primary delusional experience in Jaspers's term. The concept that this experience is immediate, direct, and irreducible, fits the rules of right hemispheric cognitive mechanism. We know that the right hemispheric representation is a continuous whole of visual situation spaces (symbolic system) simultaneously presented to consciousness.

Again, because the left hemisphere is the interpreter of the right hemisphere's experience, may this also be the case in delusions? In other words, the left hemisphere will automatically attempt, with its own logic, to make sense whenever the symbolic system of meaning changes and new meaning arises in the right hemisphere. Indeed, if we look at Jaspers's theory, we again see that his phenomenologically derived insights about the dual nature of delusions parallels our own view of the brain mechanisms of delusions. It is Jaspers who said, "Primary delusion is original experience which should be distinguished from the judgment based on it" (cited by Walker, 1991). Although Jaspers indicated that delusions are beliefs held with extraordinary conviction, a certainty unmitigated by other experience or argument, it is important to note that he emphasized that "to say simply that a delusion is a mistaken idea which is held by the patient and cannot be corrected gives only a superficial and incorrect answer to the problem" (Jaspers, 1963, p. 93; Walker, 1991). These are external characteristics of delusions. "If we want to get behind these mere external characteristics of delusions into the psychological nature of delusions, we must distinguish the original experience from the *judgment* based on it, i.e., the delusional contents as presented data from the fixed judgment" (Jaspers, 1963, p. 96; Walker, 1991). Thus, the main factor distinguishing primary delusions from other forms of belief is, according to Jaspers, their origin within the patient's experience. This experience is the core of the delusion. What Jaspers called superficial, external characteristics of the delusion is the judgment made based on this original experience, which parallels the left hemispheric interpretation.

To illustrate this complex structure of delusions, we will use an example of a patient originally presented by Mellor (1991, p. 105), and analyze a delusion from our point of view: "A patient getting off a bus at a petrol station saw a pump set apart from the others for the sale of paraffin. On the top of this pump was a sign. When he saw this sign he knew that he was damned to hell for all eternity, because God, and everyone else, erroneously believed that he had committed the unforgivable sin of sodomy with a famous figure in British public life. He said that he knew that this sign was the customary one found in such locations, and nobody had put it there specially for him, yet he knew that it had this message for him. However, when examined more closely about his perception of the sign, he said, 'I first saw a circle, and a circle has no end and so must be immortal. Inside this circle was another circle parallel to it. Parallel lines meet at infinity, so that two parallel circles having no end and meeting only at infinity must mean God. In the centre of the inner circle were bright red flames, the flames of hell, and written across these flames in capital letters was the word *Gulf.* [The oil company.] This showed me the gulf that exists between me and God.' "

The primary delusion was the immediate experience of new meaning on viewing the pump with the symbol of an encircled flame: he was damned to hell for all eternity for an (alleged) act of sodomy. Here, we think, the right hemisphere cognitive mechanism is exposed: the patient externalizes his feeling, incorporating the visual image of the pump in the new symbolic system of meaning (assigning object with meaning). The patient then interprets his experience, giving an explanation justifying why he has responded to the pump as a new sign. This justification is primarily a left hemispheric cognitive process, although here we clearly also see the exposed fragments of right hemispheric associations. These fragments reveal for us something of the mechanism of this incorporation of external objects into his symbolic system of meaning. For example, circle—immortal; two parallel circles—eternity-god. A normal person understands that one image may symbolize another but is not the same as the other; this is achieved by interaction of the right and left hemispheres in symbol formation. The patient identifies objects from different situations as belonging to one symbolic system, one indivisible whole, which shows us exposed right hemispheric consciousness without left hemispheric analysis. Then the patient uses the left hemispheric cognitive mechanism—namely, verbal logical thinking and reasoning—to make judgment and form a belief: "A circle has no end and so must be immortal ... parallel lines meet in infinity, so that two parallel circles having no end and meeting only in infinity must mean God."

In previous chapters, we proposed a cerebral mechanism for delusional misidentification syndromes (DMS), disorders involving intact recognition with misidentification of an object. Although we concentrated on the contribution of the right hemispheric cognitive mechanism in the misidentification and duplication process, DMS involves another step: the belief that the double is an impostor. We think that belief is of left hemispheric origin. It is a judgment about the double, whom the patient feels uncomfortable about and then develops a justification to explain the new meaning the object has acquired. It is an abstract concept of some perpetrator without reference to a visual image—most often, the word *impostor* is used, but it can be the Mafia, a robot, or public figure like the actor Fregoli.

7.3.4. Psychopathological and Cerebral Pathogenesis of Delusions in Schizophrenia

Examination of the development of psychosis over time reveals the structural complexity of delusions in schizophrenia. Bovett and Parnas (1993) indicated that the development of delusions in schizophrenia is preceded by a state of increasing tension, anxiety, and apprehension. Conrad called this stage *das trema*, a term taken from the world of theater and used by actors to describe the state of tension that precedes entering the stage. Authors who described schizophrenic patients with Capgras syndrome indicated that the delusion of misidentification is preceded by a diffuse paranoid state. Todd et al. (1981) indicate that the state of mind that is probably a predecessor of Capgras delusions is one of intense suspiciousness. Capgras, who first described this syndrome, also emphasized his patients' paranoid disposition (Capgras & Reboul-Lachaux, 1923). Merrin and Silverfarb (1976) spoke about increased vigilance and suspicion in patients with Capgras syndrome. Other authors who described Capgras syndrome in patients with schizophrenia indicated that there was a marked paranoid component that paved the way for the emergence of the Capgras delusion (Christodoulou, 1977; Enoch & Trethowan, 1979).

To describe the state that precedes the crystallized delusional system in schizophrenia patients, authors have used the terms *paranoid state, delusional mood, delusional atmosphere*. Mellor (1991, p. 104) discussed Jaspers's (1963) description of delusional atmosphere: "Jaspers described the experience of delusional atmosphere as an uncanny sense of something happening that is strange and suspicious. There is a feeling of distrust and distressing discomfort.... Patients find themselves in an environment that 'offers a world of new meanings,' ordinary objects seem to portend, but do not reveal, important new personal meanings.... Events are not only experienced as strange but acquire the significance of 'markers' focusing on the patient himself." Sims (1980, cited by Berner, 1991, p. 88) described delusional atmosphere: "[The] patient experiences everything around him as sinister, portentous, uncanny, peculiar, in an undefinable way ... knows he is personally involved but cannot tell how ... feels uncomfortable, often extremely perplexed and apprehensive." Conrad (1958) described this modification of the mood state in the beginning of schizophrenia as the "miniscular, scarcely remarkable experiences which impart to the situation a new and bewildering 'physiognomy.' Thus, the delusional atmosphere appears as global impression: something unknown is going on, it is in the air like an impending disaster, a general experience which indicates that one must question one's own existence" (cited from Berner, 1991, p. 88). Sims (1988, cited by Berner, 1991, p. 88) indicates that "[W]hen the delusion becomes fully formed he [the patient] often appears to accept it with a feeling of relief from the previous unbearable tension of the atmosphere."

To reiterate, the events in formation of a delusion in schizophrenia are: the experience of a changed world (new meaning); an apprehensive paranoid state (delusional atmosphere); and a belief. If we consider the last two components (the apprehensive paranoid state and belief), we have a clinical picture that is similar to the delusions observed as a result of left posterior brain injury (see section 7.3.2). As we described the psychopathological and cerebral pathogenesis of the delusion in this situation, the anxiety-apprehension-paranoia is a direct consequence of left posterior injury and the belief is an attempt at spontaneous compensation, which involves interaction between the left posterior, temporal-parietal region (affected link) and left anterior, prefrontal region (intact link). These two components are common to both the schizophrenic and organic delusion, and, we assume, reflect common brain pathways. In schizophrenia, it is the specific event of new meaning that will trigger the chain; in left brain injury, the diffuse fear or anxiety will be the precipitant.

We will give examples from two patients: the first, patient H, with the clinical manifestations of temporal lobe epilepsy, and the second, patient M, with the beginnings of schizophrenia, before the delusions were crystallized (we thank J. Sharp, Ph.D. for providing this clinical material). Patient H presented with polymorphic fragmented paranoid ideations with different content from the immediate environment. For example, on one day, she suddenly expressed fear that something would happen to her family. When asked why, she replied that while she was watching the television news about storms in Texas, she became, at that moment, very frightened about her family, thinking that something bad would happen here and her family would get hurt. Her fearful reaction to what she saw on the television is consistent with a reaction of impending doom. At another time, the patient expressed her fear that people were against her, telling lies about her. Later, she worried about the fact that patients were getting discharged and new people were coming to the ward; she was afraid of them until she got to know why they were admitted. On another occasion she expressed her fear that she would die of a heart attack. A day later, she feared that she would be raped. On exploration, the environmental stimuli that were the source of these diffuse reactions became apparent: in the case of the patient's heart attack fears, she had been put inside an MRI machine for an examination and felt increasing anxiety; in the case of her fear of rape, a new male patient with aggressive behavior had been admitted to the ward. In sum, in this patient we see fleeting paranoid ideations with changing content that reflects the environmental situation.

Patient M, in the initial stages of schizophrenia, was in a paranoid state, perplexed by the new meaning of the world. He began to "receive" thoughts that were premonitions of events that were about to occur. He began to experience strange events that he was certain had been arranged for his benefit or to convey a message to him. One morning, breakfast was not prepared on time, and he thought that he was responsible for the delay; if he did not retreat to his room, no one would be served. Later, he had a critical thought about a female patient's appearance and believed that the following fire drill was organized so that he could sit near the woman to apologize for his disparaging thoughts. The next day, he sensed that the staff standing near him was influencing him to write down the wrong date and time in the sign-out book, further evidence to him that his thoughts and actions were being controlled by others. At another time, the patient explained that when he sat down with other patients, their conversations were tailored to point out to him his previous "mistakes." When he was asked how this was happening, he explained that these patients were brought here not for their own treatment but to provide a series of lessons for him to learn.

In the first case, diffuse fear triggered the paranoid ideation. In the second case, it is the new meaning with which environment is assigned, and the accompanying anxiety and apprehension. Mellor (1991) emphasized that delusional perceptions may provide the opportunity to observe the "birth of a delusion" (p. 106).

Is delusional perception the birth of delusions in schizophrenia? Let us repeat again the initial events in delusion formation in schizophrenia, which we described in previous chapters, and then reconstruct the whole pathogenetic chain. At the heart of schizophrenia is disorder of right hemispheric self, distortion of sensations that make up the space and the boundaries of one's own body, the very core of the self. These bodily sensations are not conscious in the norm; they are fundamental and are intimately incorporated into higher levels of self. Like Cerberus, the gatekeeper, these bodily feelings (thalamic emotion) preserve the integrity of the physical self. When the sensations of body parts are distorted, the integrity is compromised. What are the consequences of distortion of bodily sensations? (1) There is a "leakage" in I-space, that is, disorder in the boundaries between environment and the self; (2) pathological sensations come to the surface of consciousness (are exposed). As we have proposed earlier, because there is no language for these unique, individual feelings, they may be expressed in visual metaphors of the symbolic self; that is, translated into visual images that are immediately included into a symbolic system of meaning. With disorder of boundaries between self and environment, the symbolic system of meaning is attributed to the outside world (new meaning of the world); the changed self is projected and perceived as a changed world. It is the symbolic self that is projected, assigning external objects with meaning, with special significance. The altered fundamental sensations of one or more body parts at the core of the visual image predispose to somatic (body) delusions ("a snake coming out of my head," the delusion of being poisoned, and so on). Leakage in I-space resulting in disorder of the boundaries between self and environment predisposes to delusions of control/being controlled.

This is when and how delusional perception is born. The anatomical base for it is, as we proposed, the system including the mediodorsal nucleus of the thalamus—right prefrontal lobe. We assume this involves some neurodevelopmental disorder in the connections and/or the regions themselves. The next event in the psychopathological picture of delusions, which happens almost simultaneously, is the paranoid state, the delusional atmosphere. Extrapolating from verifiable disorders, we assume that the anatomical base for this should be pathological activation of posterior regions of left hemisphere. The next event is interpretation of delusional perception (belief), which coincides with formation of a fixed delusional system. We assume that the anatomical base for this is the left prefrontal lobe.

References

Abler, W. L. (1976). Asymmetry in the skulls of fossil men: Evidence of lateralized brain function. *Brain, Behavior and Evolution, 13*, 2–3, 111–115.

Achutina, T. (1975). *Neurolinguistic analysis of dynamic aphasia.* Moscow: Moscow State University Press.

Albert, M., & Bear, D. (1957). A case study of word deafness with reference to the role of time in auditory comprehension. *Brain, 97*, 373–384.

Alexander, M. P., Hiltbrunner, B., & Fischer, R. S. (1989). Distributed anatomy of transcortical sensory aphasia. *Archives of Neurology, 46*, 885–892.

Alexander, M. P., & Benson, D. F. (1991). The aphasias and related disturbances. In R. J. Yoynt, Ed., *Clinical neurology*, Vol. 1 (pp. 1–58). Philadelphia: Lippincott.

Alexander, M. P., Stuss, D. T., & Benson, M. D. (1979). Capgras syndrome: A reduplicative phenomenon. *Neurology, 29*, 334–339.

Anderson, A. L. (1951). The effect of laterality localization of focal brain lesions on the Wechsler-Bellevue sbutests. *Journal of Clinical Psychology, 7*, 149–153.

Anderson, D. (1988). The delusion of inanimate doubles: Implications for understanding the Capgras phenomena. *British Journal of Psychiatry, 153*, 694–699.

Anderson, D., & Williams, E. (1994). The delusion of inanimate doubles. *Psychopathology, 27*, 220–225.

Anderson, S. W., Damasio, A. R., & Damasio, H. (1990). Troubled letters but not numbers. *Brain, 113*, 749–766.

Andreasen, N. (1996). Pieces of the schizophrenia puzzle fall into place. *Neuron, 16*, 697–700.

Andreasen, N., Arndt, S., Swayze, V., Cizadlo, T., Flaum, M., O'Leary, D., Ehrhard, J. C., & Yuh, W. T. (1994). Thalamic abnormalities in schizophrenia visualized through magnetic resonance image averaging. *Science, 266*, 294–297.

Angyal, A. (1936). The experience of the body-self in schizophrenia. *Archives of Neurology and Psychiatry, 34*, 270–279.

Anufriyev, A. K. (1979). Pathology of coenaesthesia and affective disturbances with its equivalents: Psychosomatic disorders in cyclothymic and cyclothymic-like states. R.S.F.S.R. Ministry of Health, Moscow Institute of Psychiatry Works, 87, 8–24.

Auerbach, S. H., Allard, T., Naeser, M., Alexander, M. P., & Albert, M. L. (1982). Pure word deafness.

Analysis of a case with bilateral lesions and a defect on the prephonemic level. *Brain, 105*, 271–300.

Balonov, A., Barkan, D., & Deglin, V. (1979). *Unilateral ECT*. Leningrad: Nauka.

Barbur, J., Watson, J., Frackowiak, R. S. J., & Zeki, S. (1993). Conscious visual perception without V 1. *Brain*, *116*, 1293–1302.

Bauer, R. M. (1986). The cognitive psychophysiology of prosopagnosia. In H. D. Ellis, M. A. Jeeves, M. A. Newcombe, & F. Young (Eds.), *Aspects of Face Processing*. Dordrecht: Nijhoff.

Bauer, R., & Becka, D. (1954). Intellect after cerebrovascular accidents. *Journal of Nervous and Mental Disease*, *120*, 379–384.

Bear, D. (1979). Temporal lobe epilepsy: A syndrome of sensory limbic hyperconnection. *Cortex*, *15*, 357–384.

Beauvois, M. F., Saillant, B., Meininger, V., & L'Hermitte, F. (1978). Bilateral tactile aphasia: A tacto-verbal dysfunction. *Brain*, *101*, 381–401.

Beckers, G., & Homberg, V. (1992). Cerebral visual motion blindness: Transitory akinetopsia induced by transcranial magnetic stimulation of human area V5. *Proceedings of the Royal Society of London. Series B. Biological Sciences*, *249*, 173–178.

Bein, E. S. (1961). Paraphasias in different forms of aphasia. In *Clinic and pathophysiology of aphasia* (pp. 117–139). Moscow: Medgiz.

Bench, C. J., Friston, K. J., Brown, R. G., Irackowiak, R., & Dolan, R. (1993). Regional cerebral blood flow in depression measured by positron emission tomography: The relationship with clinical dimensions. *Psychological Medicine*, *23*, 579–590.

Benes, F. M., McSparren, J., Bird E. D., et al. (1991). Deficits in small interneurons in prefrontal and cingulate cortices of schizophrenic and schizoaffective patients. *Archives of General Psychiatry*, *48*, 996–1001.

Benson, D. F. (1994). Naming disorders. In S. Kennedy (Ed.), *Psychobiology of language* (pp. 126–128). Cambridge, MA: MIT Press.

Benson, D. F. (1994). *The neurology of thinking*. New York: Oxford University Press.

Benson, D. F., & Geschwind, N. (1970). Developmental Gerstmann's syndrome. *Neurology*, *20*, 293–298.

Benson, D. F., & Geschwind, N. (1971). The aphasias and related disturbances. In *Clinical neurology* (3rd ed.). New York: Harper & Row.

Benson, D. F., & Stuss, D. T. (1990). Frontal lobe influences on delusions: A clinical perspective. *Schizophrenia Bulletin*, *16*, 403–411.

Benton, A. L. (1992). Gerstmann's syndrome. *Archives of Neurology*, *49*, 445–447.

Berndt, R. (1987). Symptom co-occurrence and dissociation in the interpretation of agrammatism. In M. Coltheart, G. Sartori, & R. Job (Eds.), *The cognitive neuropsychology of language*. Hillsdale, NJ: Erlbaum.

Berner, P. (1991). Delusional atmosphere. *British Journal of Psychiatry*, *159*(Suppl. 14), 83–93.

Bernstein, N. (1947). *On the construction of movements*. Moscow: Medgiz.

Bernstein, N. (1967). *The coordination and regulation of movements*. New York: Pergamon Press.

Bianchi, V. L. (1967). Evolution of dual function of the cerebral hemispheres. Leningrad: Leningrad University Press.

Bianchi, V. L. (1979). Individual and species hemisphere asymmetry in animals. *Journal of Higher Nervous Activity*, *29*, 2, 295–304.

Binder, J. R., Rao, S. M., Hammeke, T. A., Yerkin, T. E., Jesmanowicz, A., Bandettini, P. A, Wong, E. C., Estrowski, L. D., Goldstein, M. D., Haughton, V. M., & Hyde, J. S. (1994). Functional magnetic resonance imaging of human auditory cortex. *Annals of Neurology*, *35*, 6, 662–672.

Blinkov, S. M. (1938). Structural variability of the cerebral cortex: Middle temporal region of adult man. Moscow, Brain Institute, 314, 313–362.

Blinkov, S. M. (1955). *Temporal region of the brain in man and monkey*. Moscow: Meditsina.

Blinkov, S. M., & Glezer, I. (1964). *Human brain in numbers and tables*. Moscow: Meditsina.

Blonsky, P. D. (1935). *Memory and thinking*. Moscow–Leningrad: Ogiz.

Bloom, F. E. (1993). Advancing in neurodevelopmental origin for schizophrenia. *Archives of General Psychiatry*, *50*, 224–227.

Bogen, J. E., (1969). The other side of the brain: II. An appositional mind. III. The corpus callosum and creativity. *Bulletin of Los Angeles Neurological Societies*, *34*, 3–4, 135–162, 191–220.

Bovet, P., & Parnas, J. (1993). Schizophrenic delusions: A phenomenological approach. *Schizophrenia Bulletin*, *19*, 3, 580–597.

Brain, W. (1941). Visual disorientation with special reference to lesions of the right cerebral hemisphere. *Brain*, *64*, 244–272.

Brain, W. (1963). Some reflections on brain and mind. *Brain*, *86*, 3, 381–401.

Bradshaw, J. L., & Nettleton, N. C. (1981). The nature of hemispheric specialization in man. *Behavioral and Brain Sciences*, *4*, 51–91.

Brodmann, K. (1908). Beitrage zur histologishen Lokalisation der Grosshirnrinde: VI. Die Cortex gliederung der Menschen. *J. Psychol. Neurol.*, *10*, 231–246.

Brodmann, K. (1909). *Vergleichende Lokalisationlehre der Grosshirnrinde in ihren Prinzipien dargestellt auf Grund des Zellenbaues*. Leipzig: Barth.

Buchsbaum, M. S., Someya, T, Ying Teng, C., Abel, L., Chin, S., Najafi, A., Haier, R., Wu, J., & Bunney, W. (1996). PET and MRI of the thalamus in never-medicated patients with schizophrenia. *American Journal of Psychiatry*, *153*, 191–199.

Bunak, V. V. (1966). Speech and intellect: Their stages of development in anthropogenesis. New series, Vol. 92 (pp. 497–555). Moscow: Institute of Ethnology Academy of Science, USSR.

Butler, R. W., & Braff, D. L. (1991). Delusions: A review and integration. *Schizophrenia Bulletin*, *17*, 4, 633–647.

Campbell, A. W. (1905). Histological studies on the localization of cerebral function. London: Cambridge University Press.

Capgras, J., & Reboul-Lachaux, J. (1923). Illusion des sosies dans un delire systematise chronique. *Bulletin de la Societe Clinique de Medecine Mentale*, *2*, 6–16.

Cappa, S., Cavallotti, G., & Vignolo, L. A. (1981). Phonetic and lexical errors in fluent aphasia: Correlation with lesion site. *Neuropsychologia*, *219*, 171–177.

Caramazza, A., & Berndt, R. (1978). Semantic and syntactic processes in aphasia. *Psychological Bulletin*, *85*, 4, 898–918.

Caramazza, A., & Hills, A. E. (1991). Lexical organization of nouns and verbs in the brain. *Nature*, *349*, 788–790.

Carmon, A., & Nachshon, I. (1971). Effect of unilateral brain damage on perception of temporal order. *Cortex*, *7*, 410–418.

Chatterjee, A., Maher, L., Rothi, L., & Heilman, K. (1995). Asyntactic thematic role assignment: The use of a temporal–spatial strategy. *Brain and Language*, *49*, 125–139.

Chejf, U. L. (1975). *Meaning and structure of language*. Moscow: Progres.

Chlenoff, L. G. (1934). Body schema and its disturbances. In *New data in study of apraxia, agnosia, and aphasia* (pp. 62–91). Moscow-Leningrad.

Chomsky, N. (1957). *Syntactic structures*. The Hague: Mouton.

Christodoulou, G. N. (1991). The syndrome of Capgras. *British Journal of Psychiatry*, *130*, 556–564.

Chua, S., & McKenna, P. J. (1995). Schizophrenia—A brain disease? A critical review of structural and functional cerebral abnormality in the disorder. *British Journal of Psychiatry*, *166*, 563–582.

Conrad, K. (1958). *Die beginnende Schizpphrenie. Versuch einer Gestaltanalyse des Wahns*. Stuttgart: Thieme.

Coslett, H. B., Bowers, D., & Heilman, K. M. (1987). Reduction in cerebral activation after right hemisphere stroke. *Neurology*, *37*, 957–962.

Coughlan, A. K., & Warrington, E. K. (1978). Word-comprehension and word-retrieval in patients with localized cerebral lesions. *Brain*, *101*, 163–185.

Courbon, P., & Fail, G. (1927). Illusion de fregoli. *Bulletin de la Societe Clinique de Medecine Mentale, 15*, 121–124.

Courbon, P., & Tusques, Y. (1932). Illusions d'intermetamorphose et de charme. *Annales Medico-Psychologiques, 90*, 401–405.

Courville, C. B. (1942). Coup-contracoup mechanism of craniocerebral injuries. *Archives of Surgery, 45*, 19–43.

Critchley, M. (1950). The body image in neurology. *Lancet, 1*, 335.

Critchley, M. (1953). *The parietal lobes*. New York: Hafner.

Critchley, M. (1966). The enigma of Gerstmann's syndrome. *Brain, 89*, 183–198.

Cummings, J. L. (1985). Organic delusions: Phenomenology, anatomical correlations, and review. *British Journal of Psychiatry, 146*, 184–197.

Cutting, J. (1990). *The right cerebral hemisphere and psychiatric disorders*. Oxford: Oxford University Press.

Damasio, A. R. (1992). Aphasia. *New England Journal of Medicine, 326*, 8, 531–539.

Damasio, A. R., & Damasio, H. (1992). Brain and language. *Scientific American, 267*, 3, 89–95.

Damasio, A. R., & Tranel, D. (1993). Nouns and verbs are retrieved with differently distributed neural systems. *Proceedings of the National Academy of Sciences of the United States of America, 90*, 4957–4960.

Damasio, H., & Damasio, A. (1980). The anatomical basis of conduction aphasia. *Brain, 103*, 337–350.

Damasio, H., & Damasio, A. (1989). *Lesion analysis in neuropsychology*. New York: Oxford University Press.

Davidson, G. M. (1941). The syndrome of Capgras. *Psychiatric Quarterly, 15*, 513–521.

Delis, D. C., Robertson, L. C., & Efron, R. (1986). Hemispheric specialization of memory for visual hierarchical stimuli. *Neuropsychologia, 24*, 2, 205–214.

Delle-Sala, S., Laiagna, M., Spinler, H., & Trivelli, C. (1993). Autobiographical recollection and frontal damage. *Neuropsychologia*, 31, 8, 823–839.

Denis, M. (1979). *Les images mentales*. Paris: Presse Universitaire de France.

De Pauw, K. W. (1994). Psychodynamic approaches to the Capgras delusion: A critical historical review. *Psychopathology, 27*, 154–160.

De Renzi, E., Liotti, M., & Nichelli, P. (1987). Semantic amnesia with preservation of autobiographical memory: A case report. *Cortex, 23*, 575–597.

De Renzi, E., & Spinnler, H. (1966). Visual recognition in patients with unilateral cerebral disease. *Journal of Nervous and Mental Disease, 142*, 515–525.

Dobrochotova, T. A. (1974). *Emotional pathology in focal brain disorders*. Moscow: Meditsina.

Dobrochotova, T. A., & Bragina, N. N. (1977). *Functional asymmetry and psychopathology in focal brain disorders*. Moscow: Meditsina.

Dolan, R. J., Bench, C. J., Brown, R. J., Scott, L. C., Friston, K. J., & Frackowiak, R. S. J. (1992). Regional cerebral blood flow abnormalities in depressed patients with cognitive impairment. *Journal of Neurology, Neurosurgery and Psychiatry, 55*, 768–773.

Drews, E. (1987). Qualitatively different organizational structures of lexical knowledge in the left and right hemisphere. *Neuropsychologia, 25*, 2, 419–427.

Drevets, W. C., & Raichle, M. E. (1985). Positron emission tomographic imaging studies of human emotional disorders. In M. S. Gazzaniga, (Ed.) *The cognitive neurosciences* (pp. 1153–1164). Cambridge, MA: MIT Press.

Drevets, W. C., Spitznagel, E. L., et al. (1992a). Discriminatory capability of PET measurements of regional blood flow in familial pure depressive disease. Abstract presented at Society of Neuroscience Annual Meeting, Anaheim, CA.

Drevets, W., Videen, T., Price, J., Preskorn, S., Carmichael, S., & Raichle, M. (1992b). A functional anatomical study of unipolar depression. *Journal of Neuroscience, 12*, 3628–3641.

Drevets, W. C., Price, J. L., Simpson, J. R., Todd, T. R., Reich, T., Vannier, M., & Raichle, M. E. (1997). Subgenual prefrontal cortex abnormalities in mood disorders. *Nature, 386*, 24, 824–827.

Duffy, J. D., & Campbell, J. J. (1994). The regional prefrontal syndromes: A theoretical and clinical overview. *Journal of Neuropsychiatry and Clinical Neurosciences, 6*, 379–387.

Dupre, E. (1925). *Pathologie de l'inagination et de l'emotivite*. Paris.

Efron, R. (1963). Temporal perception, aphasia, and déjà vu. *Brain, 86*, 408–424.

Ekman, P. (1984). Expression and the nature of emotion. In K. Scherer & P. Ekman (Eds.), *Approaches to emotion* (pp. 319–344). Hillsdale, NJ: Erlbaum.

Ellis, H. D. (1994). The role of the right hemisphere in the Capgras delusion. *Psychopathology, 27*, 177–185.

Ellis, H. D., & Young, A. W. (1990). Accounting for delusional misidentifications. *British Journal of Psychiatry, 157*, 239–248.

Enoch, M. D. (1963). The Capgras syndrome. *Acta Psychiatrica Scandinavica, 39*, 439–462.

Feinberg, T. E., Schindler, R. J., Ochoa, E., Kwan, P. C., & Farah, M. J. (1994). Associative visual agnosia and alexia without prosopagnosia. *Cortex, 30*, 3, 395–411.

Flor-Henry, P. (1969). Psychosis and temporal lobe epilepsy: A controlled investigation. *Epilepsia, 10*, 363–395.

Farah, M. J., & Wallace, M. J. (1992, July). Semantically bounded anomia: Implications for the neural implementation of naming. *Neuropsychologia, 30*, 7, 609–621.

Fulford, K. W. M. (1991). Evaluative delusions: Their significence for philosophy and psychiatry. *British Journal of Psychiatry, 159* (Suppl. 14), 108–112.

Fuster, J. M. (1985). The prefrontal cortex and temporal integration. In A. Peters (Ed.), *Cerebral cortex, Vol. 4, Association and auditory cortices* (pp. 151–177). New York: Plenum Press.

Gainotti, G. (1969). Reactions "catastrophiques" et manifestations d'indifference au cours des atteintes cerebrales. *Neuropsychologia F*, 195–204.

Galaburda, A. M., Le May, M., Kemper, T. L., & Geshwind, N. (1978). Right–left asymmetries to the brain. *Science, 199*, 852–856.

Galin, D. (1974, October). Implications for psychiatry of left and right cerebral specialization. *Archives of General Psychiatry, 31*, 572–583.

Gardner, H., Ling, P., Flamm, I., & Silverman, J. (1975). Comprehension and appreciation of humorous material following brain damage. *Brain, 98*, 399–412.

Gastaut, H., and Broughton, R. (1972). *Epileptic seizures*. Springfield, IL: Charles C. Thomas.

Gazzaniga, M. S., & Le Doux, J. E. (1978). *The integrated mind*. New York: Plenum Press.

Gerstmann, J. (1940). Syndrome of finger agnosia, disorientation for right and left, agraphia and acalculia. *Archives of Neurology Psychiatry, 44*, 398–408.

Geschwind, N. (1971). Aphasia. *New England Journal of Medicine, 284*, 654–656.

Gibbon, J., Malapani, C., Dale, C. L., & Gallistel, C. (1997, April). Toward a neurobiology of temporal cognition: Advances and challenges. *Current Opinions in Neurobiology, 7*, 2, 170–184.

Glezerman, T. B. (1983). Brain dysfunctions in children. Moscow: Nauka.

Glezerman, T. B. (1986). Psychophysiological base of thinking disorder in aphasia/aphasia and intellect. Moscow: Academy of Sciences, Department of Physiology, Nauka.

Glezerman, T. B., & Dmitrova, E. D. (1989). Neuropsychological differentation of specific dyslexia in children. *Journal of Neuropathology and Psychiatry, 3*, 63–68.

Glezerman, T. B., & Novinskaja, N. L. (1983). About Gerstmann's syndrome in children. Conference of RSFSR Children Neuropathologists and Psychiatrists. Moscow, 117–119.

Glezerman, T., Balkoski, V., & Miller, A. (1996). A neuropsychiatric model of self. In *X World Congress of Psychiatry* Abstracts, Vol. 2, pp. 400–401.

Gloor, P., Oliver, A., Quesney, L., Andermann, F., & Horowitz, S. (1982). The role of the limbic system in experiential phenomena of temporal lobe epilepsy. *Annals of Neurology, 12*, 129–144.

Goldberg, E., Podell, K., & Lovell, M. (1994). Lateralization of frontal lobe functions and cognitive novelty. *Journal of Neuropsychiatry and Clinical Neurosciences, 6*, 371–378.

Goldenberg, G., Podreka, I., Steiner, M., & Willness, K. (1987). Patterns of regional cerebral blood flow related to memorizing of high and low imagery words. *Neuropsychologia, 25*, 3, 473–485.

Goldenberg, G. (1992). Loss of visual imagery and loss of visual knowledge—A case study. *Neuropsychologia, 30*, 12, 1081–1099.

Goldstein, K. (1948). *Language and language disturbances*. New York: Grune and Stratton.

Goodglass, H. (1983). Word retrieval for production. In Psychobiology of Language (pp. 129–138), Cambridge, MA: MIT Press.

Goodglass, H., & Baker, E. (1976). Semantic field, naming, and auditory comprehension in aphasia. *Brain and Language, 3*, 359–374.

Goodglass, H., & Berko, J. (1960). Agrammatism and inflectional morphology in English. *Journal of Speech and Hearing Research, 3*, 3, 33–43.

Goodglass, H., & Kaplan, E. (1972). *The assessment of aphasia and related disorders*. Philadelphia: Lea and Febiger.

Grigsby, J. P., Kemper, M. B., & Hagerman, R. (1987). Developmental Gerstmann's syndrome without aphasia in Fragile X syndrome. *Neuropsychologia, 25*, 6, 881–891.

Guariolia, C., & Antonucci, G. (1992). Personal and extrapersonal space: A case of neglect dissociation. *Neuropsychologia, 30*, II, 1001–1009.

Gruhle, H. W. (1915). Selbstschilderung und Einfühlung. *Zeitschrift für die gesamte Neurologie und Psychiatrie, 28*, 148–231.

Harris, J. C. (1995). *Developmental neuropsychiatry*. New York: Oxford University Press.

Hart, J., Berndt, R., & Caramazza, A. (1985). Category specific naming deficit following cerebral infarction. *Nature, 316*, 8, 439–440.

Hawking, S. (1988). *A brief history of time*. New York: Bantam Books.

Head, H. (1963). *Aphasia and kindred disorders*. Cambridge, England: Cambridge University Press. (Original work published 1926.)

Hebb, D. O. (1942). The effect of early and late brain injury upon test scores and the nature of normal adult intelligence. *Proceedings of the American Philosophical Society, 85*, 275–292.

Hecaen, H., & Albert, W. L. (1978). *Human neuropsychology*. New York: Wiley-Interscience.

Heilbrun, A. B. (1959). Lateralization of cerebral lesion and performance on spatial-temporal tasks. *Archives of Neurology, 1*, 282–287.

Heilman, K. M. (1973). Ideational apraxia—A redefinition. *Brain, 96*, 861–864.

Heilman, K. M., Safran, A., & Geschwind, N. (1971). Closed head trauma and aphasia. *Journal of Neurology, Neurosurgery and Psychiatry, 34*, 265–269.

Heilman, K. M., Watson, R. T., Valenstein, E. (1985). Neglect and related disorders. In K. M. Heilman & E. Valenstein (Eds.), *Clinical neuropsychology* (pp. 243–293). New York: Oxford University Press.

Heinrichs, R. W. (1993). Schizophrenia and the brain. *American Psychologist, 48*, 3, 221–233.

Hodges, J. R., Patterson, K., Oxbury, S., & Funnel, E. (1992). Semantic dementia: Progressive fluent aphasia with temporal lobe atrophy. *Brain, 115*, 1783–1806.

Hoffman, R., & McGlashan, T. (1994). Corbicocortical connectivity, autonomous networks, and schizophrenia. *Schizophrenia Bulletin, 20*, 2, 257–261.

Hofstede, B. M. T., & Kolk, H. H. J. (1994). The effects of task variation on the production of grammatical morphology in Broca's aphasia: A multiple case study. *Brain and Language, 46*, 278–328, 1994.

Horst, L. van der, & Mtschzft, L. (1932). *Psychiatry u. Neurol., 83*, 65.

Ivanov, V. V. (1978). *Asymmetry of the brain and semiotic systems*. Moscow: Sovetskkoje Radio.

Izaed, K. (1991). *The psychology of emotions*. New York: Plenum Press.

Jackson, J. H. (1958). *Selected writings of John Hughlings Jackson*. New York: Basic Books.

Jakobson, R. (1970). Linguistics in its relation to other sciences. In *Main trends of research in the social and human studies* (pp. 419–465). The Hague: Mouton.

Jakobson, R. (1971a). Pattern in linguistics. In *Selected writings of R. Jakobson. Vol. II* (pp. 223–228) The Hague: Mouton.

Jakobson, R. (1971b). Linguistic types of aphasia. In *Selected writings of R. Jakobson* (pp. 307–333). The Hague: Mouton.

Jakobson, R. (1971c). *Selected writings. Vol. II. Word and language* (pp. 23–71, 154–183). The Hague: Mouton.
Jakobson, R. (1971d). Two aspects of language and two types of aphasia disturbances. In *R. Jakobson selected writings. Vol. II. Word and language* (pp. 239–259). The Hague: Mouton.
Jakobson, R. (1976). *Six leçons sur le son et le sens*. Paris: Les Editions de Minuit.
Jakobson, R. (1980). *Brain and language*. Columbus, OH: Slavica Publishers.
Jakobson, R., & Halle, M. (1956). *Fundamentals of language*. The Hague: Mouton.
Jakobson, R., & Halle, M. (1962). Phonology and phonetics. In *R. Jakobson selected writings. Vol. I. Phonological studies* (pp. 464–504). The Hague: Mouton.
Jakobson, R., Halle, M., & Cherry, C. (1962). Toward the logical description of languages in their phonemic aspect. In *R. Jakobson selected writings. Vol. I. Phonological studies* (pp. 448–463). The Hague: Mouton.
Jaspers, K. (1963). *General psychopathology*. J. Hoening & M. W. Hamilton. Manchester: Manchester University Press.
Jones, M. R. (1976). Imagery as a mnemonic aid after left temporal lobectomy; contrast between specific and generalized memory disorders. *Neuropsychologia*, *12*, 21–30.
Jones-Gotman, M. (1979). Incidental learning of image mediated of pronounced words after right temporal lobectomy. *Cortex*, *15*, 187–198.
Jones-Gotman, M., & Milner, B. (1974). Right temporal lobe contribution to language mediated verbal learning. *Neuropsychologia*, *16*, 61–71.
Joseph, A. B., & O'Leary, D. H. (1987). Anterior cortical atrophy in Fregoli syndrome. *Journal of Clinical Psychiatry*, *48*, 409–411.
Joseph, A. B., O'Leary, D. H., & Wheeler, H. G. (1990). Bilateral atrophy of frontal and temporal lobes in schizophrenic patients with Capgras syndrome. *Journal of Clinical Psychiatry*, *51*, 322–325.
Jung, C. G. (1971). Psychological types. *The collected works of C. G. Jung. Vol. 6*. Princeton, NJ: Princeton University Press.
Kandel, E. R., & Schwartz, J. H. (1985). *Principles of neural science*. New York: Elsevier Science.
Kaplan, E. (1990). The process approach to neuropsychological assessment of psychiatric patients. *Journal of Neuropsychiatry and Clinical Neurosciences*, *2*, 1, 72–87.
Kaplan, H., Sadock, B., & Grebb, J. (1994). *Synopsis of psychiatry*, 7th ed. Baltimore: Williams and Wilkins.
Kapur, N., Turner, A., & King, C. (1988). Reduplicative paramnesia: Possible anatomical and neuropsychological mechanisms. *Journal of Neurology, Neurosurgery and Psychiatry*, *51*, 579–581.
Katznelson, S. D. (1972). *Typology of language and verbal thinking*. Leningrad: Nauka.
Katznelson, S. D. (1986). *General and typological linguistics*. Leningrad: Nauka.
Kennedy, F., & Wolf, A. (1936). The relationship of intellect to speech defects in aphasic patients. *Journal of Nervous and Mental Disease*, *84*, 125–145.
Kertesz, A., & Benson, D. F. (1970). Neologistic jargon—A clinicopathological study. *Cortex*, *6*, 362–386.
Kertesz, A., Sheppard, A., & MacKenzie, R. (1982). Localization in transcortical sensory aphasia. *Archives of Neurology*, *39*, 475–478.
Kim, E. (1991). A post-ictal variant of Capgras syndrome in a patient with a frontal meningioma. *Psychosomatics*, *32*, 448–451.
Kinsbourne, M., & Warrington, E. (1963). Developmental Gerstmann's syndrome. *Archives of Neurology*, *8*, 490.
Kinsbourne, M. (1968). Developmental Gerstmann's syndrome. *Pediatric Clinics of North America*, *15*, 771.
Kluver, H., & Bucy, P. C. (1937). Psychic blindness and other symptoms following temporal lobectomy in Rhesus monkeys. *American Journal of Physiology*, *119*, 352–353.
Knight, R., Singh, J., & Woods, D. (1989). Premovement parietal lobe input to human sensorimotor cortex. *Brain Research*, *498*, 190–194.

Knopman, D. S., Selnes, O. A., Niccum, N., & Rubens, A. B. (1984). Recovery of naming in aphasia. *Neurology, 34,* 1461–1470.

Kochetkova, V. I. (1973). *Paleoneurology.* Moscow. Moscow University Press.

Kock, E. P. (1967). *Visual agnosias.* Moscow: Meditsina.

Koestler, A. (1969). Man—One of evolution's mistakes? The place of value in a world of facts. Nobel Foundation. The New York Times Company.

Kolk, M., & Van Grunsven, M. (1985). Agrammatism as a variable phenomenom. *Cognitive Neuropsychology, 2,* 347–384.

Kononova, E. P. (1962). The frontal region of the brain. Moscow: Meditsina.

Kotrola, K. T., & Weinberger, D. R. (1995). Brain imaging and schizophrenia. *Annual Review of Medicine, 46,* 113–122.

Kretschmer, E. (1927). *Medical psychology.* Moscow: Life and Knowledge.

Kretschmer, E. (1936). *Physique and character.* New York: Harcourt Brace.

Kussmaul, A. (1877). Disturbances of speech. In M. von Ziemssien (Ed.), *Cyclopedia of the practice of medicine* (pp. 581–875). New York: William Wood.

Landis, T., Cummings, J. L., Benson, F. D., & Palmer, P. (1986). Loss of topographic familiarity—An environmental agnosia. *Archives of Neurology, 43,* 132–136.

Lebedinsky, M. S. (1941). *Aphasia, apraxia, and agnosia.* Charkov.

Lebrum, Y., & Hoops, R. (1974). *Intelligence and aphasia.* Amsterdam: Swets, Zeitlinger.

Lenneberg, E. H. (1967). *Biological foundations of language.* New York: Wiley.

Leontjev, A. A. (1969). *Language, speech, speech activity.* Moscow: "Prosveschenie."

Levin, H. S., Grossman, R. C., & Kelly, P. J. (1976). Aphasic disorder in patients with closed head injury. *Journal of Neurology, Neurosurgery and Psychiatry, 39,* 1062–1070.

Levy, J. (1968). Possible basis for the evolution of lateral specialization of the human brain. *Nature, 224,* 614–615.

Levy, J., & Trevarthen, C. (1976). Metacontrol of hemispheric function in human split-brain patients. *Journal of Experimental Psychology: Human Perception and Performance, 2,* 299–312.

Levy-Bruhl, L. (1930). Moscow: Archaic Thought. (Translated from *La Mentalite Primitive,* Paris, 1922).

Lewis, D. A., & Anderson, S. A. (1987a). The functional architecture of the prefrontal cortex and risk of schizophrenia. *Journal of Psychiatric Research, 21,* 413–421.

Lewis, S. W. (1987b). Brain imaging in a case of Capgras syndrome. *British Journal of Psychiatry, 150,* 117–121.

Lezak, M. D. (1983). *Neuropsychological assessment.* New York: Oxford University Press.

Lilly, R., Cummings, J. L., Benson, D., & Frankel, M. (1983). The human Kluver-Bucy syndrome. *Neurology, 33,* 1141–1145.

Livingstone, M., & Hubel, D. (1988). Segregation of form, color, movement, and depth: Anatomy, physiology, and perception. *Science, 240,* 740–749.

Luaute, L. P., & Bidault, E. (1994). Capgras syndrome, agnosia of identification, and delusion of reduplication. *Psychopathology, 27,* 186–193.

Lukianowicz, N. (1967). "Body image" disturbances in psychiatric disorders. *British Journal of Psychiatry, 113,* 31–47.

Luria, A. R. (1968). Small book about big memory. *The mind of a mnemonist.* Moscow: Moscow University Press.

Luria, A. R. (1970). *Traumatic aphasia: Its syndromes, psychology and treatment.* The Hague: Mouton (Original work published 1947, Moscow: USSR Academy of Sciences.)

Luria, A. R. (1980). *Higher cortical function in man.* New York: Basic Books. (Original work published 1969, Moscow: Moscow University Press.)

Lyon, M., Barr, C. Cannon, T., Mednick, S., and Shore, D. (1989). Fetal neural development and schizophrenia. *Schizophrenia Bulletin, 15,* 1, 149–160.

Maher, L., Chatterjee, A., Rothi, L., & Heilman, K. (1995). Agrammatic sentence production: The use of a temporal–spatial strategy. *Brain and Language*, *49*, 105–124.

Malloy, P. F., & Richardson, E. D. (1994). The frontal lobes and content-specific delusions. *Journal of Neuropsychiatry and Clinical Neurosciences*, *6*, 455–466.

Mandelshtam, O. E. (1921). *Word and culture. Almanac of poet's guild.* Petrograd: Dragon.

Mandler, G. (1980). Recognizing: The judgment of previous occurrence. *Psychological Review*, *87*, 252–271.

Marcuse, H. (1904). Apraktische Symptome bei einem Fail von senile Demenzr Zzentbl. *Nerv Heilk Psychiat.*, *27*, 737–751.

Markova, E. D. (1961). Clinical features and pathophysiology of amnestic aphasis. In *Problems of clinics and pathophysiology of aphasias ll* (pp. 58–95). Moscow: Medgiz.

Maslov, S. (1983). Asymmetry of cognitive mechanisms and its consequences. In *Semiotics and Informatics* (pp. 3–32). Moscow: Viniti.

Mayberg, M. S., Liotti, M., Liotti, M., Jerabek, P. A., et al. (1995). Induced sadness: A PET model of depression. *Human Brain Mapping*, *1*(suppl.), 396.

McCarthy, K. A., & Warrington, E. K. (1985). *Neuropsychologia*, *23*, 709–727.

McKenna, P., Mortimer, A., & Hodges, J. (1994). Semantic memory and schizophrenia. In A. David & J. Cutting (Eds.), *The Neuropsychology of Schizophrenia* (pp. 163–178). Hillsdale, NJ: Erlbaum.

Meck, W. H. (1996, June). Neuropharmacology of timing and time perception. *Cognitive Brain Research*, *3*, 3–4, 272–242.

Mega, M. S., & Cummings, J. L. (1994). Frontal-subcortical circuits and neuropsychiatric disorders. *Journal of Neuropsychiatry and Clinical Neurosciences*, *6*, 358–370.

Mehta, Z., Newcombe, F., & DeHaan, E. (1992, July). Selective loss of imagery in a case of visual agnosia. *Neuropsychologia*, *30*, 7, 645–655.

Mellor, C. S. (1991). Delusional perception. *British Journal of Psychiatry*, *159*(Suppl. 14), 104–107.

Mendez, M., & Benson, F. (1985). Atypical conduction aphasia. *Archives of Neurology*, *42*, 886–891.

Merrin, E. L., & Silverfarb, P. M. (1976). The Capgras phenomenon. *Archives of General Psychiatry*, *33*, 965–968.

Miceli, G., Silveri, M. C., Villa, G., & Caramazza, A. (1984). *Cortex*, *20*, 217–220.

Mishkin, M. (1972). *Cortical visual areas and their interactions.* In A. G. Karzmar & J. C. Eccles (Eds.), *Brain and human behavior* (pp. 187–208). Berlin: Springer-Verlag.

Mishkin, M., Ungerleider, L. C., & Macko, K. A. (1983). Object vision and spatial vision: Two cortical pathways. *Trends in Neurosciences*, *6*, 414–417.

Moor, J. H., & Tucker, G. J. (1979). Delusions: Analysis and criteria. *Comprehensive Psychiatry*, *20*, 388–393.

Moore, R. (1991). The suprachiasmatic nucleus and the circadian timing system. In D. Klein, R. Moore, & S. Rappaport (Eds.), *Suprachiasmatic nucleus (The mind's clock).* New York: Oxford University Press.

Morris, M. M., Luders, M., Lesser, R. P., Dinner, D. S., & Hahn, J. (1984). Transient neuropsychological abnormalities (including Gerstmann's syndrome) during cortical stimulation. *Neurology*, *34*, 877–883.

Moscovitch, M. (1976). On the representation of language in the right hemisphere of right-handed people. *Brain and Language*, *3*, 47–71.

Moscovitch, M. (1983). Stages of processing and hemispheric differences in language in the normal subjects. In M. Studdert-Kennedy (Ed.), *Psychobiology of language* (pp. 88–104). Cambridge, MA: MIT Press.

Mountcastle, V. B., Lynch, J. C., Georgupoulos, A., Sakata, H., & Acuna, C. (1975). Posterior parietal association cortex in the monkey command functions for operations with extrapersonal space. *Journal of Neurophysiology*, *38*, 871–908.

Mullan, S., & Penfield, W. (1959). Illusions of comparative intepretation and emotion. *Archives of Neurology and Psychiatry*, *81*, 269–284.

Nadeau, S., & Rothi, L. (1992). Morphologic agrammatism following a right hemisphere stroke in a mental patient. *Brain and Language*, *43*, 4, 642–667.

Nebes, R. D. (1978). Direct examination of cognitive function in the right and left hemispheres. In M. Kinsbourne (Ed.), *Asymmetrical function of the brain*. New York: Cambridge University Press.

Newcombe, F., & Russel, W. R. (1969). Dissociated visual perceptual and spatial deficits in focal lesions of the right hemisphere. *Journal of Neurology, Neurosurgery and Psychiatry*, *32*, 73–81.

Newcombe, F., Oldfield, R., Ratcliff, G., & Wingfield, A. (1971). Recognition and naming of object-drawings by men with focal brain wounds. *Journal of Neurology, Neurosurgery and Psychiatry*, *34*, 329–340.

Nielsen, J. M. (1948). *Agnosia, apraxia, and aphasia*. New York: Hafner. (Original work published 1936.)

Nikkolajenko, N. N., & Deglin, V. (1984). Semiotics of space and functional brain asymmetry. *Scientific Works of Tartu State University*, *17*, 441, 48–67.

Novinskaja, N. L., & Glezerman, T. B. (1986). Variants of Gerstmann's syndrome in children. *Journal of Neuropathology and Psychiatry*, *10*, 1467–1470.

O'Reilly, R., & Malhorta, L. (1987). Capgras syndrome—An unusual case and discussion of psychodynamic factors. *British Journal of Psychiatry*, *151*, 263–265.

Paivio, A. (1979). *Imagery and verbal processes* (2nd ed.). Hillsdale, NJ: Erlbaum.

Pakkenberg, B. (1990). Pronounced reduction of total neuron number in mediodorsal thalamic nucleus and nucleus accumbens in schizophrenics. *Archives of General Psychiatry*, *47*, 1023–1028.

Pandya, D. N., & Yeterian, E. H. (1985). Architecture and connections of cortical association areas. In A. Peters & E. G. Jones (Eds.), *Cerebral cortex: Association and auditory cortices* (pp. 3–60). New York: Plenum.

Pardo, J. V., Fox, P. T., & Raichle, M. E. (1991). Localization of a human system for sustained attention by positron emission tomography. *Nature*, *349*, 61–64.

Pardo, J. V., Pardo, P. J., & Raichle, M. E. (1993). Neural correlates of self-induced dysphoria. *American Journal of Psychiatry*, *150*, 713–719.

Paterson, A., & Zangwill, O. L. (1944). Disorders of visual space perception associated with lesions of the right cerebral hemisphere. *Brain*, *67*, 331–358.

Pearlson, G. D., Petty, R. G., Ross, C. A., & Tien, A. Y. (1996). Schizophrenia: A disease of heteromodal association cortex? *Neuropsychopharmacology*, *14*, 1–17.

Penfield, W., & Perot, P. (1963). The brain's record of auditory and visual experience. *Brain*, *86*, 595–696.

Penfield, W., & Roberts, L. (1959). *Speech and brain mechanisms*. Princeton, NJ: Princeton University Press.

Peterson, S. E., Fox, P. T., Posner, M. I., Mintin, M., & Raichle, M. E. (1988). Positron emission tomographic studies of the cortical anatomy of single-word processing. *Nature*, *331*, 585–588.

Pfefferbraum, A., & Marsh, L. (1995). Structural brain imaging in schizophrenia. *Clinical Neuroscience*, *3*, 105–111.

Piaget, J. (1977). The origins of intelligence in children. In M. E. Gruber & J. J. Veneche (Eds.), *The essential Piaget*. New York: Basic Books.

Pick, A. (1903). On reduplicative paramnesia. *Brain*, *26*, 242–267.

Pick, A. (1905). *Studien uber motorische Apraxie und ihre nahestehende Erscheinungen*. Leipzig: Deuticke.

Pick, A. (1973). *Aphasia*. Springfield, IL: Charles C. Thomas. (Original work published 1931.)

Piercy, M. (1964). The effect of cerebral lesions on intellectual function. *British Journal of Psychiatry*, *110*, 310–352.

Poljakov, R. I. (1966). About structural organization of the frontal lobe of the brain in connection with its function. In *The frontal lobes and regulation of psychical processes* (pp. 38–61). Moscow: Moscow University Press.

Posner, M. I., & Peterson, S. E. (1990). The attention system of the human brain. *Annual Review of Neuroscience*, *13*, 25–42.

Rakis, P., Lyon, M., Barr, C., Cannon, T., Mednick, S., & Shore, D. (1989). Genetic and epigenetic regulation of cortical mapping in fetal neural development and schizophrenia. *Schizophrenia Bulletin, 15*(1), 154.

Rapaport, D. (1946). *Diagnostic psychological testing*. Chicago. Year Book Publishers.

Reitan, R. M. (1959). The effects of brain lesions on adaptive abilities in human beings. Indianapolis.

Rizzo, L., Danion, J., Linden, M., & Grangé, D. (1996). Patients with schizophrenia remember that an event has occurred but not when. *British Journal of Psychiatry, 168*, 427–431.

Roeltgen, D. P. (1985). Agraphia. In K. M. Heilman & E. Valenstein (Eds.), *Clinical neuropsychology*. New York: Oxford University Press.

Roland, P. E. (1985). Cortical organization of voluntary behavior in man. *Human Neurobiology, 4*, 155–167.

Roland, P. E., Skinhoj, E., Lassen, N., & Larsen, B. (1980). Different cortical areas in man in organization of voluntary movements in extrapersonal space. *Journal of Neurophysiology, 43*, 137–150.

Roth, M. (1949). Disorders of body image caused by lesions of the right parietal lobe. *Brain, 72*, 89–111.

Rubin, S. S., & Niemeier, P. L. (1992). Nonverbal affective communication as a factor in psychotherapy. *Psychotherapy, 29*, 4, 596–602.

Sacks, O. (1987). Disembodied lady. *The man who mistook his wife for a hat* (pp. 43–54). New York: Harper & Row.

Saffran, E. M., Schwartz, M. F., & Marin, O. S. M. (1980). The word order problem on agrammatism: II. Production. *Brain and Language, 10*, 263–280.

Sala, S. D., Laiacona, M., Spinnler, H., & Trivelli, C. (1993). Autobiographical recollection and frontal damage. *Neuropsychologia, 31*, 8, 823–839.

Schmaryan, A. S. (1949). *Brain pathology and psychiatry*. Moscow: Medgiz.

Schneider, K. (1959). *Clinical psychopathology* M. W. Hamilton (trans.). New York: Grune and Stratton.

Schwartz, M. F., Saffran, E. M., & Marini, O. S. M. (1980). The word order problem in agrammatism: I. Comprehension. *Brain and Language, 10*, 249–262.

Seldon, H. L. (1982). Structure of human auditory cortex: III. Statistical analysis of dendritic trees. *Brain Research, 249*, 211–221.

Seldon, H. L. (1985). The anatomy of speech perception. In A. Peters & E. G. Jones (Eds.), *Cerebral cortex: Association and auditory cortices, Vol. 4* (pp. 273–326). New York: Plenum.

Selemon, L. D., Rajkowska, G., & Goldman-Rakic, P. S. (1995). Abnormally high neuronal density in the schizophrenic cortex: A morphometric analysis of prefrontal area and occipital area 17. *Archives of General Psychiatry, 52*, 805–820.

Semmes, J. (1968). Hemisphere specialization: A possible clue to mechanism. *Neuropsychologia, 6*, 11–26.

Shenton, M. E., Kikinis, R., Jolesz, F. A., et al. (1992). Abnormalities of the left temporal lobe and thought disorder in schizophrenia. *New England Journal of Medicine, 327*, 604–612.

Sidtis, J. J., Volpe, B. T., Holtzman, J. D., Wilson, D. H., & Gazzaniga, M. S. (1981). Cognitive interaction after staged callosal section: Evidence for transfer of semantic activation. *Science, 212*, 17, 344–346.

Signer, S. F. (1994). Self-substitution as a variant of the Capgras syndrome. *Psychopathology, 27*, 232–239.

Signer, S. F. (1994). Localization and lateralization in the delusion of substitution. *Psychopathology, 27*, 168–176.

Silva, J. A., Leong, G. B., Shaner, A. L., & Chang, C. Y. (1989, May–June). Syndrome of intermetamorphosis: A new perspective. *Comprehensive Psychiatry, 30*, 3, 209–213.

Silva, J. A., Leong, G. B., Weinstock, R., Sharama, K., & Klein, R. L. (1994). Delusional misidentification syndromes and dangerousness. *Psychopathology, 27*, 215–219.

Simernitskaya, E. G. (1978). *Cerebral dominance*. Moscow: Moscow University Press.

Slotinzeva, T. V. (1979). Dynamics of voluntary attention reflected in EEG parameters in norm and in patients with the focal brain lesions. In *Neurophysiological mechanisms of attention*. Moscow: Moscow University Press.

Smirnov, V. (1976). *Stereotaxical neurology*. Moscow: Meditsina.

Smith, A. (1965). Verbal and nonverbal test performance of patients with acute lateralized brain lesions. *Journal of Nervous and Mental Disease*, *741*, 517–523.

Smith, V. W. (1991). Lyric poems of eight Russian poets (Pushkin plus ...), edited and annotated with parallel verses translation by V. W. Smith. Palo Alto, CA: Zapizclat.

Sno, H. N. (1994). A continuum of misidentification symptoms. *Psychopathology*, *27*, 14–147.

Springer, S., & Deutsch, G. (1989). *Left brain, right brain* (3rd ed.). New York: W. H. Freeman.

Stankevich, I. A., & Shevchenko, Ju. T. (1935). Inferior parietal region in adult human. In *Variability of brain cortical formations*. Works of Moscow Brain Institute, Vol. 1, pp. 119–172.

Stern, K., & McNaughton, D. (1945). Capgras syndrome. *Psychiatric Quarterly*, *19*, 139–163.

Terzian, M. (1964). Behavioral and EEG effects of intracorotid sodium amytal injections. *Acta Neurochirurgica (Wien)*, *12*, 230–240.

Thomsen, I. V. (1975). Evaluation and outcome of aphasia in patients with severe closed head trauma. *Journal of Neurology, Neurosurgery and Psychiatry*, *38*, 713–718.

Thompson, M. I., Silk, K. R., & Hover, G. L. (1980). Misidentification of a city: Delimiting criteria for Capgras syndrome. *American Journal of Psychiatry*, *137*, 10, 1270–1271.

Todd, J. (1957). The syndrome of Capgras. *Psychiatric Quarterly*, *31*, 250–265.

Todd, J., Dewhurst, R., & Walls, G. (1981). The syndrome of Capgras. *British Journal of Psychiatry*, *139*, 319–327.

Traugott, N. N. (1973). *Mechanisms of memory disorders*. Leningrad: Nauka.

Trubetskoj, N. C. (1960). *Basics of phonology*. Moscow.

Tulving, E. (1983). *Elements of episodic memory*. Oxford: Clarendon/Oxford University Press.

Tulving, E., Kapur, S., Markowitsch, H. J., Cravik, I. M., Habib, R., & Houle, S. (1994). Neuroanatomical corrolates of retrieval in episodic memory. *Proceedings of the National Academy of Sciences of the United States of America*, *15*, 91, 6, 2012–2015.

Ullman, S. (1995). The visual analysis of shape and form. In *The cognitive neuroscience* (pp. 339–350). Cambridge, MA: MIT Press.

Ungerleider, L. G., & Mishkin, M. (1982). Two cortical visual systems. In D. J. Ingle, M. A. Goodale, & R. Y. W. Mansfield (Eds.), *Advances in the analysis of visual behavior* (pp. 549–586). Cambridge, MA: MIT Press.

Valdman, A. V., Evartau, E. E., & Kozlovskaya, M. M. (1976). *Psychopharmacology of emotions*. Moscow: Meditzina.

Vinarskaya, E. N. (1971). *Clinical problems of aphasia*. Moscow: Meditsina.

Vinarskaya, E. N. (1973). Modern state of the problems of dysarthria. In *Dysarthria and its topical-diagnostic meaning in the clinic of focal brain damage* (pp. 9–105). Tashkent: Meditsina.

Vogt, C., & Vogt, O. (1919). Allgemeinere Ergebnisse unserer Hirnforschung. *J. Psychol. Neurol.*, *25*, 279–461.

von Economo, C., & Koskinas, G. N. (1925). *Die Cytoarchitektonic der Kirnrinde des erwachsenen Menchen*. Berlin: Springer.

Vygotsky, L. S. (1960). *Development of higher psychological functions*. Moscow: Akad Ped Nauk.

Vygotsky, L. S. (1962). *Thought and language*. E. Hanfmann & G. Vaker (trans.). Cambridge, MA: MIT Press. (Original work published 1934.)

Waddington, J. L. (1993). Schizophrenia: Developmental neuroscience and pathobiology. *Lancet*, *341*, 531–536.

Walker, C. (1991). Delusion: What did Jaspers really say? *British Journal of Psychiatry*, *159*(Suppl. 14), 94–103.

Wapner, W., Hamby, S., & Gardner, H. (1981). The role of the right hemisphere in the apprehension of complex linguistic materials. *Brain and Language*, *14*, 15–33.

Warren, R. M., Obusek, C. J., Farmer, R. M., & Warren, R. P. (1969). Auditory sequence: Confusion of patterns other than speech or music. *Science*, *164*, 586–587.

Warrington, E. (1975). The selective impairment of semantic memory. *Quarterly Journal of Experimental Psychology*, *72*, 635–657.

Warrington, E., & McCarthy, R. (1983). Category-specific access dysphasia. *Brain*, *106*, 859–878.

Warrington, E., & Shallice, T. (1984). Category-specific semantic impairments. *Brain*, *106*, 859–878.

Watson, J. D. G., Myers, R. S. Frackowiak, R. S. J., Hajnal, J. V., Woods, R. P., Mazziotta, J. C., Shopp, S., & Zeki, S. (1993). Area V5 of the human brain: Evidence from a combined study using PET and MRI. *Cerebral Cortex*, *3*, 79–94.

Weinberger, D. R. (1987). Implications of normal brain development for the pathogenesis of schizophreina. *Archives of General Psychiatry*, *44*, 660–669.

Weinberger, D. R. (1996). On the plausibility of "the neurodevelopmental hypothesis" of schizophrenia. *Neuropsychopharmacology*, *14*(Suppl.), 1–11.

Weinrich, N., Steele, R., Kleczewska, M., Carlson, G. S., Baker, E., & Wertz, R. (1989). Representations of "verbs" in computerized visual communication system. *Aphasiology*, *3*, 6, 501–512.

Weinstein, E. A. (1994). The classification of delusional misidentification syndromes. *Psychopathology*, *27*, 130–135.

Weinstein, E. A., & Kahn, R. L. (1952). Nonaphasic misnaming in organic brain disease. *Archives of Neurology and Psychiatry*, *67*, 72–79.

Weinstein, E. A., Kahn, R. L., Malitz, S., & Rozanski, J. (1954). Delusional reduplication of parts of the body. *Brain*, *77*, 45–60.

Weisenberg, T., Roe, A., & McBride, K. (1936). *Intelligence: A psychological study of test performance*. New York.

Winner, E., & Gardner, H. (1977). The comprehension of metaphor in brain damaged patients. *Brain*, *100*, 717–729.

Young, A. W., Leafhead, K. M., & Szulecka, T. K. (1994). The Capgras and Cotard delusions. *Psychopathology*, *27*, 226–231.

Zeki, S., & Shipp, S. (1988). The functional logic of cortical connections. *Nature*, *335*, 311–317.

Zeki, S. (1993). *A vision of the brain*. Oxford: Blackwell Scientific.

Zeki, S., & Lamb, M. (1994). The neurology of kinetic art. *Brain*, *117*, 607–636.

Zingeser, L. B., & Berndt, R. S. (1990). *Brain and Language*, *39*, 14–32.

Index